Current Topics in Behavioral Neurosciences

Series Editors:
Mark Geyer, La Jolla, CA, USA
Bart Ellenbroek, Wellington, New Zealand
Charles Marsden, Nottingham, UK

About this series

Current Topics in Behavioral Neurosciences provides critical and comprehensive discussions of the most significant areas of behavioral neuroscience research, written by leading international authorities. Each volume offers an informative and contemporary account of its subject, making it an unrivalled reference source. Titles in this series are available in both print and electronic formats.

With the development of new methodologies for brain imaging, genetic and genomic analyses, molecular engineering of mutant animals, novel routes for drug delivery, and sophisticated cross-species behavioral assessments, it is now possible to study behavior relevant to psychiatric and neurological diseases and disorders on the physiological level. The *Behavioral Neurosciences* series focuses on "translational medicine" and cutting-edge technologies. Preclinical and clinical trials for the development of new diagnostics and therapeutics as well as prevention efforts are covered whenever possible.

Jo C. Neill • Jayashri Kulkarni
Editors

Biological Basis of Sex Differences in Psychopharmacology

Springer

Editors
Prof. Dr. Jo C. Neill
School of Pharmacy
University of Bradford
Bradford BD7 1DP
United Kingdom
j.c.neill@bradford.ac.uk

Prof. Dr. Jayashri Kulkarni
Monash Alfred Psychiatry Research Centre
The Alfred Hospital and Monash University
Commercial Road
Prahran, Victoria
Australia
j.kulkarni@alfred.org.au

ISSN 1866-3370 e-ISSN 1866-3389
ISBN 978-3-642-20005-2 e-ISBN 978-3-642-20006-9
DOI 10.1007/978-3-642-20006-9
Springer Heidelberg Dordrecht London New York

Library of Congress Control Number: 2011931783

Cover illustration: Artistic representation of oscillatory synchrony and timing of neurons in networks by Gyorgy Buzsaki

Cover design: deblik, Berlin

Printed on acid-free paper

Springer is part of Springer Science+Business Media (www.springer.com)

Preface

This volume attempts to answer the question of how sex influences brain and behaviour. It brings together experts in this field, psychiatrists and other mental health care professionals with preclinical researchers to review the latest work in this area and give a thorough overview of how males and females are different in terms of brain function and behaviour. This is followed by a clinical perspective, applying brain biology to explain why some illnesses are gender specific and how gonadal steroids are involved in the aetiology and symptomatology of psychiatric diseases and may be modulated to provide new therapeutic approaches to mental illnesses. Appreciation and improved understanding of sex differences will certainly lead to improvements in the diagnosis, tailored treatment approaches and hence outcomes for people suffering with mental disorders. This volume will be essential reading for all health care professionals.

Animal behavioural models of efficacy studies in pharmacology generally use male rodents. However, the presentation of sex differences in behaviour underscores the need to include female animals in basic research studies. Similarly, gender differences in the presentation, treatment and outcomes of mental illnesses are often overlooked. The biological basis of sex differences and its application in clinical disorders are important issues that are highlighted by the research discussed in this volume.

The volume is divided into two sections, the first deals with the importance of recognising and studying sex differences in brain function and behaviour in animal models. The second section is dedicated to the consideration of biological sex differences in the presentation of aspects of mental illnesses such as schizophrenia, bipolar disorder, depression and anorexia nervosa. The two sections are interrelated and provide an integrated approach with animal research informing the human application in considering the biological basis of sex differences in psychopharmacology.

The volume starts with an overview by Professor Kay Marshall (a reproductive endocrinologist who has been persuaded to study the brain!) This chapter gives an important introduction into the mechanisms by which gonadal steroids produce their effects which explains how sex differences come about. The second chapter by Berend Olivier and colleagues explores how sex matters for rats. They provide an overview of rodent sexual behaviour and how to measure this in the laboratory. The chapter has a particular emphasis on the role of serotonin (5-HT) on sexual activity and on sexual dimorphism in response to serotonergic agents. This is clearly of much relevance as drugs such as SSRIs cause sexual dysfunction in the clinic, and it is essential to model this appropriately in the laboratory. In a later clinical chapter on the impact of sex on antidepressants, John Sramek and colleagues detail clinical trial work on this important area. An investigation into sex differences and the effect of gonadal steroids on cognitive function in rodents is provided by Jane Sutcliffe, which is of particular importance as cognitive dysfunction occurs in many psychiatric illnesses such as depression, ADHD and PTSD. In schizophrenia most notably this remains an unmet clinical need, with emphasis on the development of new therapies for cognitive and other symptoms of this illness. Implications for the aetiology of schizophrenia are explored by Veena Kumari in her chapter dealing with human sensorimotor gating. Chapter 9 written by Anita Riecher-Rossler and Jayashri Kulkarni, and Chapter 10 by Angelika Wieck present the very important role that oestrogen plays as a key neuroprotective agent, and its impact on the timing and gender differences in illness presentation. The possibility of using hormone modulation as a new treatment approach is also discussed with respect to psychotic disorders.

Chapter 4 by Dai Mitsushima illustrates that in rodents neurotransmitters show sexual dimorphism, and that neurotransmitter release is affected by gonadectomy with a focus on acetylcholine, again of particular importance for cognition. A subsequent chapter by Justin Anker and Marilyn Carroll deals with the very important topic of drug dependence. They show that females are more sensitive than males to the reinforcing effects, and less sensitive to withdrawal effects, of certain drugs of abuse, making them more vulnerable to drug dependence which can be effectively modelled in animals. The translation from animals to humans here is impressive with female rats showing greater propensity for drug self-administration and relapse in animal models. The authors go on to demonstrate how these effects in females may be mediated by gonadal steroid hormones, in particular oestrogen and progesterone (which is important, as it is not all about oestrogen, as Kay Marshall explains in her opening chapter). They discuss possible mechanisms including oestrogen receptors and their interaction with the mesolimbic dopamine system with the emphasis on addiction to psychostimulants such as cocaine. Applying this framework, a novel proposal for the noted sex differences in anorexia nervosa is described in a chapter by Charlotte Keating, with applicability for new thinking about the aetiology of this severe and female dominant eating disorder.

Stress is of course an important feature of human lives, including our response to drugs of abuse and Christina Dalla and her colleagues cover this topic in some depth in their chapter. Men and women differ in their vulnerability to stress and stress-related psychiatric disorders such as depression. The authors explore sex differences in the response to acute and chronic stress in several animal models in some detail. Male and female rodents differ in their reactivity and adaptation to various stressors, and the authors demonstrate the link between this and differences in the neuroendocrine system and its interaction with neurotransmitter systems, such as serotonin and dopamine. The final preclinical chapter provides an elegant review by Elizabeth Tunbridge and Paul Harrison into sex differences in the catechol-*O*-methyltransferase (COMT) gene. The gene encodes an enzyme that metabolises catechol compounds including dopamine, and the authors explain how sexual dimorphism in this gene and its interaction with oestrogen impacts on psychiatric disease states. Many of the preclinical studies suggest that sex-specific interventions may be a beneficial approach when treating patients, and understanding the sex differences in developmental disorders experienced early in life is an important area detailed by Bruce Tonge and colleagues in a clinical chapter that also proposes treatment strategies for early psychiatric presentations.

In summary, sex differences are observed in humans and animals in brain function and behaviour and in the response to illness. Men and women have different advantages in many aspects of behaviour particularly cognitive function which is a key component of many psychiatric disorders. Indeed, there is both clinical and preclinical evidence to support a role for the sex steroids in modulating performance in certain cognitive domains. At present, these interactions are complex and the underlying mechanisms have yet to be elucidated. Once these relationships are understood, there is potential for more effective therapeutic exploitation. This volume covers in some depth many illnesses in which sex differences and gonadal steroids are important in terms of aetiology, symptomatology, progression and treatment. It is the first volume to successfully achieve this and will be of considerable importance to workers in all aspects of mental illness.

Bradford, UK — Jo C. Neill
Melbourne, Australia — Jayashri Kulkarni

Contents

Contributors

Justin J. Anker
Department of Psychiatry, University of Minnesota, MMC 392, MN 55455 Minneapolis, USA, anke0022@umn.edu

Marilyn E. Carroll
Department of Psychiatry, University of Minnesota, MMC 392, Minneapolis, MN 55455 USA, anke0022@umn.edu

Johnny S. W. Chan
Division of Pharmacology, Utrecht Institute for Pharmaceutical Sciences and Rudolf Magnus Institute of Neuroscience, Utrecht University, Sorbonnelaan 16, 3584 CA Utrecht, The Netherlands, j.s.w.chan@uu.nl

Kim M. Cornish
Centre for Developmental Psychiatry and Psychology, School of Psychology and Psychiatry, Faculty of Medicine, Nursing and Health Sciences, Melbourne, VIC, Australia

Neal R. Cutler
Worldwide Clinical Trials, Inc., 401 N. Maple Drive, Beverly Hills, CA 90210 USA, neal.cutler@wwctrials.com

Christina Dalla
Department of Pharmacology, Medical School, University of Athens, Mikras Asias 75, Goudi, 11527 Athens, Greece, cdalla@med.uoa.gr

Paul J. Harrison
Department of Psychiatry, University of Oxford, Neurosciences Building, Warneford Hospital, Oxford, OX3 7JX, UK, paul.harrison@psych.ox.ac.uk

Charlotte Keating
Monash Alfred Psychiatry Research Centre (MAPrc), The Alfred Hospital, 1st floor, Old Baker Building, Commercial Road, Prahran, VIC 3181 Australia, charlottekeating1@gmail.com

Nikolaos Kokras
Department of Pharmacology, Medical School, University of Athens, Mikras Asias 75, Goudi, 11527 Athens, Greece, NKokras@med.uoa.gr

Jayashri Kulkarni
Monash Alfred Psychiatry Research Centre, The Alfred Hospital and School of Psychology and Psychiatry, Monash University, Commercial Road, Melbourne, VIC 3004 Australia, jayashri.kulkarni@med.monash.edu.au

Veena Kumari
Department of Psychology, Institute of Psychiatry, King's College London, De Crespigny Park, London SE5 8AF, UK, veena.kumari@kcl.ac.uk

Kay M. Marshall
School of Pharmacy, University of Bradford, Bradford BD7 1DP, UK, k.m.marshall@bradford.ac.uk

Dai Mitsushima
Department of Physiology, Yokohama City University Graduate School of Medicine, 3-9 Fukuura Kanazawaku, Yokohama 236-0004, Japan, dm650314@med.yokohama-cu.ac.jp

Berend Olivier
Division of Pharmacology, Utrecht Institute for Pharmaceutical Sciences and Rudolf Magnus Institute of Neuroscience, Utrecht University, Sorbonnelaan 16, 3584 CA Utrecht, The Netherlands, b.olivier@uu.nl

Jocelien D. A. Olivier
Department Clinical Neuroscience, Division of Psychiatry, Karolinska Institutet, KFC Novum Level 6, Exp 617 Lab 614, 14157 Huddinge, Sweden, jocelien.olivier@ki.se

Ronald S. Oosting
Division of Pharmacology, Utrecht Institute for Pharmaceutical Sciences and Rudolf Magnus Institute of Neuroscience, Utrecht University, Sorbonnelaan 16, 3584 CA Utrecht, The Netherlands, r.s.oosting@uu.nl

Zeta Papadopoulou-Daifoti
Department of Pharmacology, Medical School, University of Athens, Mikras Asias 75, Goudi, 11527 Athens, Greece, zdaifoti@med.uoa.gr

Pothitos M. Pitychoutis
Department of Pharmacology, Medical School, University of Athens, Mikras Asias 75, Goudi, 11527 Athens, Greece, PPitychoutis@med.uoa.gr

Anita Riecher-Rössler
Psychiatric University Clinic Basel, University Psychiatric Outpatient Department, c/o University Hospital Basel, Petersgraben 4, Basel, 4031 Switzerland, anita.riecher@upkbs.ch

Nicole J. Rinehart
Centre for Developmental Psychiatry and Psychology, School of Psychology and Psychology, Faculty of Medicine, Nursing and Health Sciences, Monash University, Melbourne, VIC Australia

Eelke M. Snoeren
Division of Pharmacology, Utrecht Institute for Pharmaceutical Sciences and Rudolf Magnus Institute of Neuroscience, Utrecht University, Sorbonnelaan 16, 3584 CA Utrecht, The Netherlands, e.m.snoeren@uu.nl

John J. Sramek
Worldwide Clinical Trials, Inc., 401 N. Maple Drive, Beverly Hills, CA 90210 USA, jsramek@wwctrials.com

Jane Suzanne Sutcliffe
Maccine Pte Ltd, 10, Science Park Road, 01-05 The Alpha, Singapore Science Park II, Singapore 117684 Singapore, Jane.Sutcliffe@Maccine.com

Bruce J. Tonge
Centre for Developmental Psychiatry and Psychology, School of Psychology and Psychology, Faculty of Medicine, Nursing and Health Sciences, Monash University, Melbourne, VIC, Australia, bruce.tonge@monash.edu

Elizabeth M. Tunbridge
Department of Psychiatry, University of Oxford, Neurosciences Building, Warneford Hospital, Oxford, OX3 7JX, UK, elizabeth.tunbridge@psych.ox.ac.uk

Jan G. Veening
Division of Pharmacology, Utrecht Institute for Pharmaceutical Sciences and Rudolf Magnus Institute of Neuroscience, Utrecht University, Sorbonnelaan 16, 3584 CA Utrecht, The Netherlands, j.g.veening@uu.nl

Christiaan H. Vinkers
Division of Pharmacology, Utrecht Institute for Pharmaceutical Sciences and Rudolf Magnus Institute of Neuroscience, Utrecht University, Sorbonnelaan 16, 3584 CA Utrecht, The Netherlands, c.h.vinkers@uu.nl

Marcel D. Waldinger
Division of Pharmacology, Utrecht Institute for Pharmaceutical Sciences and Rudolf Magnus Institute of Neuroscience, Utrecht University, Sorbonnelaan 16, 3584 CA Utrecht, The Netherlands, md@demon.waldinger.nl

Angelika Wieck
Laureate House, Wythenshawe Hospital, Manchester Mental Health and Social Care Trust, University of Manchester, Southmoor Road, Manchester M239LT, UK, angelika.wieck@manchester.ac.uk

Introduction to the Interaction Between Gonadal Steroids and the Central Nervous System

Kay M. Marshall

Contents

Abstract The sex steroids are frequently referred to as the gonadal steroids and are erroneously assumed to be exclusively linked to the ovaries in women or the testes in men and the functions of the reproductive tract. This chapter will provide an overview of some of the extragonadal effects of these hormones, focusing on the central nervous system, and the mechanisms of hormone action. Hormone synthesis and metabolism within the CNS will be discussed with particular focus on the role of aromatase. Sex steroids exert many of their effects via intracellular receptors and these genomic responses tend to be slow in onset, however, some responses to steroids occur more quickly and are mediated via membrane receptors and involve interactions with many different transduction pathways to produce a diverse array of responses. These complexities do pose challenges but also offer opportunity for novel approaches for therapeutic exploitation as the pharmacological tools with which to modulate systems become increasingly available.

Keywords Androgen · Mechanism of action · Oestrogen · Progesterone

---

K.M. Marshall
School of Pharmacy, University of Bradford, Bradford BD7 1DP, UK
e-mail: k.m.marshall@bradford.ac.uk

J.C. Neill and J. Kulkarni (eds.), *Biological Basis of Sex Differences in Psychopharmacology*, Current Topics in Behavioral Neurosciences 8, DOI 10.1007/7854_2011_136

The sex steroids are commonly referred to as the gonadal steroids and are erroneously assumed to be exclusively linked to the ovaries in women or to the testes in men and to the functions of the reproductive tract (Chabbert Buffet et al. 1998; Genazzani et al. 2002; Mooradian et al. 1987). The sex steroids themselves are not gender specific; for example, testosterone has several key functions in women; testosterone imbalance has been linked to depression in women (Rohr 2002) and oestrogen promotes hippocampal neurogenesis in male rats (Bowers et al. 2010). These steroids can be synthesised at extragonadal sites and have effects that are outside those obviously linked with reproductive processes. This chapter will provide an overview of some of the extragonadal effects of these hormones, focusing on the central nervous system (CNS), and the mechanisms of hormone action. The examples cited will be somewhat oestrogenocentric as this reflects the perhaps disproportionate amount of research time spent on oestrogen (a simple Medline search for 2010 suggests that twice as many papers were published on oestrogen than on testosterone). However, progesterone and testosterone and their related steroidal products should not be overlooked as it is often the balance of these hormones and/or their metabolites with one another that will determine the overall response. For example, progesterone can reverse the effects of oestrogen on dendritic spine density and on brain-derived neurotrophic factor (Murphy and Segal 2000; Aguirre et al. 2010). In addition, all responses mediated by sex steroids will in turn be influenced by chronobiology beginning in utero (Pilgrim and Hutchison 1994) and including the lifting of the hypothalamic–pituitary block that facilitates the onset of puberty. The reactivation of the hypothalamic gonadotrophin-releasing hormone (GnRH) secretory system (which can be stimulated by noradrenaline and glutamate and inhibited by GABA) results in the establishment of reproductive rhythms. In women, these can be disrupted by pregnancy, where the magnitude of the hormonal changes far exceed those occurring during the menstrual cycle (for detailed review, see Brunton and Russell 2010), and end at menopause or in the male, the increasingly recognised, although less clearly demarked andropause (Keenan and Veldhuis 2009). The timing of these biological life events has considerable impact on, and importance for, long-term health.

1 Hormone Levels in the CNS and Role of Aromatase

For a recent review of steroid hormone biosynthesis, see Gilep et al. (2011). The extent to which peripheral levels of either endogenous or exogenous circulating hormones reflect levels in the CNS is still subject to scrutiny. It is known that the sex steroids can be synthesised in the brain; for example, progesterone is synthesised by glial cells (Garcia-Segura and Melcangi 2006). A recent study by Caruso et al. (2010) measured levels of sex steroids in plasma and in the CNS (cerebellum, cerebral cortex and spinal cord) in intact and gonadectomised male and female rats and found that after gonadectomy, changes in the CNS did not necessarily reflect the situation in the plasma. In addition to this, there is the further complication of identifying which hormone is the active moiety. For example, there are several forms of endogenous oestrogen namely: 17β-estradiol (E2), which is the most active and has the highest receptor affinity; estrone (E1), which is a less active product of oxidation of E2; estriol (E3), which can be produced from either of the former or from the androgen androstenedione. E3 is produced in abundance during pregnancy [it has been postulated that this happens to protect the developing CNS in the foetus (Reyes-Romero 2001)] but after the menopause levels do not really change and are similar to those found in men. 17α-estradiol, another form of endogenous oestrogen, has lower receptor affinity, but it is known to be synthesised locally in rodent brain (Simpkins et al. 1997; Levin-Allerhand et al. 2002). However, Nguyen et al. (2011) have recently reported measuring all of these oestrogens after derivitisation (using liquid chromatography separation, electrospray ionisation and tandem mass spectrometry, ESI-MS/MS) in human cerebrospinal fluid taken from trauma patients and E2 was found to predominate. This may represent assay or sample limitation; alternatively it could reflect another species difference and steroidogenesis does differ between species (Gilep et al. 2011). The synthetic pathways and resultant products (including metabolites, some of which retain biological activity) are determined by the types of steroidogenic cytochromeP450s (CYPs) and the dehydrogenase enzymes. In addition to inter-species variation, there is also some intra-species variation as the CYPs are subject to polymorphisms which may influence hormone as well as drug metabolism. Hormone metabolism may also explain some of the apparently paradoxical responses observed; for example, when the endogenous metabolite 2-methoxyestradiol is formed, it may possess some of the antioxidant properties of E2 but unlike E2 it cannot protect hippocampal neurones from insult by kainic acid (Picazo et al. 2003). Furthermore, E2 metabolism can lead to redox cycling and free radical formation and neural damage can ensue (Liehr and Roy 1990; Picazo et al. 2003).

In terms of variation between peripheral versus central steroid levels, dehydroepianstrosterone (DHEA) is a good example as DHEA can be detected in brain tissue of many species, including humans, at levels that exceed those in the periphery (Baulieu and Robel 1996; Tunbridge and Harrison 2011). DHEA is a substrate common to both oestrogen and androgen biosynthesis, and it appears that astrocytes from different brain areas can metabolise DHEA differentially; for example, hypothalamic cells are more active in producing E2 than similar cells from the cerebral cortex (Zwain and Yen 1999). However, there are some metabolic consistencies, as characterisation of the 5α-reductase-3α-hydroxysteroid dehydrogenase complex (which is key to androgen metabolism) in human brain samples has indicated no sex-specific differences and no differences over time (Steckelbroeck et al. 2001), which may suggest a non-reproductive role such as catabolism of neurotoxic steroids. Oestrogens can also be metabolised in the brain as indicated by high levels of 2- and 4-hydroxyoestrogens (the catecholoestrogens) which are metabolised further by catecholeamine-*O*-methyltransferases, an enzyme which itself is subject to oestrogenic influence and polymorphisms (Harrison and Tunbridge 2008).

The role of cytochrome P450 aromatase which is the protein product of gene *Cyp 19* also needs to be taken into account when considering the level of steroid hormones in the brain. This enzyme is responsible for converting androgens (C19 products) to oestrogens (C18 products), namely testosterone to E2 and androstenedione to E1; thus it plays a key role in regulating the androgen–oestrogen balance. Aromatase is present at extragonadal sites including the breast (where its role and inhibition, using aromatase inhibitors or AIs, have been most extensively studied; for more details, see Furr 2006) and the brain where it appears to be concentrated in the preoptic area, ventro-medial hypothalamus and the bed nucleus striae terminalis (Balthazart et al. 2003). In terms of clinical relevance, an example would be the expression of aromatase by astrocytes following injury (Garcia-Segura et al. 2003). Aromatase is subject to different regulatory controls; for example, in the gonads cAMP is stimulatory but in neural tissues it is inhibitory (Lephart 1996). Interestingly in rat brain, androgen is not only the substrate for aromatase, but dihydrotestosterone can, like oestradiol, also induce its activity (Roselli and Resko 1993). This regulation can be slow (as would be expected if control is via genomic mechanisms) or fast (suggesting non-genomic mechanisms). The presence of phosphorylation consensus sites on the enzyme would correlate with the latter mechanism (Balthazart and Ball 2006). Neurotransmitters such as dopamine and glutamate can inhibit the activity of aromatase as can kainate and NMDA (Balthazart et al. 2003).

Other considerations when attempting to correlate levels of sex steroids with function could arise from the fact that only free, that is hormone unbound to plasma proteins such as albumin or more specifically in the case of the sex hormones, sex hormone-binding globulin (SHBG), hormone is active (the presence or absence of this globulin should also be considered when using animal models and extrapolating to the human). However, SHBG seems to have biological activity via a membrane receptor coupled to a cAMP pathway (Nakhla et al. 2009), and it may even be a marker for the integrity of the blood–brain barrier (Gustafson et al. 2007).

2 Mechanisms of Action

Sex steroids can act as ligand-activated transcription factors when they exert their effects via a genomic mechanism, and the response is slow in onset to allow for the eventual translation of message (see Fig. 1).

2.1 Oestrogen

In the case of oestrogen, these receptors are known as ERα and ERβ (Green et al. 1986; Kuiper et al. 1996), and for progesterone there are also two forms namely PR-A and PR-B. These receptors are part of the nuclear/steroid receptor superfamily (see Mangelsdorf et al. 1995). Oestrogen is also acknowledged to exert some of its actions via membrane receptors as discussed below.

Responses mediated by these receptors tend to be slow in onset to allow for subsequent modulation of gene transcription (Tsai and O'Malley 1994). ERα and ERβ share considerable sequence homology at the DNA (96%) and ligand (56%) (Weiser et al. 2008) binding domains and, in vitro at least, ERα is a stronger

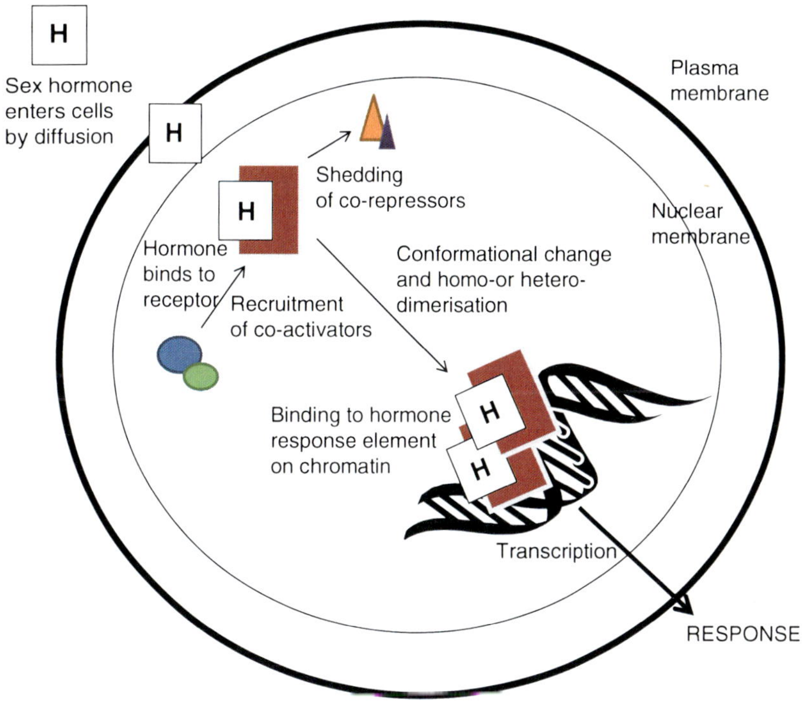

Fig. 1 Diagramatic representation of a sex hormone entering a target cell and binding to its receptor. The sex hormone (H) diffuses into the cell and binds to its intranuclear receptor, co-activator and co-repressor proteins are then recruited or shed (respectively) from the complex as it dimerises (this can either be with an identical complex, eg another ERα to form a homo-dimer or with a simlar complex eg an ERβ to form a hetero-dimer). This results in the remodelling or activation of the complex to facilitate its binding to the hormone response element of the chromatin which will in turn induce transcription and the response to the hormone (Beato and Sanchez-Pacheco 1996)

transcriptional activator (Delaunay et al. 2000). Effective transcription does require the recruitment of steroid receptor co-activators (SRCs) as shown in Fig. 1. These molecules influence several processes that follow ligand receptor interaction such as phosphorylation and thus modulate transcription (O'Malley 2006). They may also be involved in the aetiology of disease including neurological disorders (Lonard et al. 2007). The distribution of ERα and ERβ has been extensively studied in different species (see Österlund and Hurd 2001; Weiser et al. 2008).

ERα and ERβ differ not only in their distribution in the brain but also in their ligand-binding ability (Damdimopoulos et al. 2008) and roles (Bodo and Rissman 2006). A pertinent example could be oestrogenic protection against insult by NMDA in the hippocampus, an activity which is mediated by ERβ and induction of BDNF, a response inhibited by progesterone (Aguirre et al. 2010).

The potential for selective agonism or antagonism at receptor subtypes has been exploited therapeutically by, for example, the selective oestrogen receptor modulators (SERMs) and the selective oestrogen receptor down-regulators (SERDs) such as tamoxifen and faslodex respectively, which have been used extensively in the management of breast cancer. Indeed, since the publication of the findings from the Women's Health Initiative (WHI) in the USA and in the UK, the Million Women Study (MWS) (for more information and resultant publications go to http://www.nhlbi.nih.gov/whi/references.htm and http://www.millionwomenstudy.org/publications/ respectively), which in many respects were controversial not least with respect to the findings in relation to neuroprotection, the use of agents like the SERMs or even more selective compounds is likely to increase as hormone replacement regimens with better side-effect profiles are sought to manage the consequences of the menopause as women can now expect to live approximately a third of their lives in an oestrogen-deficient state. The central effects of the SERMS have yet to be fully evaluated, but they do have potential as tools and therapies (Arevalo et al. 2011) as summarised in Fig. 2.

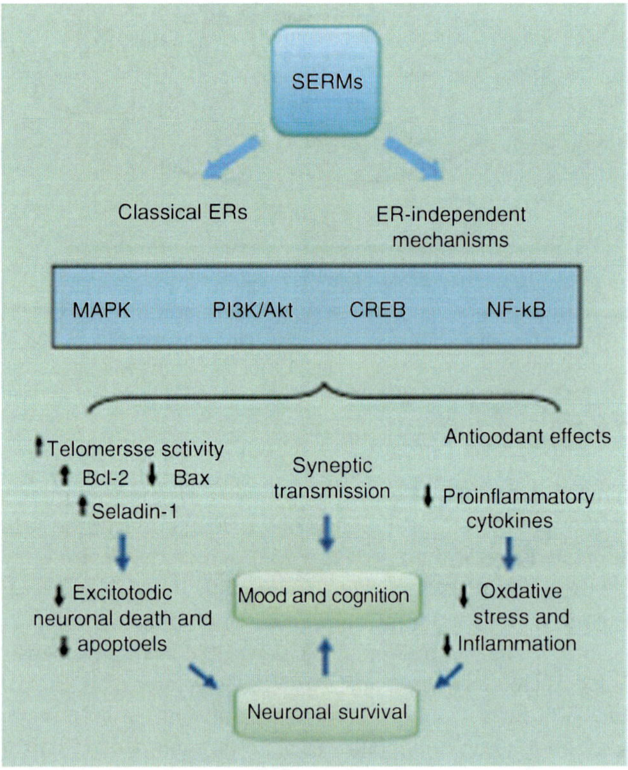

Fig. 2 Summary of the molecular mechanisms involved in the neuroprotective effects of SERMs. SERMs act in the nervous system through classical ERs or by ER-independent mechanisms and activate a variety of signalling molecules, including MAPK, PI3K, Akt, CREB and NF-κB. These molecules, in turn, trigger different coordinated mechanisms to promote neuronal survival and regulate mood and cognition (Journal of Molecular Endocrinology (2011) 46, 111–119)

In addition to the intracellular receptors, ERα and ERβ, there are also membrane receptors which are as yet less well defined. One such receptor is called GPR30 and is a 7 transmembrane domain G-protein-coupled structure (Lappano et al. 2010). Most of the work on this receptor has been done in relation to breast cancer (for review, see Maggiolini and Picard 2010), but GPR30 is expressed in the hypothalamic– pituitary axis, hippocampus and substantia nigra (Brailoiu et al. 2007). In the neo-cortex, a membrane-associated ER (ER-X) has been reported which mediates the activation of mitogen-activated protein kinase (MAPK), and 17αE2 appears to be the preferred ligand (Toran-Allerand 2004). The MAPK pathway can evoke the transcription of cAMP response element (CREB). 17βE2 can also interact with the phosphotidylinositol-3-kinase (P13K) pathway resulting in Akt (protein kinase B) activation which inhibits proapoptotic proteins.

E2 can also affect the monoaminergic systems as it can influence synthesis and degradation of dopamine, noradrenaline and 5-hydroxytryptamine as well as regulating monoamine reuptake transporters and second messengers. Findings from a study in female rats by Lubbers et al. (2010) indicate that SERMs may have the potential to manipulate monoamines selectively allowing some modulation of cognition and affective function. Oestrogens have several other central properties including (Amantea et al. 2005): an antioxidant effect principally due to direct free radical scavenging; anti-inflammatory action via suppression of interleukin-1β induction of COX-2 (for review and information on structure–activity relationships, see Prokal and Simpkins 2007). Oestrogen can also rapidly induce nitric oxide synthase activity in cortical neurones including in the hippocampus (Mannella et al. 2009).

2.2 *Progesterone*

The progesterone receptor also has two known subtypes namely PR-A and PR-B, the former being a truncated form of PR-B. PR-A dominantly represses the transcriptional activity mediated by PR-B (Giangrande and McDonnell 1999). These receptors are distributed in the human brain (Bezdickova et al. 2007) including in the hippocampus and frontal cortex (for review of form and function, see Brinton et al. 2008). Increased expression of PR-A can increase responsiveness to oestrogen via ERα (Mesiano 2001). Membrane receptors for progesterone, coupled to G-proteins, have also been identified that are associated with more rapid actions of the hormone (Thomas 2008; Dressing et al. 2011). Interestingly, unlike with PR-A and PR-B, many of the synthetic progestogens (see below) do not bind to the membrane receptor.

Progesterone, like oestrogen, has effects beyond those well recognised in the hypothalamus. Progesterone can also activate the MAPK and ERK signalling pathways which, can via CREB, lead to up-regulation of *bcl-2* in hippocampal neurones (Nilsen and Brinton 2002). Progesterone has been found to up-regulate BDNF in murine models of cerebral ischaemia (Coughlan et al. 2009) and may play a role in maintenance of the integrity of the blood–brain barrier (Ishrat et al. 2010).

A small clinical trial involving patients with traumatic brain injuries showed that the use of progesterone was not harmful and it may indeed have some beneficial effects (Wright et al. 2007). Phase III trials are now underway.

Selective agents for progesterone receptors have also been used experimentally to investigate further the role of progesterone. Therapeutic application of these agents will follow, and one agent, ulipristal acetate (ellaOne), was licenced in May 2009 in the UK for emergency contraception within 120 h (5 days) of unprotected intercourse or contraceptive failure. Ulipristal is an orally active synthetic selective progesterone receptor modulator (SPRM) with high affinity for PR-A. However, in animal studies it also has affinity for the glucocorticoid receptor, and in animals antiglucocorticoid effects have been seen. However, these antiglucocorticoid effects have not been seen in humans to-date and this may be indicative of real species differences.

A variety of synthetic progestogens are used in hormone replacement therapy and contraception, as progesterone itself has poor oral bioavailability. Not all progestogens have the same pharmacological profile (Hapgood et al. 2004), and these differences have implications for their usage. Two of the most widely used synthetic progestogens are medroxyprogesterone acetate and norethisterone. These are used as the progestogenic component of an HRT regimen in combination with oestrogen but have been shown to increase the risk of breast cancer in long-term HRT users (Million Women Study 2003; Women's Health Initiative 2002). Structurally, medroxyprogesterone acetate is more similar to natural progesterone than norethisterone. The metabolism of these two compounds is also different, as medroxyprogesterone acetate is the major progestogenic compound rather than its metabolites. In contrast, the metabolites of norethisterone (a first generation progestagen) exhibit significant activity in addition to a wide range of non-progestogenic actions. Norethisterone also binds to SHBG, whereas medroxyprogesterone acetate does not. The most notable difference in steroid receptor-binding affinity between the two synthetic progestogens and endogenous progesterone is that, although all the compounds have affinity for the mineralocorticoid receptor, only the natural compound has antagonist activity. MPA is the most potent of the three in terms of glucocorticoid activity and correspondingly it has the highest affinity for the glucocorticoid receptor, but this relationship is complicated further by the fact that the resultant effect appears to be dependent on glucocorticoid receptor density (Hapgood et al. 2004). Unlike endogenous progesterone, both MPA and the 19-nortestosterone derivatives also have affinity for the androgen receptor (but MPA has less intrinsic activity), although this is decreased in the third-generation compounds such as desogestrel. Such off-target pharmacology influences the side-effect profiles of these synthetic analogues.

In addition, natural progesterone and its metabolites allopregnanolone and pregnanolone have high affinity for $GABA_A$ receptors (Paul and Purdy 1992), and progesterone can decrease glutamic acid decarboxylase (GAD) and so attenuate GABA synthesis (Wallis and Luttge 1980). The GABAergic system can also be influenced by the androgen metabolite 5α-androstane-3α,17β-diol (3α-Adiol), which is a metabolite of the potent androgen dihydrotestosterone (DHT) which is formed from testosterone by the action of 5α-reductase, and can modulate $GABA_A$ receptors (Reddy 2004). 3α-Adiol is in turn metabolised by CYP7B1 which is expressed in the hippocampus, and this metabolic route may, therefore, indirectly influence $GABA_A$ pathways (Pettersson et al. 2009).

2.3 *Androgen*

As with oestrogen and progesterone, it is recognised that androgens may also contribute to CNS physiology and pathology outside the control of reproduction. The two most important androgens are testosterone which is transformed within the CNS by 5α-reductase to 5α-dihydrotestosterone (DHT) (testosterone can also undergo aromatisation to estradiol as discussed earlier), and they mediate their effects, like oestrogen and progesterone, via a ligand-activated transcription factor, namely the androgen receptor (AR). Much of the knowledge of the AR has been derived from work done in the field of prostate cancer (Powell et al. 2004). In the CNS, mapping of AR distribution has revealed that there is a high level of co-localisation with ERs (Patchev et al. 2004). Overlap also occurs with respect to AR expression as in, for example, the male rat forebrain oestrogens up-regulate expression. Androgen withdrawal (via castration) can increase expression in some brain areas (preoptic nucleus) but have no effect in others (amygdala); these effects are also influenced by age (Kumar and Thakur 2004), and that perinatal or even prenatal androgen exposure may be important (Goren and Kruijver 2002). Therefore, there appears to be age-dependent, hormone and tissue-specific control of the AR. The influence of androgens outside reproduction is less clear than for oestrogen and progesterone; however, high levels of AR mRNA have been detected in human hippocampus (Beyenberg et al. 2000).

There is evidence to suggest that testosterone and DHT can, like oestrogen, influence neuronal plasticity (Matsumoto and Prins 2002) and dendritic spine density (Leranth et al. 2004). Androgens may also reduce neuronal cell death after exposure to various insults, such as β-amyloid (Pike 2001) and oxidative stress (Ahlbom et al. 2001). These effects could be mediated via AR activation of MAPK/ERK signalling pathway (Nguyen et al. 2005). Androgens may also be involved in the regulation of several proteins involved in axonal regeneration such as neuritin and tubulin, and there is evidence that androgen action is potentiated by BDNF (Fargo et al. 2009). The complexities of these mechanisms may allow for selective activation of signalling pathways, and selective androgen receptor modulators (SARMs) are being developed to specifically manipulate certain effects; as yet these are limited to the periphery to improve outcomes in the management of prostatic disease, osteoporosis and muscle wasting (Gao and Dalton 2007).

3 Conclusion

The effects of the sex steroids in the brain, begin in utero (Swaab 2007), are complex, and the final response is dependent not only on receptor up- or down-regulation but also on a series of coordinated metabolic events and cross-talk between receptor signalling pathways. The importance of each step may vary depending on the surrounding milieu, the tissue, the species and the gender of the species. Gender differences are important and should not be discounted in the interest of sample homogeneity. Findings from research using males may not always be valid when extrapolated to cover the female population (Beery and Zucker 2011), and a case could be made for sex-specific medicines (Gillies and McArthur 2010). These complexities do pose challenges but also offer opportunity for novel approaches for therapeutic exploitation and the pharmacological tools with which to modulate systems are becoming available from SERMs to SERDs.

References

Aguirre C, Jayaraman A, Pike C, Baudry M (2010) Progesterone inhibits estrogen-mediated neuroprotection against excitotoxicity by down-regulating estrogen receptor-β. J Neurochem 115:1277–1287

Ahlbom E, Prins GS, Ceccatelli S (2001) Testosterone protects cerebellar granule cells from oxidative stress-induced cell death through a receptor mediated mechanism. Brain Res 892:255–262

Amantea D, Russo R, Bagetta G, Corasaniti MT (2005) From clinical evidence to molecular mechanisms underlying neuroprotection afforded by estrogens. Pharmacol Res 52:119–132

Arevalo MA, Santos-Galindo M, Lagunas N, Azcoitia I, Garcia-Segura LM (2011) Selective estrogen receptor modulators as brain therapeutic agents. J Mol Endocrinol 46:R1–R9

Balthazart J, Ball GF (2006) Is brain estradiol a hormone or a neurotransmitter? Trends Neurosci 29:241–249

Balthazart J, Baillien M, Charlier TD, Cornil CA, Ball GF (2003) Multiple mechanisms control brain aromatase activity at the genomic and non-genomic level. J Steroid Biochem Mol Biol 86:367–379

Baulieu EE, Robel P (1996) Dehydroepiandrosterone and dehydroepiandrosterone sulfate as neuroactive neurosteroids. J Endocrinol (Suppl)150:S221–S239

Beato M, Sanchez-Pacheco A (1996) Interaction of steroid hormone receptors with the transcription initiation complex. Endocr Rev 17:587–609

Beery AK, Zucker I (2011) Sex bias in neuroscience and biomedical research. Neurosci Biobehav Rev 35:565–572

Beyenberg S, Watzka M, Clusmann H, Blumcke I, Bidlingmaier F, Elger CE, Stoffel-Wagner B (2000) Androgen receptor mRNA expression in the human hippocampus. Neurosci Lett 94:25–28

Bezdickova M, Molikova R, Bebarova L, Kolar Z (2007) Distribution of nuclear receptor for steroid hormones in the human brain: a preliminary study. Biomedical Papers of the Medical Faculty of The University Palacky, Olomouc 151:69–71

Bodo C, Rissman EF (2006) New roles for estrogen receptor beta in behaviour and neuroendocrinology. Front Neuroendocrinol 27:217–232

Bowers JM, Waddell J, McCarthy MM (2010) A developmental sex difference in hippocampal neurogenesis is mediated by endogenous oestradiol. Biol Sex Differ 1:1–8

Brailoiu E, Dun SL, Brailoiu GC, Mizuo K, Sklar LA, Oprea TI, Prossnitz ER, Dun NJ (2007) Distribution and characterization of estrogen receptor G protein-coupled receptor 30 in the rat central nervous system. J Endocrinol 193:311–321

Brinton RD, Thompson RF, Foy MR, Baudry M, Wang J, Finch CE, Morgan TE, Pike CJ, Mack WJ, Stanczyk FZ, Nilsen J (2008) Progesterone receptors: form and function in brain. Front Neuroendocrinol 29:313–339

Brunton PJ, Russell JA (2010) Endocrine induced changes in brain function during pregnancy. Brain Res 1364:198–215

Caruso D, Pesaresi M, Maschi O, Garcia-Segura LM, Melcangi RC (2010) Effect of short- and long-term gonadectomy on neuroactive steroid levels in the central and peripheral nervous system of male and female rats. J Neuroendocrinol 22:1137–1147

Chabbert Buffet N, Djakoure C, Maitre SC, Bouchard P (1998) Regulation of the human menstrual cycle. Front Neuroendocrinol 19:151–186

Coughlan T, Gibson C, Murphy S (2009) Progesterone, BDNF and neuroprotection in the injured CNS. Int J Neurosci 119:1718–1740

Damdimopoulos AE, Spyrou G, Gustafsson JA (2000) Ligands differentially modify the nuclear ability of estrogen receptors alpha and beta. Endocrinology 149:339–345

Delaunay F, Pettersson K, Tujague M, Gustafsson JA (2000) Functional differences between thye amino-terminals of estrogen receptors alpha and beta. Mol Pharmacol 58:584–590

Dressing GE, Goldberg JE, Charles NJ, Schwertfeger KL, Lange CA (2011) Membrane progesterone receptor expression in mammalian tissues: A review of regulation and physiological implications. Steroids 76:11–17

Fargo KN, Foecking EM, Jones KJ, Sengelaub DR (2009) Neuroprotective actions of androgens on motoneurons. Front Neuroendocrinol 30:130–141

Freedman LP (1999) Multimeric coactivator complexes for steroid/nuclear receptors. Trends Endocrinol Metab 10:403–407

Furr BJA (2006) Aromatase inhibitors. Birkhäuser Verlag, Basel-Boston-Berlin

Gao W, Dalton JT (2007) Expanding the therapeutic use of androgens via selective androgen receptor modulators (SARMs). Drug Discov Today 12:241–248

Garcia-Segura LM, Melcangi RC (2006) Steroids and glial cell function. Glia 54:485–498

Garcia-Segura LM, Veiga S, Sierra A, Melcangi RC, Azcoitia I (2003) Aromatase: a neuroprotective enzyme. Prog Neurobiol 71:31–41

Genazzani AR, Monteleone P, Gambacciani M (2002) Hormonal influence on the central nervous system. Maturitas 43:S11–S17

Giangrande PH, McDonnell DP (1999) The A and B isoforms of the human progesterone receptor: two functionally different transcription factors encodes by a single gene. Recent Prog Horm Res 54:291–313

Gilep AA, Sushko TA, Usanov SA (2011) At the cross-roads of steroid hormone biosynthesis: the role, substrate specificity and evolutionary development of CYP17. Biochim Biophys Acta 1814.200–209

Gillies GE, McArthur S (2010) Estrogen actions in the brain and the basis for differential action in men and women: a case for sex-specific medicines. Pharmacol Rev 62:155–198

Goren LJG, Kruijver FPM (2002) Androgens and male behavior. Mol Cell Endocrinol 198:31–40

Green S, Walter P, Kumar V, Krust A, Bornert JM, Argos P, Chambon P (1986) Human oestrogen receptor cDNA: sequence, expression and homology to v-erb-A. Nature 320:134–139

Gustafson DR, Karlsson C, Skoog I, Rosengren L, Lissner L, Blennow K (2007) Mid-life adiposity factors relate to blood-brain barrier integrity in late life. J Intern Med 262:643–650

Hapgood JP, Koubovec D, Louw A, Africander D (2004) Not all progestins are the same: implications for usage. Trends Pharmacol Sci 25:554–557

Harrison PJ, Tunbridge EM (2008) Catechol-*O*-Methyltransferase (COMT): a gene contributing to sex differences in brain function, and to sexual dimorphism in the predisposition to psychiatric disorders. Neuropsychopharmacology 33:3037–3045

Ishrat T, Sayeed I, Atif F, Hua F, Stein DG (2010) Progesterone and allopregnanolone attenuate blood-brain barrier dysfunction following permanent focal ischemia by regulating the expression of matrix metalloproteinases. Exp Neurol 226:183–190

Keenan DM, Veldhuis JD (2009) Age-dependent regression analysis of male gonadal axis. Am J Physiol Regul Integr Comp Physiol 297:R1215–R1227

Kuiper GG, Enmark E, Pelto-Huikko M, Nilsson S, Gustafsson JA (1996) Cloning of a novel receptor expressed in rat prostate and ovary. Proc Natl Acad Sci USA 93:5925–5930

Kumar RC, Thakur MK (2004) Androgen receptor mRNA is inversely regulated by testosterone and estradiol in adult mouse brain. Neurobiol Aging 25:925–933
Lappano R, Rosano C, De Marco P, De Francesco EM, Pezzi V, Maggiolini M (2010) Estriol acts as a GPR30 antagonist in estrogen receptor-negative breast cancer cells. Mol Cell Endocrinol 320:162–170
Lephart E (1996) A review of brain aromatase cytochrome P450. Brain Res Rev 22:1–26
Leranth C, Hajszan T, MacLusky NJ (2004) Androgens increase spine synapse density in the CA1 hippocampal subfield of ovariectomized female rats. J Neurosci 24:485–499
Levin-Allerhand JA, Lominska CE, Wang J, Smith JD (2002) 17Alpha-estradiol and 17beta-estradiol treatment are effective in lowering cerebral amyloid-beta levels in AbetaPPSWE transgenic mice. J Alzheimers Dis 4:449–457
Liehr JG, Roy D (1990) Free radical generation by redox cycling of estrogens. Free Radic Biol Med 8:415–423
Lonard DM, Lanz RB, O'Malley BW (2007) Nuclear receptor coregulators and human disease. Endocr Rev 28:575–587
Lubbers LS, Zafian PT, Gautreaux C, Gordon M, Alves SE, Correa L, Lorrain DS, Hickey GJ, Luine V (2010) Estrogen receptor (ER) subtype agonists alter monoamine levels in the female rat brain. J Steroid Biochem Mol Biol 122:310–317
Maggiolini M, Picard D (2010) The unfolding stories of GPR30, a new membrane-bound estrogen receptor. J Endocrinol 204:105–114
Mangelsdorf DJ, Thummel C, Beato M, Herrlich P, Schutz G, Umesono K, Blumberg B, Mark M, Chambon P, Evans RM (1995) The nuclear receptor superfamily: the second decade. Cell 83:835–839
Mannella P, Sanchez AM, Giretti MS, Genazzani AR, Simoncini T (2009) Oestrogen and progestins differently prevent glutamate toxicity in cortical neurons depending on prior hormonal exposure via the induction of neural nitric oxide synthase. Steroids 74:650–656
Matsumoto A, Prins G (2002) Androgenic regulation of expression of androgen receptor protein in peripheral motoneurons of aged male rats. J Comp Neurol 443:383–387
Mesiano S (2001) Roles of estrogen and progesterone in human parturition. Front Horm Res 27:86–104
Million Women Study Collaborators (2003) Breast cancer and hormone replacement therapy in the Million Women Study. Lancet 362:419–427
Mooradian AD, Morley JE, Korenman SG (1987) Biological actions of androgens. Endocr Rev 8:1–28
Murphy DD, Segal M (2000) Progesterone prevents estradiol-induced dendritic spine density formation in hippocamp0la neurons. Neuroendocrinology 72:133–143
Nakhla AM, Hryb DJ, Hryb DJ, Rosner W, Romas NA, Xiang Z, Kahn SM (2009) Human sex hormone-binding globulin gene expression- multiple promoters and complex alternative splicing. BMC Mol Biol 10:37–55
Nguyen T-V, Yao M, Pike CJ (2005) Androgens activate mitogen-activated protein kinase signaling: Role in neuroprotection. J Neurochem 94:1639–1651
Nguyen HP, Li L, Gatson JW, Maass D, Wigginton JG, Simpkins JW, Schug KA (2011) Simultaneous quantification of four native estrogen hormones at trace levels in human cerebrospinal fluid using liquid chromatography–tandem mass spectrometry. J Pharm Biomed Anal 54:830–837
Nilsen J, Brinton RD (2002) Impact of progestins on estrogen-induced neuroprotection: synergy by progesterone and 19-norprogesterone and antagonism by medroxyprogesterone acetate. Endocrinology 143:205–212
O'Malley BW (2006) Little molecules with big goals. Science 313:1749–1750
Österlund MK, Hurd YL (2001) Estrogen receptors in the human forebrain and the relation to neuropsychiatric disorders. Prog Neurobiol 64(3):251–267
Patchev VK, Schroeder J, Goetz F, Rohde W, Patchev AV (2004) Neurotropic action of androgens: principles, mechanisms and novel targets. Exp Gerontol 39:1651–1660
Paul SM, Purdy RH (1992) Neuroactive steroids. FASEB J 6:2311–2322

Pettersson H, Lundqvist J, Oliw E, Norlin M (2009) CYP7B1-mediated metabolism of 5α-androstane-3α,17β-diol (3α-Adiol):A novel pathway for potential regulation of the cellular levels of androgensand neurosteroids. Biochim Biophys Acta 1791:1206–1215

Picazo O, Azcoitia I, Garcia-Segura LM (2003) Neuroprotective and neurotoxic effect of estrogens. Brain Res 990:220–227

Pike CJ (2001) Testosterone attenuates β-amyloid toxicity in cultured hippocampal neurons. Brain Res 919:160–165

Pilgrim C, Hutchison JB (1994) Developmental Regulation of Sex Differences in the Brain: Can the role of the gonadal steroids be redefined. Neuroscience 604:843–855

Powell SM, Christiaens V, Voulgaraki D, Waxman J, Claessens F, Bevan CL (2004) Mechanisms of androgen receptor signalling via steroid receptor coactivator-1 in prostate. Endocr Relat-Cancer 11:117–130

Prokal L, Simpkins JW (2007) Structure-non-genomic neuroprotection relationship of estrogens and estrogen-derived compounds. Pharmacol Ther 114:1–12

Reddy DS (2004) Testosterone modulation of seizure susceptibility is mediated by neurosteroids 3α androstanediol and 17β-estradiol. Neuroscience 129:195

Reyes-Romero MA (2001) The physiological role of estriol during human fetal development is to act as antioxidant at lipophilic milieus of the central nervous system. Med Hypotheses 56:107–109

Rohr UW (2002) The impact of testosterone imbalance on depression and women's health. Maturitas 41:S25–S46

Roselli CE, Resko JA (1993) Aromatase activity in the brain:hormonal regulation and sex differences. J Steroid Biochem Mol Biol 61:365–374

Simpkins JW, Rajakumar G, Zhang YQ, Simpkins CE, Grenwald D, Yu CJ, Bodor N, Day LJ (1997) Estrogens may reduce mortality and ischemic damage caused by middle cerebral artery occlusion in the female rat. Neurosurgery 87:724–730

Steckelbroeck S, Watzka M, Reichelt R, Hans VHJ, Stoffel-Wagner B, Heidrich DD, Schramm J, Bidlingmaier F, Klingmüller D (2001) Characterization of the 5α-reductase-3α-hydroxysteroid dehydrogenase complex in the human brain. J Clin Endocrinol Metab 86:1324–1331

Swaab DF (2007) Sexual differentiation of the brain and behavior. Best Practice & Research Clinical Endocrinology and Metabolism 21:431–444

Thomas P (2008) Characteristics of membrane progestin receptor alpha (mPRalpha) and progesterone membrane receptor component 1 (PGMRC1) and their roles in mediating rapid progestin actions. Front Neuroendocrinol 29:292–312

Toran-Allerand CD (2004) Estrogen and the brain: beyond ER-alpha and ER-beta. Exp Gerontol 39:1579–1586

Tsai MJ, O'Malley BW (1994) Molecular mechanisms of action of steroid/thyroid receptor superfamily members. Annu Rev Biochem 63:451–486

Tunbridge EM, Harrison PJ (2011) Importance of the COMT gene for sex differences in brain function and predisposition to psychiatric disorders. Curr Topics Behav Neurosci DOI: 10.1007/7854_2010_97.

Wallis CJ, Luttge WG (1980) Influence of estrogen and progesterone on gluatamic acid decarboxylase activity in discrete regions of rat brain. J Neurochem 34:609–613

Weiser MJ, Foradori CD, Handa RJ (2008) Estrogen receptor beta in the brain: from form to function. Brain Res Rev 57:309–320

WHI Investigators (2003) Influence of Estrogen Plus Progestin on Breast Cancer and Mammography in Healthy Postmenopausal Women. JAMA 289(24):3243–3253

Wright DW, Kellermann AL, Hertzberg VS, Clark PL, Frankel M, Goldstein FC, Salomone JP, Dent LL, Harris OA, Ander DS, Lowery DW, Patel MM, Denson DD, Gordon AB, Wald MM, Gupta S, Hoffman SW, Stein DG (2007) ProTECT: a randomized clinical trial of progesterone for acute traumatic brain injury. Ann Emerg Med 49:391–402

Yore MA, Im D, Webb LK, Zhao Y, Chadwick JG, Molenda-Figueira HA, Haidacher SJ, Denner L, Tetel MJ (2010) Steroid receptor coactivator-2 expression in brain and physical associations with steroid receptors. Neuroscience 169:1017–1028

Zwain IH, Yen SSC (1999) Dehydroepiandrosterone: biosynthesis and metabolism in the brain. Endocrinology 140:880–887

Differences in Sexual Behaviour in Male and Female Rodents: Role of Serotonin

Berend Olivier, Johnny S.W. Chan, Eelke M. Snoeren, Jocelien D.A. Olivier, Jan G. Veening, Christiaan H. Vinkers, Marcel D. Waldinger, and Ronald S. Oosting

Contents

Abstract Serotonin plays an important role in both male and female sexual behaviour. In general, reduction of 5-HT function facilitates, whereas enhancement inhibits sexual behaviour. Most fundamental research on the involvement of 5-HT in sex has been performed in rats. Selective serotonin reuptake inhibitors (SSRIs)

B. Olivier (✉), J.S.W. Chan, E.M. Snoeren, J.G. Veening, C.H. Vinkers, M.D. Waldinger, and R.S. Oosting
Division of Pharmacology, Utrecht Institute for Pharmaceutical Sciences and Rudolf Magnus Institute of Neuroscience, Utrecht University, Sorbonnelaan 16, 3584 CA Utrecht, The Netherlands
e-mail: b.olivier@uu.nl; j.s.w.chan@uu.nl; j.g.veening@uu.nl; c.h.vinkers@uu.nl; md@demon.waldinger.nl; r.s.oosting@uu.nl

J.D.A. Olivier
Division of Psychiatry Dept. Clinical Neuroscience, Karolinska Institutet, KFC Novum Level 6, Exp 617 Lab 614, SE - 14157 Huddinge, Sweden
e-mail: jocelien.olivier@ki.se

J.C. Neill and J. Kulkarni (eds.), *Biological Basis of Sex Differences in Psychopharmacology*, Current Topics in Behavioral Neurosciences 8, DOI 10.1007/7854_2010_116 have comparable effects on male and female sexual behaviour in rats; they inhibit it but only after chronic administration. Activation of the 5-HT_{1A} receptor facilitates sexual behaviour in male rats but inhibits sexual behaviour in female rats, suggesting a differential role for 5-HT_{1A} receptors in male and female rats. Research on sexual behaviour in rats with null mutations in the serotonin transporter (SERT) indicated also a differential role for 5-HT_{1A} receptors in male and female sexual behaviour. Evidence exists that different pools of 5-HT_{1A} receptors have differential roles in various parts of the cascade of sexual events occurring during sexual interactions. Roles for other 5-HT receptors are less well defined although 5-HT_{1B}, 5-$HT_{2A/B}$ and 5-HT_7 receptors seem to be involved. Identification of putative differential or comparable roles in female and male sexual activities requires more research.

Keywords 5-HT · 5-HT_{1A} receptor · 5-HTT · 8-OH-DPAT · Gender · Hypersexual behaviour · Hyposexual behaviour · Paroxetine · Premature ejaculation · Retarded ejaculation · Serotonin · Serotonin receptor knockout rat · Serotonin transporter · SERT polymorphism · Sexual behaviour · SSRI · WAY100635

1 Introduction

Sexual behaviour in rodents (and we strictly focus on the rat) happens when animals reach adulthood and engage in behaviours that result in the joining of a male and female, ending in copulation, with the intent to reproduce. The female rat's sexual behaviour is dependent on the reproductive cycle, whereas the male's sexual behaviour is not. The female's sexual behaviour is strongly dependent on peripheral gonadal steroids that have both peripheral and central nervous system (CNS) effects. Steroids act on the brain to induce sexual receptivity and all associated behaviours (proceptive, receptive and pacing behaviours). Quite some work has been performed to delineate the neural circuitry and neurochemistry of female behaviour, especially from lordosis, a behaviour that is evoked by external stimuli, normally, a male rat. Lordosis behaviour is only observed when the female is hormonally (or naturally) primed (oestradiol + progesterone) and the circuitry involves sensory, brain, spinal cord and motoric activation. In the CNS, the ventromedial nucleus of the hypothalamus (VMH), the preoptic area (POA), the midbrain central gray (MCG) and two areas in the spinal cord (cervical and lumbar) are the key structures. All structures contain oestrogen receptors that seem essential for the final integrative performance of full sexual behaviour. Many neurotransmitter systems in the CNS regulate or modulate (aspects of) sexual behaviour, including serotonin. There is strong evidence that the serotonergic modulation of sexual behaviours mainly occurs at the level of the VMH and POA.

In male rodents (rat), testosterone (T) acts during development to promote genital development and organization of the CNS neural circuitry. In adulthood, the neural circuitry along with the appropriate sensory and motoric systems controls the male's sexual motivation and performance. Male rat's sexual behaviour includes penile erection, sexual motivation and mating behaviour. All can be studied while observing the mating behaviour of a male rat in direct interaction with a receptive female. In such an interaction, the male approaches the female, sniffs her and starts mounting (the female displays lordosis). The male displays a series of mounts and intromissions that end in ejaculation. After an ejaculation, the male displays for some time sexual quiescence followed by the next series of mounts and intromissions, leading again to ejaculation, and so on. In the male rat, the testicular secretion of T occurs throughout the year, although pulsatile patterning occurs over the day. Seasonal variations in behavioural responsiveness to T of male rats have not been found, making the male (and female) rat ideal experimental animals to study the neural mechanisms of, and neurotransmitter involvement in, sexual behaviour.

The neural systems involved in male sexual behaviour seem to involve many structures that are also involved in female sexual behaviour, although clear differences are also notable. The POA and the bed nucleus of the stria terminalis (BNST) are core structures via which T acts to activate male sexual behaviour. In particular, the POA seems an integrative structure in coordinating the actions of T on both motivational and consummatory aspects of male sexual behaviour. Several neurochemical systems, including peptidergic, dopaminergic and serotonergic systems, play a role in mediating sexual behaviour.

2 Serotonin and Sexual Behaviour

The focus of this chapter is the role of serotonin in male and female sexual behaviour in the rat. There is hardly any research performed on gender differences in the development and adult functioning of the 5-HT system in the brain and spinal cord. Seeing the overlap, but also the divergence of the various neural structures and hormonal receptor systems in the male and female rat CNS, it may be difficult to predict the effects of psychopharmacological treatment with serotonergic ligands on male and female sexual behaviour.

Serotonergic psychopharmacology in humans is rather limited, only the selective serotonin reuptake inhibitors (SSRI) are selective serotonergic drugs extensively used in patients, whereas most other drugs with some serotonergic profile exert inherently other mechanisms like dopamine D_2 receptor antagonism (olanzapine, risperidone, buspirone). In the latter case, it is often impossible to purely deduct the specific contribution of the serotonergic component on the putative effects on sexual behaviour or sexual dysfunctions induced. SSRIs are widely used to treat depression both in human males and females and are notoriously implicated (Zemishlany and Weizman 2008) in inducing sexual disturbances (Kennedy and Rizvi 2009; Balon 2006). However, a complicating factor is that major depression per se is often (if not always) associated with sexual disturbances (e.g. in libido, motivation, erection: Kendurkar and Kaur 2008; Kennedy and Rizvi 2009). SSRIs enhance serotonergic neurotransmission which is generally believed to inhibit sexual behaviour, both in males and females (Zemishlany and Weizman 2008; Williams et al. 2006; Kennedy and Rizvi 2009; Kendurkar and Kaur 2008). This is confirmed by various studies showing that SSRI antidepressants induce sexual disturbances, in addition to already present dysfunctions due to the underlying depression, in both males and females (Cyranowski et al. 2004; Regitz-Zagrosek et al. 2008). No studies in humans have looked into the brain mechanisms underlying the SSRI-induced sexual dysfunction and putative gender differences. While it is still assumed that high extracellular 5-HT levels (e.g. after SSRI treatment) are needed to promote antidepressant activity, the disadvantage is the directly associated decrease in sexual behaviours. The emerging pattern seems to indicate that SSRIs, which enhance serotonergic neurotransmission in the brain, have similar inhibitory effects in human males and females. In line with the latter notion is the finding (Sugden et al. 2009) that gene expression for 5 serotonergic genes (including 5-HTT) did not differ between genders in postmortem human brains.

3 Serotonin, Serotonergic Receptors and Male Sexual Behaviour

3.1 Introduction

The importance of 5-HT in male sexual behaviour has been demonstrated by numerous studies showing that, for instance, lesions of the brainstem raphé nuclei (Albinsson et al. 1996) and 5-HT depletion (Tagliamonte et al. 1969) facilitate sexual behaviour. On the other hand, administration of 5-hydroxytryptophan, the direct precursor of 5-HT, 5-HT itself and 5-HT releasers such as MDMA and fenfluramine, inhibits sexual behaviour (Ahlenius et al. 1980; Dornan et al. 1991; Foreman et al. 1992; Gonzales et al. 1982). Altogether these findings suggest that a decrease in 5-HT neurotransmission may be involved in facilitation, whereas an increase in 5-HT neurotransmission may result in inhibition of male sexual behaviour.

3.2 SSRIs and Male Sexual Behaviour

The frequently reported sexual effects of SSRIs in men demonstrate an important role of 5-HT in human ejaculatory behaviour. In several human studies we and others have demonstrated that SSRIs including paroxetine, sertraline and fluoxetine are able to delay ejaculation in premature ejaculation (for review, see Waldinger 2002; De Jong et al. 2006). Moreover, these studies show that SSRIs exert only a minimal ejaculation delay in the first week that is often not clinically relevant. A clinically relevant ejaculation delay occurs gradually after 2–3 weeks of daily treatment. Interestingly, despite the putative similar underlying mechanism of action of SSRIs – briefly, preventing the reuptake of 5-HT, thereby elevating 5-HT levels – not all SSRIs delay ejaculation to the same extent. In humans, the tricyclic antidepressant, clomipramine and the SSRI, paroxetine have stronger ejaculation-delaying effects after 4–6 weeks of daily treatment than other SSRIs (Waldinger et al. 1998, 2001a, b).

3.3 *Acute and Chronic SSRI Administration in Male Rats*

Analogous to the human situation, in male rats a distinction can be made between the effects of acute and chronic SSRI administration on ejaculation. Acute administration of various SSRIs, such as citalopram, paroxetine, sertraline, fluoxetine and fluvoxamine, did not or marginally delay ejaculation (Mos et al. 1999; Ahlenius and Larsson 1999; Matuszcyk et al. 1998). On the other hand, chronic administration of fluoxetine (Matuszcyk et al. 1998; Cantor et al. 1999; Frank et al. 2000) and paroxetine (Waldinger et al. 2001a, b) delayed ejaculation in male rats. Nonetheless, as in humans, not all SSRIs potently delay ejaculation after chronic administration in male rats: fluvoxamine slightly affected some aspects of copulatory behaviour, but did not affect ejaculation (Waldinger et al. 2001a, b; De Jong et al. 2005a). It is unclear why the various SSRIs differ in their ability to delay ejaculation after chronic administration. The delay in onset of the therapeutic effect of SSRIs in depression and anxiety disorders has been related to adaptive changes of serotonergic autoreceptors (Haddjeri et al. 1998; Le Poul et al. 2000), and it is conceivable that the ejaculation-delaying effects of various SSRIs are due to differential adaptive changes of 5-HT receptors.

An example of the effects of an SSRI antidepressant (paroxetine) in male rat sexual behaviour is shown in Fig. 1. The effect is clearly seen in the number of ejaculations per 30-min test in sexually trained animals. Acutely (Day 1: 30 min after injection) paroxetine does not inhibit sexual behaviour whereas after 7 (subchronic; 5 and 10 mg/kg) or 14 days treatment (chronic; 2.5, 5.0 and 10.0 mg/kg)

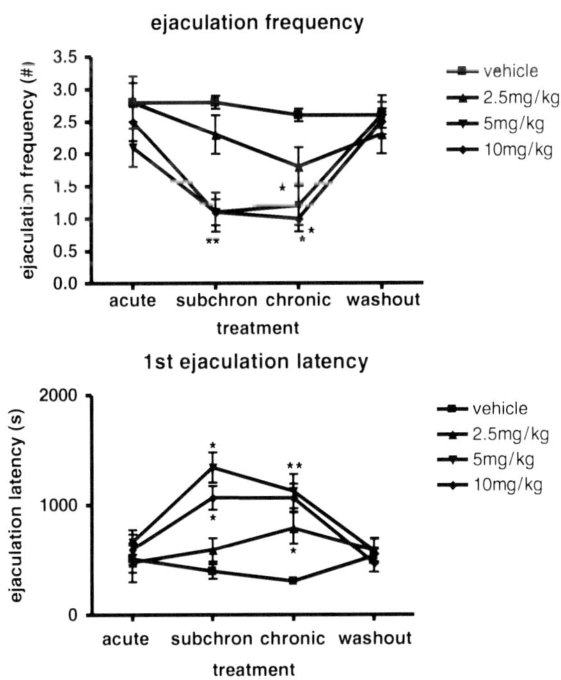

Fig. 1 The mean number of ejaculations ± SEM of male rat groups treated with vehicle or different doses of the SSRI paroxetine (2.5, 5.0 and 10.0 mg/kg IP) is given after acute (30 min pre-treatment), sub-chronic (7 days; once daily) and chronic (14 days: once daily) treatment. One week after cessation of treatment (washout), sexual behaviour was again measured but now without any treatment. Sexual behaviour tests were run on days 1, 7, 14 and 21 and consisted of a 30-min test in which a male rat had free access to a female that was hormonally brought into oestrus (method: Chan et al. 2010). $*p < 0.05$ compared to vehicle

paroxetine strongly (and dose dependently) reduces sexual behaviour. The effect is reversible as animals return to their pre-testing level 1 week after cessation of treatment. A similar picture emerges for the first ejaculation latency that is not affected acutely, but is dose-dependently enhanced after 7 days and 14 days of treatment, and returns to baseline 1 week after cessation of treatment.

Ahlenius and Larsson (1999) have studied the mechanism of SSRI-induced delay of ejaculation in more detail and showed that acute treatment with citalopram did not affect ejaculatory behaviour. Co-administration of the 5-HT_{1A} receptor antagonist WAY-100635 with citalopram strongly delayed ejaculation latencies, suggesting 5-HT_{1A} receptor involvement in the effect of citalopram on ejaculation. De Jong et al. (2005a, b) also showed that citalopram, acutely or chronically, while not inhibiting sexual behaviour itself, when combined with a sexually inactive dose of WAY100635 completely abolished sexual behaviour.

We studied this phenomenon further and confirmed earlier findings (Looney et al. 2005) that a dose as low as 0.01 mg/kg of WAY100635 facilitated the behaviourally inactive acute 10 mg/kg paroxetine dose and led to strong inhibition of male sexual behaviour (Fig. 2). The data suggest that the inhibitory action of SSRIs after (sub) chronic treatment are related to changes at certain 5-HT_{1A} receptors after long-term treatment.

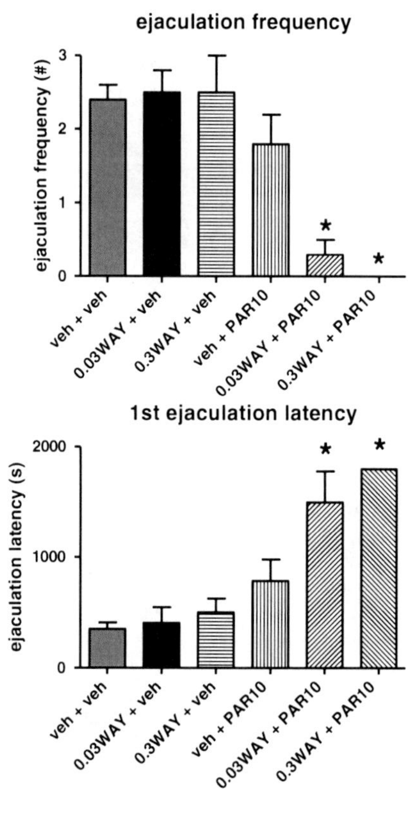

Fig. 2 Sexually trained male rats were acutely injected with saline or 10 mg/kg paroxetine (IP; 30 min before testing) immediately followed by an injection of either saline or a dose (0.03 and 0.3 mg/kg IP) of the 5-HT_{1A} receptor antagonist WAY100635. During an ensuing sexual behaviour test of 30 min, the sexual behaviour of the male was scored. In the figure, the mean number of ejaculations $\pm$ SEM is given. *PAR* paroxetine, *WAY* WAY100635, *VEH* vehicle. $*p < 0.05$ compared to vehicle

Subsequently, it was found that the ejaculation-delaying effects of the combination of citalopram and WAY100635 could be fully blocked by a selective 5-HT_{1B} receptor antagonist, suggesting a role for this receptor subtype in the delay of ejaculation (Hillegaart and Ahlenius 1998). Interestingly, a previous study from the same laboratory also suggested a role of the 5-HT_{1B} receptor in the delay of ejaculation. In this study, it was shown that the 5-HT_{1B} receptor agonist anpirtoline dose-dependently delayed ejaculation in rats (Hillegaart and Ahlenius 1998).

3.4 SERT-KO Rats and Male Sexual Behaviour

In humans, the SERT plays a prominent role in the homeostasis of serotonergic neurotransmission. Polymorphisms in the promoter region of the SERT influence the activity of SERT, and the two length alleles (S and L allele) have functional consequences for the function of the 5-HT system (Murphy and Lesch et al. 2008). L and S (LL > LS > SS) generate allele-dependent 5-HT activity with associated functional consequences (Lesch et al. 2008). Rats do not possess such promoter length polymorphisms but genetic knockout of the SERT gene might generate rat models of the S-allele versions of the human SERT. Therefore, $SERT^{-/-}$ and $SERT^{+/-}$ can be compared to wild-type ($SERT^{+/+}$) male rats and their sexual behaviour studied (Chan et al. 2011). It was expected, in analogy to treatment with chronic SSRI treatment, that $SERT^{-/-}$ and $SERT^{+/-}$ rats would display a lowered sexual behaviour compared to $SERT^{+/+}$ rats.

All rats (30 per genotype) were trained up to seven times (once weekly a test of 30 min) and gene knockout rats indeed showed lower sexual performance than wild-type rats. On average the mean number of ejaculations at week 7 was 1.6 for $SERT^{+/+}$, 1.1 for $SERT^{+/-}$ and 0.7 for $SERT^{-/-}$ rats (Fig. 3), a significant decrease

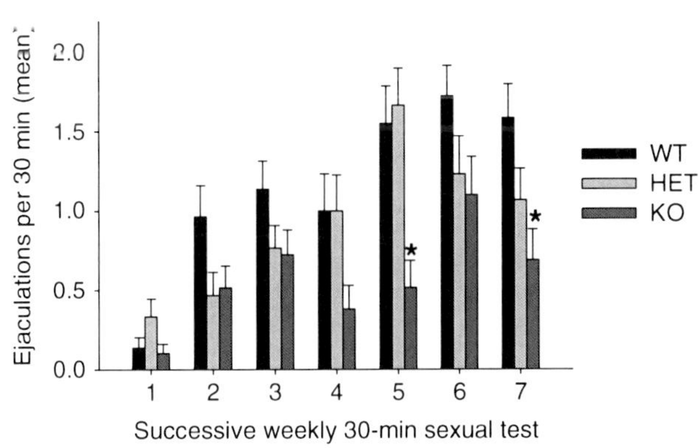

Fig. 3 Development of sexual behaviour (mean number of ejaculations/test) in male wild-type ($SERT^{+/+}$, WT), heterozygous ($SERT^{+/-}$, HET) and homozygous ($SERT^{-/-}$, KO) rats tested weekly over 7 weeks in a sexual behaviour test of 30 min with an oestrus female. *$p < 0.05$ compared to WT animals

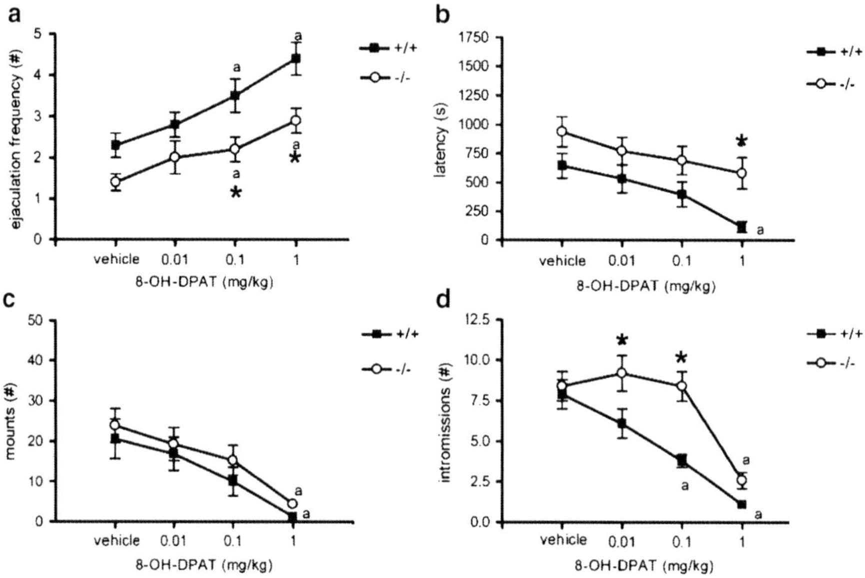

Fig. 4 Effects of the 5-HT_{1A} receptor agonist 8-OH-DPAT (s.c.) on ejaculation frequency over a 30-min test (**a**), latency to first ejaculation (**b**), first ejaculatory series mounts (**c**) and first ejaculatory series intromissions (**d**) of $SERT^{+/+}$ (+/+) and $SERT^{-/-}$ (−/−) animals. *$p < 0.05$ compared to wild type (+/+); *a*: $p < 0.05$ compared to vehicle treatment

for the homozygote gene knockout. The heterozygote KO was not different from the wild type.

Next, the 5-$HT_{1A/7}$ receptor agonist +/−8-OH-DPAT was tested. 5-HT_{1A} stimulation has pro-sexual activities in rats which also occur in the three genotypes. Although the basal level of sexual behaviour (number of ejaculations, ejaculation latency, post-ejaculatory latency) in the $SERT^{-/-}$ is lower than in the other two genotypes (Fig. 4), the stimulant effect of 8-OH-DPAT in all three genotypes is similar, indicating that 5-HT_{1A} receptors mediating this effect have not changed [(de)sensitized].

The 5-HT_{1A} receptor antagonist WAY100635 had no effects in the WT and heterozygote rats but had a dose-dependent inhibitory effect in the SERT-KO (Fig. 5), suggesting that a different pool of 5-HT_{1A} receptors is involved in its action and that these receptors appear sensitized in the SERT-KO. Remarkably, the heterozygous $SERT^{+/-}$ rats did in no way differ from the WT rats. Heterozygous SERT-KO rats have intermediate enhanced extracellular 5-HT levels compared to WT and SERT-KO ($SERT^{-/-} > SERT^{+/-} > SERT^{+/+}$). Apparently, like the effective dose of SSRIs that need to occupy at least 80% of the SERTs before antidepressant efficacy is observed (Kugaya et al. 2003), the $SERT^{+/-}$ still has sufficient SERT capacity (50%) left to show undisturbed sexual behaviour.

To summarize, the sexual side effects of SSRIs are still not fully understood. Nevertheless, some recent findings and genetic evidence suggest that adaptive changes in the 5-HT system and probably its interactions with neuroendocrine systems (De Jong et al. 2007) may be responsible for their sexual effects.

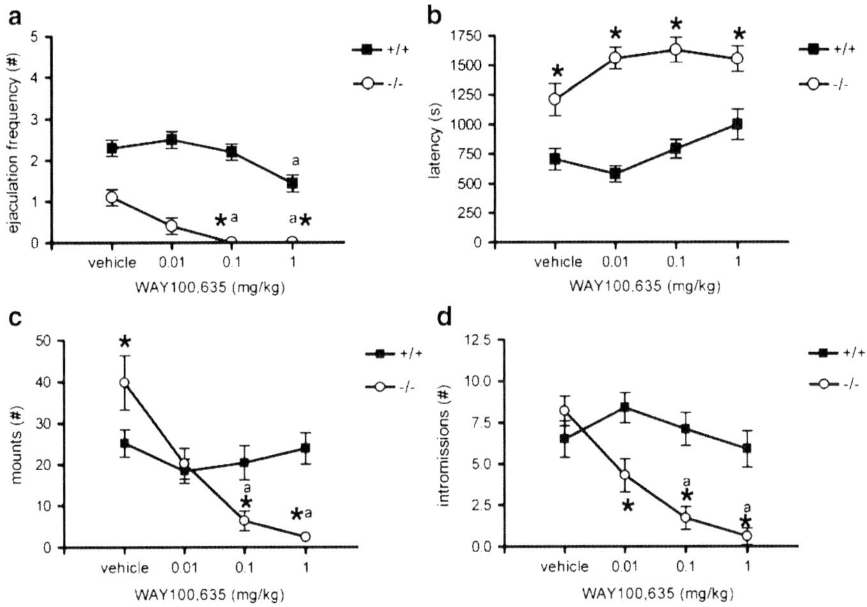

Fig. 5 Effects of WAY100635 (IP) on ejaculation frequency over 30-min test (**a**), latency to first ejaculation (**b**), first ejaculatory series mounts (**c**) and first ejaculatory series intromissions (**d**) of wild-type (+/+) and serotonin transporter knockout (−/−) animals. *$p < 0.05$ compared to wild type (+/+); *a*: $p < 0.05$ compared to vehicle treatment

3.5 *Serotonin Receptor Agonists and Antagonists and Ejaculation in Male Rats*

As described above, activation of 5-HT_{1B} receptors has been associated with delaying ejaculation in male rats. 5-HT_2 receptors are also implicated in modulation of sexual activity, e.g. shown by the 5-$HT_{2A/2C}$ receptor agonist DOI-induced inhibition of sexual behaviour (Klint and Larsson 1995). On the other hand, several other studies have shown that 5-$HT_{2A/2C}$ receptor agonists generally inhibit sexual behaviour by decreasing the number of animals that initiated copulation, but do not affect ejaculation latencies in animals that do initiate copulation (Ahlenius and Larsson 1998; Klint et al. 1992; Watson and Gorzalka 1991). Thus, it appears that 5-HT_2 receptors in general inhibit sexual behaviour, but their precise role in the regulation of ejaculation is not entirely clear.

A facilitatory role on ejaculation has been ascribed to activation of 5-HT_{1A} receptors, and various selective agonists for this receptor, such as 8-OH-DPAT (Ahlenius and Larsson 1990), FG-5893 (Andersson and Larsson 1994) and flesinoxan (Haensel and Slob 1997; Mos et al. 1991), potently facilitate sexual behaviour and decrease ejaculation latencies. Nevertheless, the underlying mechanisms of the facilitatory effects of 5-HT_{1A} receptor agonists are still unclear. A possibility for the mechanism of action may be activation of presynaptic 5-HT_{1A} receptors that

Table 1 Mean number of ejaculations, mounts and intromissions and ejaculation latency (in seconds) for sexually naïve male rats during a 15-min test with a sexually active, oestrus female

Drug (route)	Dose (mg/kg)	EF	MF	IF	EL (s)
8-OH-DPAT (SC)	0	0.1	10.5	8.3	869
	0.1	1.5[a]	6.5[a]	7.6	351[a]
	0.2	1.9[a]	3.5[a]	5.1	187[a]
	0.4	1.7[a]	1.1[a]	1.1[a]	238[a]
Flesinoxan (IP)	0	0.3	13.9	12.2	854
	0.1	1.0[a]	7.9	13.4	636[a]
	0.3	1.3[a]	5.5[a]	9.8	459[a]
	1.0	1.8[a]	3.3[a]	8.2	281[a]
Buspirone (IP)	0	0.3	10.3	9.9	860
	3.0	1.2[a]	7.0	11.3	502[a]
	10.0	0.1	0.3[a]	1.8[a]	849
Ipsapirone (IP)	3.0	0.9([a])	7.9	11.3	502[a]
	10.0	1.5[a]	10.9	12.2	636[a]

All data are depicted as means
EF ejaculation frequency, *MF* mount frequency, *IF* intromission frequency, *EL* ejaculation latency
[a]Significantly different ($p < 0.05$) from the corresponding vehicle (0 mg/kg) dose

will lead to an inhibition of 5-HT neuronal firing and consequently results in facilitation of sexual behaviour as described above. Alternatively, activation of postsynaptic 5-HT$_{1A}$ receptors may result in facilitation of sexual behaviour. Evidence for a postsynaptic mechanism of action is provided by studies demonstrating that injection of 8-OH-DPAT directly into the medial preoptic area potently facilitated sexual behaviour and lowered ejaculatory threshold (Matuszewich et al. 1999). Administration of 5-HT$_{1A}$ receptor antagonists does not lead to any change in sexual behaviour (Ahlenius and Larsson 1999; De Jong et al. 2005a; Sura et al. 2001). Moreover, the effects of 5-HT$_{1A}$ receptor agonists can be antagonized by 5-HT$_{1A}$ receptor antagonists. When 5-HT$_{1A}$ receptor antagonists are combined with SSRIs (after acute or chronic administration), the inhibitory action of SSRIs is facilitated indicating a role for the 5-HT$_{1A}$ receptor in the inhibitory action of SSRIs in male sexual behaviour (De Jong et al. 2005a, b; Table 1)

3.6 Animal Models of Premature and Retarded Ejaculation

Most of our current understanding of the anatomy and neurobiology of sexual behaviour is based on animal studies using rats that are sexually experienced and display normal sexual behaviour. Interestingly, the comparable ejaculation-delaying effects of SSRIs in humans and rats suggest high translational validity with regard to the regulation of ejaculation. Nevertheless, face validity is low when one tries to extend results obtained in rats that display normal sexual behaviour to dysfunction such as premature and retarded or even (an)-ejaculation. Over the last decades, several groups have studied rats that display hyposexual behaviour and are referred to, by different investigators, as sexually inactive, sluggish, impotent or

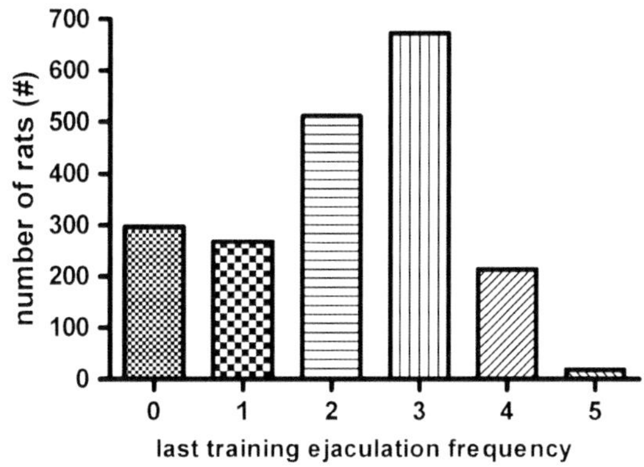

Fig. 6 More than 1,900 male rats were tested over a period of 5 years and trained weekly for 4 weeks in a sex test of 30 min against a female rat brought into behavioural oestrus. The graph represents the number of animals that displayed: 0, 1, 2, 3, 4 or 5 ejaculations during the last training test. Animals with 0 or 1 ejaculations/test were depicted as "slow" or "sluggish"; animals with two to three ejaculations/test as "normal" and animals with more than three ejaculations/test as "fast"

non-copulating rats. Recent findings suggest the presence of neurobiological differences associated with the hyposexual behaviour that these rats display. On the other hand, hypersexual behaviour can also be provoked pharmacologically. However, there are only few studies that have studied rats that are hypersexual by nature. Thus, investigating animals that do not display normal sexual behaviour may help understanding of the underlying neurobiological mechanisms and hopefully will provide further insight in the aetiology of ejaculatory dysfunction.

In our laboratory, we have found (Pattij et al. 2005; Olivier et al. 2005) that male outbred Wistar rats display sexual "endophenotypes". In subsequent cohorts of 100–120 male rats, we consistently found rats that display a very low (0–1), normal (2–3) or high (4–5) number of ejaculations in 30-min tests with a receptive female even after four to eight training tests. The behaviour of these males seems very stable, and we suggest the low performing animals as putative model for delayed ejaculation in humans and the high performing rats as model for premature ejaculation (Pattij et al. 2005; Olivier et al. 2006). Figure 6 shows the distribution of these "endophenotypic" sexual phenotypes in 1,982 male rats we tested thus far.

These various endophenotypes are now the subject of pharmacological studies.

3.7 *Studies with Rats Displaying Hyposexual Behaviour*

It was already demonstrated in early experiments in the 1940s that rats reared in isolation are either not capable to achieve ejaculation or remain sexually inactive, after repeated exposure to a receptive female (Beach 1942). In contrast, rats that were reared in groups with either same-sex or hetero-sex cage mates did not show these clear deficits in copulatory behaviour. Importantly, in most but not all of the isolation-reared males, sexual performance gradually improved with experience. These early findings suggest that experience and learning play an important role in rat copulatory performance, but apparently do not exclusively determine the ability to successfully copulate until ejaculation. In early studies focussing on rats displaying different levels of sexual performance, in our laboratory we have tried to create hyposexual behaviour in male rats by manipulating the level of sexual experience (Mos et al. 1990). To this end, we have studied the sexual behaviour of 278 sexually naïve male Wistar rats in 15-min tests with an oestrus female. From those 278 males, 23 showed no sexual activity at all, i.e. no intromissions and maximally one mount was scored during the test. From the remaining 255 rats, 211 displayed sexual activity, but failed to ejaculate during the test. The average ejaculation latency of the 44 ejaculating males was 620 $\pm$ 28 s. If sexually naïve male rats were treated with 5-HT_{1A} receptor agonists, these males performed quite well (Table 1). In particular, the two full 5-HT_{1A} receptor agonists ($\pm$)-8-OH-DPAT and flesinoxan enhanced sexual behaviour to the level of sexually experienced male rats. The partial 5-HT_{1A} receptor agonists buspirone and ipsapirone also facilitated sexual activity. These findings indicate that naïve male rats are able to perform sexual activities reminiscent of sexually "experienced" rats in a very short time interval. Apparently, sexually naïve rats may be influenced by certain factors that can be overcome by treatment with psychoactive drugs, at least 5-HT_{1A} receptor agonists and (not shown here) α_2-adrenoceptor antagonists like yohimbine and idazoxan (Mos et al. 1990, 1991).

These pharmacological studies strongly suggest that neurobiological mechanisms underlie the differences observed in basal sexual behaviour.

3.8 Studies with Rats Displaying Hypersexual Behaviour

In contrast to studies focussing on rats that are hyposexual by nature, reports of rats that are hypersexual by nature are scarce. Nevertheless, numerous studies have indicated that a variety of selective pharmacological compounds, neurotransmitters and neuropeptides may facilitate sexual behaviour (Bitran and Hull 1987; Argiolas 1999). Most interesting are those studies in which male rat sexual behaviour is potently facilitated and in which the behaviour shares some of the characteristics of human premature ejaculation. Indeed, some of the clinical symptoms of premature ejaculation can be evoked pharmacologically in male rats. For instance, various selective 5-HT_{1A} receptor agonists have been shown to potently decrease ejaculation latencies and intromission and mount frequencies. Apart from selective 5-HT_{1A} receptor agonists, a selective dopamine D_2 receptor agonist SND-919 (Ferrari and Giuliani 1994) has also been shown to decrease ejaculation latencies in rats, although its effects were much less pronounced compared to the effects of 5-HT_{1A} receptor agonists.

Not only can pharmacological manipulations facilitate ejaculatory behaviour, but "tactile" stimulation, such as shock and tail-pinching (Barfield and Sachs 1968; Wang and Hull 1980), also facilitate ejaculatory behaviour. Presumably these facilitatory effects are mediated by activation of the brain dopaminergic system (Leyton and Stewart 1996).

3.9 Conclusion: Serotonin and Male Sexual Behaviour

Research in humans and rats has indicated that modulating 5-HT levels in the CNS changes ejaculatory thresholds and associated sexual behaviour. Activation of 5-HT_{1A} receptors and blockade of 5-HT_{2C} receptors facilitates sexual behaviour, whereas activation of 5-HT_{1B} and 5-HT_{2A} receptors inhibits it. SSRIs, which facilitate serotonin neurotransmission, inhibit sexual behaviour but only after chronic administration or genetic inactivation of the SERT gene. There is a paucity of data on the putative role of other 5-HT receptors in the modulation of male sexual behaviour.

4 Serotonin, Serotonergic Receptors and Female Sexual Behaviour

4.1 Introduction

The pharmacology of sexual behaviour in females is rather restricted compared to males. The majority of work has focused on one aspect of it: the lordosis reflex. Female sexual behaviour consists of attractivity, proceptivity and receptivity. Attractivity reflects behaviour, smell and sounds by the female that attract the male and most often leads to proceptive behaviour of the female, including solicitation, hopping and darting. Receptivity is reflected in the lordosis reflex required for successful copulation. Beach (1948) introduced the lordosis quotient (LQ = lordosis to mount ratio X 100) reflecting the lordotic response of the female to a mounting male. The LQ is the most frequently used parameter when studying effects of hormones and drugs on female sexual behaviour (cf. Uphouse 2000; Uphouse and Guptarak 2010). The lordosis reflex (arching of the back, elevation of the rump, dorsoflexion of the tail and extension of the neck) is a very stereotyped posture in response to a mounting male (Pfaff 1999). The tactile stimulation stimulates cutaneous receptors in the flank, rump, tail base and perineum, which feed their information to the brain where primarily areas in the hypothalamus (notably the VMH) are crucial in the control of lordosis. Oestrogen (Er_{α}) receptor activation is required to induce the lordosis reflex, and there is a minimum amount of circulating oestrogen needed to reach a certain lordosis threshold. Moreover, a latent period (minimally 16 h) is needed for receptivity development. Normally, both oestrogen and progesterone are used to optimally organize the libido reflex, but progesterone is not needed if the oestrogen dose is extra high. Adding progesterone reduces the amount of oestrogen needed to induce lordosis behaviour.

Pharmacological studies often use submaximal oestrogen (or progesterone) doses in ovariectomized females which produce submaximal lordosis quotients and generate a model that can be pharmacologically manipulated. Early studies showed that reduction of monoamine levels in the brain (e.g. by pCPA or reserpine) activated lordosis in suboptimally oestrogen-primed ovariectomized rats, while activation of 5-HT function inhibits it (for review, see Uphouse 2000; Uphouse and Guptarak 2010). With the emerging availability of selective 5-HT receptor ligands more specific studies could be performed, but still serotonergic psychopharmacology has been mainly restricted to 5-HT_{1A} and 5-HT_2 receptors.

Activation of 5-HT_{1A} receptors leads to inhibition of the lordosis reflex in hormonally suboptimally and optimally primed female rats (Ahlenius et al. 1986; Mendelson and Gorzalka 1986). Work from Uphouse's group (Uphouse 2000) has found that the underlying mechanism of this inhibition is mediated via postsynaptic 5-HT_{1A} receptors in the hypothalamus, specifically, although not exclusively, in the VMH. Blocking of these 5-HT_{1A} receptors, however, did not lead to facilitation of the lordosis reflex which also does not happen after systemic administration of 5-HT_{1A} receptor antagonists (Uphouse 2000), a finding we confirmed in our laboratory (see SERT-KO data later).

The role of 5-HT_{1B} receptors in lordosis is somewhat disputed (Uphouse and Guptarak 2010). Notwithstanding the limited evidence and lack of selective agonists, data suggest that activation of presynaptic 5-HT_{1B} receptors facilitates lordosis (Mendelson 1992), whereas blockade of 5-HT_{1B} receptors inhibits it (Uphouse et al. 2009).

Activation of 5-$HT_{2A/2C}$ receptors (e.g. by DOI) facilitates lordosis in suboptimally primed rats (Mendelson and Gorzalka 1990), whereas 5-$HT_{2A/2C}$ receptor antagonists inhibit it (Hunter et al. 1985; Mendelson and Gorzalka 1985). These effects seem also to be mediated in the hypothalamus probably in close interaction with those mediated by 5-HT_{1A} receptors (Uphouse 2000; Uphouse and Guptarak 2010).

5-HT_3 receptors do not play an important role in female sexual behaviour; the few studies reported (for overview, see Uphouse and Guptarak 2010) do not point to central 5-HT_3 receptors as a primary target. Similarly, an inhibitory role in lordosis of 5-HT_7 receptors has been suggested (Siddiqui et al. 2007), but these data are much linked to 5-HT_{1A} receptor modulation and research involving selective 5-HT_7 receptor agonists is required.

As SSRIs are reported to induce a high incidence of sexual disturbance in human females (Balon 2006; Montgomery et al. 2002), it is relatively surprising that only a few studies have been performed in rats. Acute treatment with SSRIs reduces lordosis in hormonally primed ovariectomized rats (Frye et al. 2003; Sarkar et al. 2008). Because sexual side effects of SSRIs in humans are particularly disturbing after chronic administration, animal studies using chronic SSRIs are particularly relevant. Matuszcyk et al. (1998) found that chronic fluoxetine reduced sexual behaviour in female rats. This and other studies (Maswood et al. 2008; Uphouse and Guptarak 2010) are complicated by the fact that natural cycling females were used and fluoxetine affected the cycle, at least in a large number of the animals. A better strategy would be to chronically treat ovariectomized female rats with an SSRI, prime them with a dose of oestrogen and progesterone to induce lordosis and to test the effects of the SSRI in this model. Sarkar et al. (2008) found, using this paradigm, that fluoxetine acutely reduced lordosis but this effect was attenuated after sub-chronic fluoxetine administration, suggesting that some tolerance for the sexual inhibitory effect of the SSRI had occurred.

4.2 SERT-KO Rats and Female Sexual Behaviour

An alternative way to study the role of the SERT in female sexual behaviour is using genetically modified animals, in this case the SERT-KO rat made by ENU mutagenesis (Smits et al. 2006). Female Wistar intact rats were tested in a paced mating design where sexually experienced males were restricted to one side of a cage, whereas the female (brought into behavioural oestrus by a high dose of oestradiol) could spend time on both sides of a divider which allowed passage of the female (but not the male) through a couple of openings in the divider. Figure 7 shows that mutant SERT genotypes ($SERT^{+/-}$ and $SERT^{-/-}$) were not different from wild types ($SERT^{+/+}$) in any aspect of proceptive or receptive behaviour over three consecutive tests of 30 min. This indicates that permanent absence of the serotonin transporter has no influence on female sexual behaviour under normal conditions. Treatment with a 5-HT_{1A} receptor agonist (+/− 8 OH-DPAT) dose-dependently reduced proceptive behaviours (b) in all three genotypes, but in the

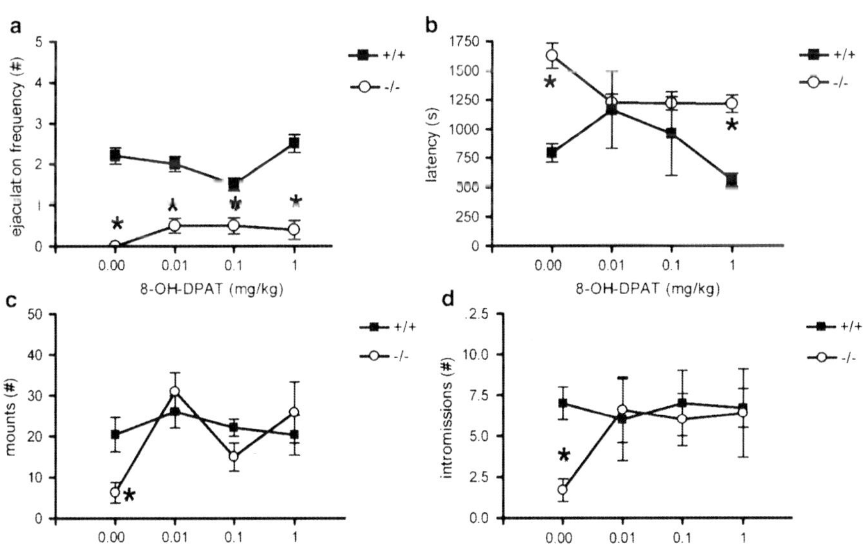

Fig. 7 Effects of three doses of 8-OH-DPAT (0.01, 0.1 and 1 mg/kg, SC) and one dose of WAY100635 (0.1 mg/kg, IP) on ejaculation frequency over 30-min test (**a**), latency to first ejaculation (**b**), first ejaculatory series mounts (**c**) and first ejaculatory series intromissions (**d**) of $SERT^{+/+}$ (+/+) and $SERT^{-/-}$ (−/−) animals. *$p < 0.05$ compared to wild type (+/+)

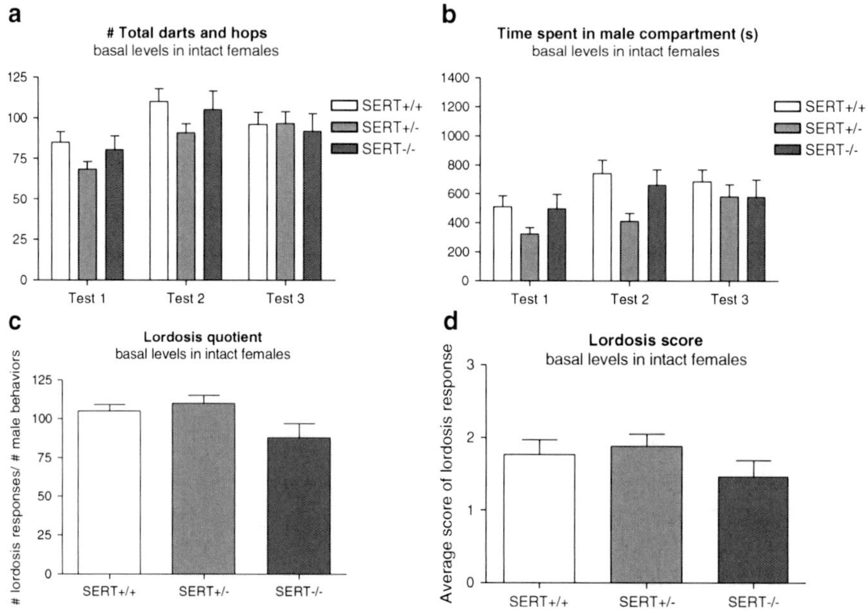

Fig. 8 In a paced mating situation (Snoeren et al. 2010) female wild-type ($SERT^{+/+}$), heterozygous ($SERT^{+/-}$) and homozygous ($SERT^{-/-}$) rats were brought into behavioural oestrus by hormonal priming and tested against a sexually experienced male rat. Females could pace the behaviour and stay in- or outside the male compartment. The number of proceptive [hopping and darting (**a**)] and receptive behaviours [Lordosis quotient and Lordosis score (**c** and **d**)] and the time spent in the male compartment (**b**) were measured

SERT-KO the dose–response curve clearly shifted to the right, indicative of a desensitized 5-HT_{1A} receptor (Fig. 8). However, time spent with the male was not affected (a), showing that the decreased proceptive behaviour was not caused by a diminished interaction with the male. Treatment with a 5-HT_{1A} receptor antagonist (WAY100635) did not affect any behaviour alone [(c) and (d)], whereas a selected dose of WAY100635 (0.1 mg/kg IP) was able to antagonize the 8-OH-DPAT-induced reduction in proceptive behaviour (f). Apparently, normal female sexual behaviour is not dependent on the functional status of 5-HT_{1A} receptors, but when challenged 5-HT_{1A} receptors appear desensitized in homozygous, but not heterozygous SERT-KO rats (Fig. 9).

5 Conclusions

The neurotransmitter serotonin clearly plays a role in male and female sexual behaviour (Table 2). Lowering serotonergic function seems to facilitate and enhancing it to inhibit sexual behaviour. The availability of blockers of the

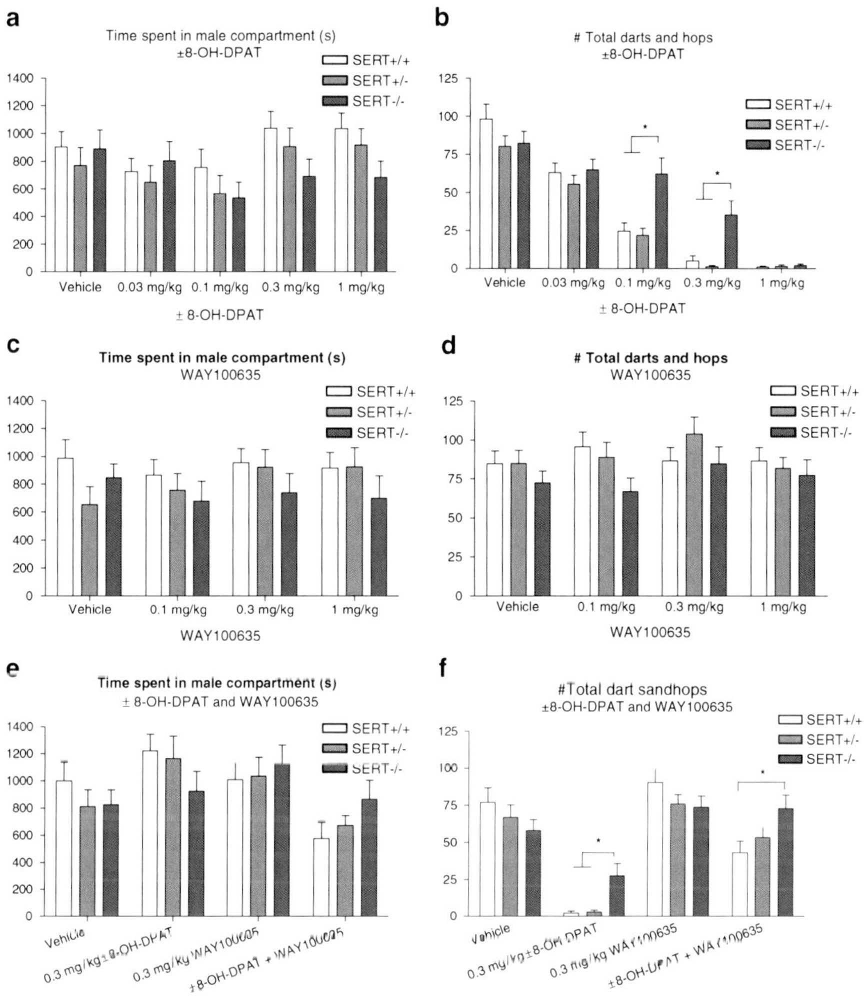

Fig. 9 Female wild-type ($SERT^{+/+}$), heterozygous serotonin transporter knockout ($SERT^{+/-}$) and homozygous serotonin transporter knockout ($SERT^{-/-}$) rats brought into behavioural oestrus were treated with the $5\text{-}HT_{1A}$ receptor agonist +/−8-OH-DPAT (**a**); the $5\text{-}HT_{1A}$ receptor antagonist WAY100639 (**b**) or a combination of selected doses of 8-OH-DPAT (0.3 mg/kg) and WAY100639 (0.3 mg/kg) (**c**). The *left part* of each figure shows the time spent by the female in the male compartment; the *right part* the number of proceptive behaviours (hopping and darting) performed by the female during the test. The test was performed using a paced mating design in which the male and female were separated by a perforated wall that could be crossed by the female but not by the male. The female decides whether she wants to spend time with the male and receive mounts, intromissions and ejaculations. *$p < 0.05$ compared to wild type

Table 2 Summary of the effects of various serotonergic ligands on male and female sexual behaviour in rats after acute or chronic treatment

Target/ligand	Treatment	Male sexual behaviour	Female sexual behaviour
SERT/SSRI	Acute	=	=
SERT/SSRI	Chronic	↓	nd
5-HT_{1A} R agonist	Acute	↑	↓
5-HT_{1A} R agonist	Chronic	↑	nd
5-HT_{1A} R antagonist	Acute	=	=
5-HT_{1A} R antagonist	Chronic	nd	nd
5-HT_{1B} R agonist	Acute	↓	↑
5-HT_{1B} R agonist	Chronic	nd	nd
5-HT_{1B} R antagonist	Acute	=	↓
5-HT_{1B} R antagonist	Chronic	nd	nd
5-$HT_{2A/C}$ R agonist	Acute	↓	↑
5-$HT_{2A/C}$ R agonist	Chronic	nd	nd
5-$HT_{2A/C}$ R antagonist	Acute	↓	↓
5-$HT_{2A/C}$ R antagonist	Chronic	nd	nd
5-HT_7 R agonist	Acute	nd	↓
5-HT_7 R agonist	Chronic	=	nd
5-HT_7 R antagonist	Acute	=	=
5-HT_7 R antagonist	Chronic	=	nd

= not affected, *nd* not determined, ↑ enhanced, ↓ lowered, *R* receptor, *SSRI* selective serotonin reuptake inhibitor, *SERT* serotonin transporter

serotonin transporter and ligands for various serotonergic receptors has led to studies on male and female rat sexual behaviour that shed light on the contributions of individual receptors/transporter in male and female sexual function. SSRIs, blocking the SERT, generally lead to inhibition (after chronic treatment) of male and female sexual behaviour in agreement with the theory that enhancement of serotonergic function inhibits sexual behaviour. 5-HT_{1A} receptor activation facilitates male ejaculatory behaviour but inhibits female lordosis behaviour, suggesting an opposing role for this receptor in males and females. Clear-cut roles for other serotonergic receptors are less developed and need considerable research efforts.

Genetic manipulation of the SERT in rats indicated a differential influence of the absence of the SERT in male and female sexual behaviour; KO males, but not females, had lower baseline sexual activities. 5-HT_{1A} receptors were not desensitized in male KO, but were desensitized in females, indicating a differential role of various 5-HT_{1A} receptor pools in male and female sexual behaviour.

References

Ahlenius S, Larsson K (1990) In: Rodgers RJ, Cooper SJ (eds) 5-HT_{1A} agonists, 5-HT_3 antagonists and benzodiazepines: their comparative behavioural pharmacology. Wiley, Chichester, pp 281–301

Ahlenius S, Larsson K (1998) Evidence for an involvement of 5-HT$_{1B}$ receptors in the inhibition of male rat ejaculatory behaviour produced by 5-HTP. Psychopharmacology 137:374–382
Ahlenius S, Larsson K (1999) Synergistic actions of the 5-HT$_{1A}$ antagonist WAY-100635 and citalopram on male rat ejaculatory behaviour. Eur J Pharmacol 379:1–6
Ahlenius S, Larsson K, Svensson L (1980) Further evidence for an inhibitory role of central 5-HT in male rat sexual behaviour. Psychopharmacology 68:217–220
Ahlenius S, Fernandez-Guasti A, Hjorth S, Larsson K (1986) Suppression of lordosis behaviour by the putative 5-HT receptor agonist 8-OH-DPAT in the rat. Eur J Pharmacol 124:361–363
Albinsson A, Andersson G, Andersson K, VegaMatuszczyk J, Larsson K (1996) The effects of lesions in the mesencephalic raphe systems on male rat sexual behaviour and locomotor activity. Behav Brain Res 80:57–63
Andersson G, Larsson K (1994) Effects of FG-5893, a new compound with 5-HT$_{1A}$ receptor agonistic and 5-HT$_2$ antagonistic properties, on male-rat sexual-behaviour. Eur J Pharmacol 255:131–137
Argiolas A (1999) Neuropeptides and sexual behaviour. Neurosci Biobehav Rev 23:1127–1142
Balon R (2006) SSRI-associated sexual dysfunction. Am J psychiatry 163:1504–1509
Barfield RJ, Sachs BD (1968) Sexual behaviour: stimulation by painful electrical shock to skin in male rats. Science 161:392–395
Beach FA (1942) Comparison of copulatory behaviour of male rats raised in isolation, cohabitation, and segregation. J Genet Psychol 60:3–13
Beach FA (1948) Hormones and behaviour. Hoeber Harber, New York
Bitran D, Hull EM (1987) Pharmacological analysis of male rat sexual behaviour. Neurosci Biobehav Rev 11:365–389
Cantor J, Binik I, Pfaus JG (1999) Chronic fluoxetine inhibits sexual behaviour in the male rat: reversal with oxytocin. Psychopharmacology 144:355–362
Chan JSW, Waldinger MD, Olivier B, Oosting RS (2010) Drug-induced sexual dysfunction in rats. Curr protoc Neurosci 53:9.34.1–9.34.11
Chan JSW, Snoeren EMS, Cuppen E, Waldinger MD, Olivier B, Oosting RS (2011) The serotonin transporter plays an important role in male sexual behavior: a study in serotonin transporter knockout rats. J Sex Medicine 8:97–108
Cyranowski JM, Bromberger J, Youk A, Matthews K, Kravitz HM, Powell LH (2004) Lifetime depression history and sexual function in women at midlife. Arch Sex Behav 33:539–548
De Jong TR, Pattij T, Veening JG, Dederen PFJ, Waldinger MD, Cools AR, Olivier B (2005a) Citalopram combined with WAY100635 inhibits ejaculation and ejaculation-related Fos immunoreactivity. Eur J Pharmacol 509:49–59
De Jong TR, Pattij T, Veening JG, Waldinger MD, Cools AR, Olivier B (2005b) Effects of chronic selective serotonin reuptake inhibitors on 8-OH-DPAT induced facilitation of ejaculation in rats: comparison of fluvoxamine and paroxetine. Psychopharmacology 179:509–515
De Jong TR, Veening JG, Waldinger MD, Cools AR, Olivier B (2006) Serotonin and the neurobiology of the ejaculation threshold. Neurosci Biobehav Rev 30:893–907
De Jong TR, Veening JG, Olivier B, Waldinger MD (2007) Oxytocin involvement in SSRI-induced delayed ejaculation: a review of animal studies. J Sex Med 4:14–28
Dornan WA, Katz JL, Ricaurte GA (1991) The effects of repeated administration of MDMA on the expression of sexual behaviour in the male rat. Pharmacol Biochem Behav 39:813–816
Ferrari F, Giuliani D (1994) The selective D-2 dopamine-receptor antagonist eticlopride counteracts the ejaculatio-praecox induced by the selective D-2-dopamine agonist SND-919 in the rat. Life Sci 55:1155–1162
Foreman MM, Hall JL, Love RL (1992) Effects of fenfluramine and para-chloroamphetamine on sexual behaviour of male rats. Psychopharmacology 107:327–330
Frank JL, Hendricks SE, Olson CH (2000) Multiple ejaculations and chronic fluoxetine: effects on male rat copulatory behaviour. Pharmacol Biochem Behav 66:337–342
Frye CA, Petralia SM, Rhodes ME, Stein B (2003) Fluoxetine may influence lordosis of rats through effects on midbrain 3 alpha, 5 alpha-THP concentrations. Ann RNY Acad Sci 1007:37–41

Gonzales G, Mendoza L, Ruiz J, Torrejon J (1982) A demonstration that 5-hydroxytryptamine administered peripherally can affect sexual behaviour in male rats. Life Sci 31:2775–2781

Haddjeri N, Blier P, De Montigny C (1998) Long-term antidepressant treatments result in a tonic activation of forebrain 5-HT_{1A} receptors. J Neurosci 18:10150–10156

Haensel SM, Slob AK (1997) Flesinoxan: a prosexual drug for male rats. Eur J Pharmacol 330:1–9

Hillegaart V, Ahlenius S (1998) Facilitation and inhibition of male rat ejaculatory behaviour by the respective 5-HT_{1A} and 5-HT_{1B} receptor agonists 8-OH-DPAT and anpirtoline, as evidenced by use of the corresponding new and selective receptor antagonists NAD-299 and NAS-181. Br J Pharmacol 125:1733–1743

Hunter AJ, Hole DR, Wilson CA (1985) Studies into the dual effects of serotonergic pharmacological agents on female sexual behaviour in the rat: preliminary evidence that endogenous 5-HT is stimulatory. Pharmacol Biochem Behav 22:5–13

Kendurkar A, Kaur B (2008) Major depressive disorder, obsessive-compulsive disorder, and generalized anxiety disorder: do the sexual dysfunctions differ? Primary Care Companion J Clin Psychiatry 10:299–305

Kennedy SH, Rizvi S (2009) Sexual dysfunction, depression, and the impact of antidepressants. J Clin Psychopharmacol 29:157–164

Klint T, Larsson K (1995) Clozapine acts as a 5-HT_2 antagonist by attenuating DOI-induced inhibition of male rat sexual behaviour. Psychopharmacology 119:291–294

Klint T, Dahlgren IL, Larsson K (1992) The selective 5-HT2 receptor antagonist amperozide attenuates 1-(2, 5-dimethoxy-4-iodophenyl)-2-aminopropane-induced inhibition of male rat sexual behaviour. Eur J Pharmacol 212:241–246

Kugaya A, Seneca NM, Snyder PJ, Williams SA, Malison RT, Baldwin RM, Seibyl JP, Innis RB (2003) Changes in human in vivo serotonin and dopamine transporter availabilities during chronic antidepressant administration. Neuropsychopharmacology 28:413–420

Le Poul E, Boni C, Hanoun N, Laporte AM, Laaris N, Chauveau J, Hamon M, Lanfumey L (2000) Differential adaptation of brain 5-HT_{1A} and 5-HT_{1B} receptors and 5-HT transporter in rats treated chronically with fluoxetine. Neuropharmacology 39:110–122

Leyton M, Stewart J (1996) Acute and repeated activation of male sexual behaviour by tail pinch: opioid and dopaminergic mechanisms. Physiol Behav 60:77–85

Looney C, Thor KB, Ricca D, Marson L (2005) Differential effects of simultaneous or sequential administration of paroxetine and WAY-100, 635 on ejaculatory behaviour. Pharmacol Biochem Behav 82:427–433

Maswood N, Sarkar J, Uphouse L (2008) Modest effects of repeated fluoxetine on estrous cyclicity and sexual behaviour in Sprague Dawley female rats. Brain Res 1745:53–60

Matuszcyk JV, Larsson K, Eriksson E (1998) The selective serotonin reuptake inhibitor fluoxetine reduces sexual motivation in male rats. Pharmacol Biochem Behav 60:527–532

Matuszewich L, Lorrain DS, Trujillo R, Dominguez J, Putnam SK, Hull EM (1999) Partial antagonism of 8-OH-DPAT's effects on male rat sexual behaviour with a D_2, but not a 5-HT_{1A} antagonist. Brain Res 820:55–62

Melis MR, Argiolas A (1995) Dopamine and sexual behaviour. Neurosci Biobehav Rev 19:19–38

Mendelson SD (1992) A review and reevaluation of the role of serotonin in the modulation of lordosis behaviour in the female rat. Neurosci Biobehav Rev 16:309–350

Mendelson SD, Gorzalka BB (1985) A facilitatory role for serotonin in the sexual behaviour of the female rat. Pharmacol Biochem Behav 22:1025–1033

Mendelson SD, Gorzalka BB (1986) 5-HT_{1A} receptors: differential involvement in female and male sexual behaviour in the rat. Eur J Pharmacol 132:323–326

Mendelson SD, Gorzalka BB (1990) Sex differences in the effects of 1-(m-trifluoromethylphenyl) piperazine an 1-(m-chlorophenyl) piperazine on copulatory behaviour in the rat. Neuropharmacology 29:783–786

Montgomery SA, Baldwin DS, Riley A (2002) Antidepressant medications: a review of the evidence for drug-induced sexual dysfunction. J Affect Disord 69(1–3):119–140

Mos J, Olivier B, Bloetjes K, Poth M (1990) Drug-induced facilitation of sexual behaviour in the male rat: behavioural and pharmacological aspects. In: Slob AK, Baum MJ (eds) Psychoneuroendocrinology of growth and development. Medicom Publishers, Rotterdam, pp 221–232

Mos J, Van Logten J, Bloetjes K, Olivier B (1991) The effects of idazoxan and 8-OH-DPAT on sexual behaviour and associated ultrasonic vocalizations in the rat. Neurosci Biobehav Rev 15:505–515

Mos J, Mollet I, Tolboom JTBM, Waldinger MD, Olivier B (1999) A comparison of the effects of different serotonin reuptake blockers on sexual behaviour of the male rat. Eur Neuropsychopharmacol 9:123–135

Murphy DL, Lesch KP (2008) Targeting the murine serotonin transporter insights into human neurobiology. Nat Rev Neurosci 9:85–96

Olivier J, Cools A, Ellenbroek B, Cuppen E, Homberg J (2010) The serotonin transporter knockout rat: a review. In Kalueff AV, LaPorte JL (eds) Experimental models in serotonin transporter research. Cambridge University Press, Cambridge, PP170–213

Pattij T, de Jong TR, Uitterdijk A, Waldinger MD, Veening JG, Cools AR, van der Graaf PH, Olivier B (2005) Individual differences in male rat ejaculatory behaviour: searching for models to study ejaculation disorders. Eur J Neurosci 22:724–734

Regitz-Zagrosek V, Schubert C, Krüger S (2008) Gesclechterunterschiede in der neuropsychiatrischen Pharmakotherapie. Der Internist 12:1516–1523

Sarkar J, Hiegel C, Ginis E, Hilbun E, Uphouse L (2008) Subchronic treatment with fluoxetine attenuates effects of acute fluoxetine on female rat sexual behaviour. Brain Res 1190:58–64

Siddiqui A, Niazi A, Shahariar S, Wilson CA (2007) The 5-Ht(7) receptor is involved in the regulation of female sexual behaviour in the rat. Pharmacol Biochem Behave 87:386–392

Smits BM, Mudde JB, van de Belt J, Verheul M, Olivier JD, Homberg J, Gurvey V, Cools AR, Ellenbroek BA, Cuppen E (2006) Generation of gene knockouts and mutant models in the laboratory rat by ENU-driven target selected mutagenesis. Pharmacogenet Genomics 16:159–169

Snoeren EMS, Chan JSW, De Jong TR, Cuppen E, Waldinger MD, Olivier B, Oosting RS (2010) Serotonin transporter null mutation and sexual behaviour in female rats: 5-HT_{1A} receptor desensitization. J Sex Med 7:2424–2434

Sugden K, Tichopad A, Khan N, Craig IW, D'Souza UM (2009) Genes within the serotonergic system are differentially expressed I the human brain. BMC Neurosci 10:50

Sura A, Overstreet DH, Marson L (2001) Selectively bred male rat lines differ in naïve and experienced sexual behaviour. Physiol Behav 72:13–20

Tagliamonte A, Tagliamonte P, Gessa GL, Brodie BB (1969) Compulsive sexual activity induced by p-chlorophenylalanine in normal and pinealectomized male rats. Science 166:1433–1435

Uphouse L (2000) Female gonadal hormones, serotonin and sexual receptivity. Brain Res Rev 33:242–257

Uphouse L, Guptarak J (2010) Serotonin and sexual behaviour. In: Müller C, Jacobs B (eds) Handbook of behavioural neurobiology of serotonin. Elsevier, Amsterdam, pp 347–365

Uphouse L, Hiegel C, Guptarak J, Maswood N (2009) Progesterone reduces the effect of the serotonin 1B/1D receptor antagonist GR127935, on lordosis behaviour. Horm Behav 55(1):169–174

Waldinger MD (2002) The neurobiological approach to premature ejaculation (review article). J Urol 168:2359–2367

Waldinger MD, Hengeveld MW, Zwinderman AH, Olivier B (1998) Effect of SSRI antidepressants on ejaculation: a double-blind, randomized, placebo-controlled study with fluoxetine, fluvoxamine, paroxetine, and sertraline. J Clin Psychopharmacol 18:274–281

Waldinger MD, Plas A vd, Pattij T, Oorschot R v, Coolen LM, Veening JG, Olivier B (2001a) The SSRIs fluvoxamine and paroxetine differ in sexual inhibitory effects after chronic treatment. Psychopharmacology 160:283–289

Waldinger MD, Zwinderman AH, Olivier B (2001b) SSRIs and ejaculation: a double-blind, randomized, fixed-dose study with paroxetine and citalopram. J Clin Psychopharmacol 21:556–560

Wang L, Hull EM (1980) Tail pinch induces sexual behaviour in olfactory bulbectomized male rats. Physiol Behav 24:211–215

Watson NV, Gorzalka BB (1991) DOI-induced inhibition of copulatory behaviour in male rats: reversal by 5-HT_2 antagonists. Pharmacol Biochem Behav 39:605–621

Williams VSL, Baldwin DS, Hogue SL, Fehnel SE, Hollis KA, Edin HM (2006) Estimating the prevalence and impact of antidepressant-induced sexual dysfunction in 2 European countries< a cross-sectional patient survey. J Clin Psychiatry 67:204–210

Zemishlany Z, Weizman A (2008) The impact of mental illness on sexual dysfunction. Adv Psychosom Med 29:89–106

Female Rats Are Smarter than Males: Influence of Test, Oestrogen Receptor Subtypes and Glutamate

Jane Suzanne Sutcliffe

Contents

Abstract Interest in the influence of sex hormones within the central nervous system is a rapidly expanding area of research. A considerable amount of evidence has recently been obtained to support an important role of the gonadal steroids in cognitive processing. Not only are distinct and complementary behavioural phenotypes evident for each gender, in the case of the female but they are also reliant upon hormonal status. Gender influences and hormonal status are thus paramount and should encourage the development of more hypothesis-driven research strategies to understand gender differences in both normal behaviour and where this is altered in neuropsychiatric disorders.

Keywords Cognition · In vivo · Oestrogen · Rat · Sex differences

J.S. Sutcliffe
Maccine Pte Ltd, 10 Science Park Road, #01-05 The Alpha, Singapore Science Park II, Singapore 117684, Singapore
e-mail: Jane.Sutcliffe@Maccine.com

J.C. Neill and J. Kulkarni (eds.), *Biological Basis of Sex Differences in Psychopharmacology*, Current Topics in Behavioral Neurosciences 8, DOI 10.1007/7854_2011_120,

1 Introduction

Men and women differ not only in their physical attributes and reproductive function but also in many other characteristics including cognitive abilities and intellectual problem solving skills. For the past few decades, it has been ideologically fashionable to insist that these behavioural differences are minimal; research into sex differences in cognition has often been neglected and for many years, the male animal has been the standard gender used for behavioural research. Historically, female subjects have not been studied due to perceived problems inherent with oestrous cycle monitoring and also since it is assumed that their behaviour will be similar to those of a male. Strangely enough, evidence from human studies illustrates that men are actually more variable on tests of cognitive score and intellectual performance (Sahay and Hen 2007). Sex differences cannot be considered irrelevant as there are indisputable, and well-established gender differences in metabolism, neuroanatomy, endocrinology, biochemistry and behaviour (Cahill 2006). Moreover, sex differences in many behavioural paradigms have been reported since the early days of the last century (Baker 1987; Corey 1930). One of the most important discoveries regarding gender differences is concerned with the innate differences between the anatomy and neurophysiology of the male and female brain. Differences begin during early development due to a combination of genetic and hormonal events and persist throughout the lifespan of an individual [for a comprehensive review on the origins of gender differences, see (Wilson and Davies 2007)]. The imprinting of the female brain is not dependent upon androgen-oestrogen influence but for male differentiation, androgens must be present (Pilgrim and Reisert 1992).

Research during the past decade confirms that gonadal steroids, such as oestrogen, progesterone and testosterone, have the ability to influence the structural properties of the brain regions that sub-serve learning and memory. One of the most comprehensively characterised regions of the brain is the hippocampus, which has been shown to display different structural changes in response to different gonadal hormones. Evaluation of brain structure, function and chemistry over the course of the menstrual cycle as well as across the life span in women is critical to understanding sex differences in both normal and aberrant behaviour. This chapter provides an overview of common cognitive paradigms with respect to inherent sex-specific abilities and the impact of cyclic or manipulated hormonal changes on learning and memory in male and female rats. Mechanisms by which these gender specific abilities are purported to arise are also discussed with particular relevance to oestrogen receptors (ERs) and glutamate.

2 In vivo Evidence for Gender Differences in Cognition

The relationship between gonadal hormones and cognition remains immensely complex and depends on many factors, which require unique consideration such as the target brain structures and the recruited memory systems for that particular cognitive challenge. Effects of the gonadal hormones and inherent sex differences on cognitive behaviour are not immediately obvious since gender differences are investigated across multiple laboratories, under different environmental conditions, with different strains of animals and sex differences are rarely discussed. Throughout this section, gender differences in popular cognitive paradigms are discussed and concisely compiled in Table 1.

2.1 *Object Recognition Tasks*

During the past few years, the ethologically relevant novel object recognition (NOR) and object displacement (OD) paradigms have enjoyed much scientific interest. These paradigms depend entirely on the rat's natural preference for novelty (objects and location), there is no requirement for rule specific learning and thus, there is a logistical advantage whereby the inherent variability during rule acquisition and undue stress is avoided. There is also good evidence to suggest that the NOR task is more sensitive to recognition memory impairments than other tasks such as the Delayed Non-Matching to Sample (DNMS) task (Nemanic et al. 2004; Pascalis et al. 2004). The NOR consists of a familiarisation phase, where two identical objects are presented to the subject, an inter-trial delay (which can be manipulated) and a test phase where one familiar and one novel object are presented to the subject (Ennaceur and Delacour 1988). Similar to the NOR task, the OD task consists of a familiarisation phase with presentation of two identical objects, which progresses, after an inter-trial delay, onto the test phase where one of these objects now occupies a new location with respect to the previous trial. Compared to incremental learning tasks using multiple learning trials, the NOR and OD tasks allow the investigation of drug effects on different stages of memory formation and recollection. To assess the influence of a compound on object encoding, it may be administered prior to the sample trial. The compound in question can also be administered intermediately (during the inter-trial interval) to assess the influence of drug effects on consolidation of object information. Thus, object recognition tasks are fast becoming a powerful ethologically relevant, scientific tool.

Recent research using object recognition tasks has provided clear evidence of gender-specific abilities (Sutcliffe et al. 2007). Both genders show a progressive decline when longer inter-trial intervals are experienced for both the NOR and the OD tasks; however, gender-specific cognitive abilities are evident. Female hooded-Lister rats have been shown to display a sustained object recognition memory when compared to male rats during this task when the length of memory retention is challenged. Recent studies (Sutcliffe et al. 2007) have demonstrated a clear female advantage during the NOR paradigm, where female hooded-Lister rats exhibit memory retention at inter-trial intervals of up to 3 h compared with 30 min in their male counterparts. The converse is true for the OD task, where male hooded-Lister rats exhibit a preference for the displaced object at an inter-trial interval of 3 h compared to only 30 min when compared to female hooded-Lister rats

Table 1 Summary of gender differences in common behavioural paradigms

Cognitive paradigm	Gender bias summary	Oestrous cycle influence	References
NOR	Females → Males Inconsistencies on actual length of memory retention across research groups	Cyclic performance, improved performance with high oestrogen and detrimental influence of ovariectomy which is reversed by hormone replacement	Sutcliffe et al. (2007), King et al. (2004), Ghi et al. (1999), Frye et al. (2007), Walf et al. (2006), Inagaki et al. (2010) and Aubele et al. (2008)
OD	Females ← Males	Cyclic performance, improved performance with high oestrogen	Sutcliffe et al. (2007) and Frye et al. (2007)
RAM/MWM	Small male advantage but inconsistent findings, often no sex differences observed	Negative effect of rising oestrogen across the oestrous cycle	Jonasson (2005), Bucci et al. (1995) and Faraji et al. (2010)
ZT	Females ← Males	Not reported	Faraji et al. (2010)
Social recognition	Young; Females → Males Aged; Females = Males	High oestrogen at pro-oestrous results in extended memory retention. Detrimental influence of ovariectomy, which is reversed by hormone replacement	Markham and Juraska (2007), Sánchez-Andrade and Kendrick (2011) and Hlinäck (1993)
Eyeblink conditioning	Females → Males during acquisition only. Evidence suggests more females reach experimental criterion and retain the knowledge of the task longer than males	Detrimental influence of ovariectomy, which is reversed by supraphysiological levels of hormone replacement	Dalla et al. (2009), Wood and Shors (1998) and Leuner et al. (2004)
Fear conditioning	Females ← Males	Ovariectomy results in male-like performance ability	Maren et al. (1994), Pryce et al. (1999) and Gupta et al. (2001)
FR1	Females → Male during acquisition	No effect of ovariectomy	Dalla et al. (2008), Shors et al. (2007), Beatty and Beatty (1970) and Van Oyen et al. (1981)

(*continued*)

Table 1 (continued)

Cognitive paradigm	Gender bias summary	Oestrous cycle influence	References
FR2	Females → Male during acquisition	No effect of ovariectomy	Dalla et al. (2008), Shors et al. (2007), Beatty and Beatty (1970) and Van Oyen et al. (1981)
DRL/Lever pressing	Simple tasks Females ← Males ↑ Complexity Females → Male	Ovariectomy results in male-like performance ability	van Haaren et al. (1990), van Hest et al. (1987), Beatty (1973), Roth et al. (2004) and Lynch et al. (2002)
Self-administration	Faster acquisition in females	Ovariectomy results in male-like performance ability	Fattore et al. (2007, 2009), Roth et al. (2004) and Lynch et al. (2002)

(Fig. 1a–d) (Sutcliffe et al. 2007). Conversely, other research groups have demonstrated male hooded-Lister rats retain the preference for the novel object up to a 2 h inter-trial interval (King et al. 2004). Gender differences within the NOR paradigm have also been confirmed in the Wistar rat, females again exhibit superior performance during longer inter-trial intervals (90 min) when compared to their male counterparts (60 min) (Ghi et al. 1999).

Cyclic variations in sex hormones also exert a significant impact on performance during object recognition tasks. Female rats demonstrated a distinct variability in cognitive performance during the OD task throughout the oestrous cycle with enhanced object recognition during pro-oestrous or oestrous when compared to di-oestrous when oestrogen levels are at their lowest (Sutcliffe et al. 2007; Frye et al. 2007). No cyclic alteration in performance was observed during the NOR paradigm at a 1 h inter trial interval (Sutcliffe et al. 2007), but a significant influence of the oestrous cycle during the NOR has been reported in female Long-Evans rats when the inter-trial interval is increased to a 4 h inter-trial interval, which was concomitant with increases in serum estradiol, progesterone and 3 alpha-hydroxy-5 alpha-pregnan-20-one (Walf et al. 2006). Further supporting the hypothesis that the sex steroids play an important role in cognition, ovariectomy produces a robust deficit in the NOR and OD paradigms, which is reversed with ovarian steroid replacement regimens (Inagaki et al. 2010; Frye et al. 2007. The cognitive deficit due to ovariectomy is a robust finding across research groups, in our laboratory sexually mature, ovariectomised animals were shown to significantly identify the novel from familiar object after an inter-trial interval of 1 h on weeks 1 and 2 but not weeks 3–6 and week 40 following surgery in comparison with sham-operated animals who retained recognition ability. Furthermore, ovariectomised animals receiving sub-cutaneous hormone replacement consisting of oestrogen alone or oestrogen in combination with a progestin (450 μg/kg 17-β-estradiol propionate once per week plus medroxyprogesterone 17-acetate or levonorgestrel at 15 mg/kg

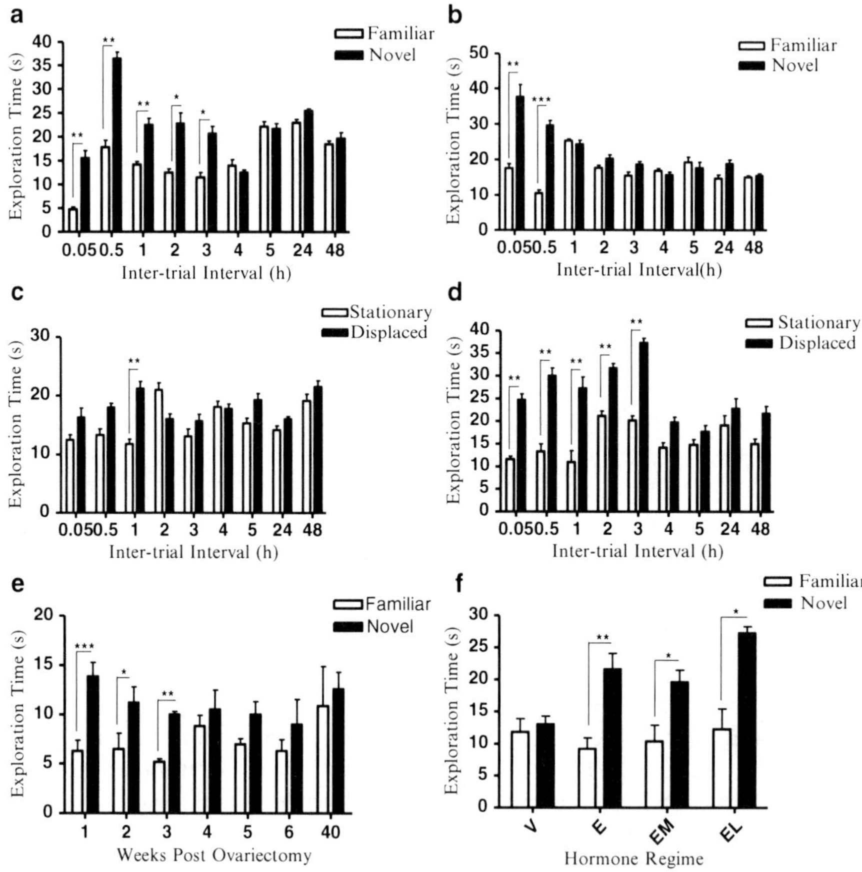

Fig. 1 Exploration times of gonadally intact mature female (**a**, **c**) and male (**b**, **d**) hooded-Lister rats during the retention trial in the NOR task (where one familiar and one novel object are presented to the subjects, **a** and **b**) and during the OD task (where two familiar, identical objects are presented to the subject but one object is displaced compared to the acquisition trial, **c** and **d**). Figures 1e and f illustrate the robust impact of ovariectomy (**e**) and continuous 16 week hormone replacement (**f**) – V (vehicle 1 ml/kg s.c), E (450 μg/kg 17-β-estradiol propionate s.c once per week), EM (E plus 15 mg/kg medroxyprogesterone 17-acetate s.c once every second week) and EL (E plus 15 mg/kg levonorgestrel s.c once every second week) during the NOR task with a 1h inter-trial interval. $*p < 0.05$, $**p < 0.01$ represent significant differences between the time spent exploring the familiar compared to the novel object. $**p < 0.05$, $**p < 0.01$, $***p < 0.001$ represent significant differences between the time spent exploring the displaced object compared to the stationary object. Data are expressed as the mean ± S.E.M. and were analysed at each time point using paired sample's t-tests ($n = 6$)

once every second week) for 15 weeks immediately after ovariectomy sustained NOR but not OD performance at a 1 h inter-trial interval. Delayed (initiation of hormone replacement at 13 weeks post-ovariectomy) and intermittent (initiation of treatment on weeks 1–6 resuming on 13–18) hormone replacement regimes highlighted the benefit of initiating hormone replacement immediately following ovariectomy using a 1 h inter-trial interval in the NOR task. However, when the inter-trial interval was increased to 3 h, only the oestrogen alone intermittent group was able to significantly differentiate between the novel and familiar object (Fig. 1e, f. Sutcliffe et al., unpublished findings). Interestingly, gonadectomised (GDX) mature male Sprague-Dawley rats explore a novel and familiar object equally during the NOR task, and these adverse effects of gonadectomy are attenuated by supplementing GDX animals with testosterone-propionate but not 17-β-estradiol (Aubele et al. 2008). Endogenous or exogenous oestrogen replacement is consistently reported to be beneficial during the NOR task for females, a finding which is not reported in males, suggesting innate activational differences by which oestrogen acts within the brain.

2.2 Spatial Maze Paradigms

The existence of gender differences in rodent models of spatial learning and memory is a prominent yet controversial and often contested topic in the literature. A variety of studies have revealed a male superiority during performance on spatial tasks such as the Radial Arm Maze (RAM) developed by Olton and Samuelson (Olton and Samuelson 1976) and the Morris Water Maze (MWM) developed by Roger Morris (Morris 1981). In the MWM task, subjects will swim in a pool of water to identify/locate a hidden platform employing the topographical relationships among the distal visual cues, pool wall and goal location. Similar to the MWM, subjects completing the RAM must find the locations of food rewards at the end of some of the maze arms and retrieve this information for successive trials. The RAM has primarily been adopted to assess spatial working memory but is also adaptable to recruit working and reference memory simultaneously. During both the standard 8 arm version and the 17-arm version of this task, gender differences have not always been reliably demonstrated with some inconsistent findings (Jonasson 2005).

A direct comparison between male and female Long Evans rats at 6 months of age in the MWM task has concluded that there are no sex differences in place learning ability. Furthermore, search accuracy on probe trials, when the platform was unavailable, was also equivalent for the male and female groups. A recent review illustrates that, while there is a distinct male advantage on spatial tasks there is a lack of reliable replication across different laboratories which may be due to the strain of animal used, variations in stage of the oestrous cycle and stress levels whilst under test (Jonasson 2005). Fisher, Long Evans and Sprague-Dawley rats have been shown to yield the largest and most robust male advantage during spatial cognition tasks, while the same tasks employed in Wistar rats denote only a small male advantage when compared to female performance (Bucci et al. 1995). Interestingly, the promising development of a new paradigm – the dry-land ziggurat task (ZT) may provide the research community with more robust and consistent gender-related results. The ZT consists of an open field containing 16 identical ziggurats (pyramid-shaped towers) positioned equal distances apart. One ziggurat is baited with a food reward and the rat must navigate through the open field to retrieve the food reward using a combination of distal and/or proximal cues. The ZT relies on the ability of the test subject to acquire and recall the location of the baited ziggurat and this is tested in consecutive training sessions of eight trials per day for 10 days. The location of the baited ziggurat is changed every second day, requiring the rats to learn a total of five different locations. Indices of learning and memory are based on several parameters, including latency to find the target, distance travelled, the number of visits to non-baited ziggurats (errors) and the number of returns. A recent study directly compared the performance of male and female Long-Evans rats in the wet-land MWT with the dry-land ZT. While males and females did not display significant differences in the traditional measures of spatial navigation within the MWT, they displayed a robust, male biased, sex difference in all measures of the ZT indicating task-specific gender differences in spatial performance. Taken together these findings suggest that males and females may employ different learning strategies in the MWT and ZT and that the latter task provides a more favourable task for assessing gender differences in spatial memory in rats (Faraji et al. 2010).

Fluctuations in sex hormones (e.g., oestrogen, progesterone and testosterone) will undoubtedly cause a shift in male and female performance. Varying levels of oestrogen results in variations in spatial learning and memory so that, when tested across the oestrous cycle females perform as well as males on days of low oestrogen but poorly when levels rise (Frye 1995; Warren and Juraska 1997). When female Long Evans rats are tested at a single point during the oestrous cycle, females in oestrous outperform those during the pro-oestrous phase (Warren and Juraska 1997). Conversely, other studies have found limited or no cyclical variations in performance (Healy et al. 1999; Bucci et al. 1995; Berry et al. 1997), indicating that retention for spatial information may be preserved despite morphological alterations in hippocampal dendritic spine density in the normally cycling female rat. Interestingly, administration of chronic oestrogen to ovariectomised rats enhanced spatial memory during the RAM task (Luine et al. 1998) and age-associated decline in sex steroid levels have been found to more profoundly impair spatial working memory in female rats in comparison with their male counterparts (Markowska 1999). Taken together, this evidence suggests a pivotal role for oestrogen for mnemonic abilities.

Variations in stress levels of the animals whilst under testing conditions may also play a critical role in gender-specific abilities during spatial paradigms. A growing body of literature demonstrates that certain aspects of spatial mnemonic function are dependent upon a stress response pattern, which is shown to be sexually dimorphic. The deleterious influence of chronic stress evident in male subjects in numerous spatial tasks, displayed as an impaired performance is not evident in female rats and this effect is thought to be mediated by oestrogen (to be discussed in more detail later in this chapter, Sect. 3).

2.3 Social Recognition

Social recognition memory, vitally important for social interaction and the establishment of dominance hierarchies, is the ability to discriminate between unfamiliar and familiar conspecifics. In rodents, the task utilises chemosensory cues present in the anogenital region, which composes an "olfactory signature". Similar to the NOR task, the amount of time one individual spends investigating another may be evaluated in the laboratory. Young (3–5 months) adult female rats have been shown to discriminate between novel and familiar juveniles for longer intervals than males (a 120 min compared to a 90 min interval), although aged male and female rats (16.5–19.5 months) have been shown to display the same social discrimination memory abilities (Markham and Juraska 2007). The superior memory ability of females was not found to be preserved during ageing, discrimination between a novel and familiar juvenile was abolished after a 120 min interval.

More recently, the influence of gender and the oestrous cycle has been evaluated on the formation and long-term (24 h) maintenance of social recognition memory in mice with a focus on the respective involvement of α- and β-oestrogen receptors. Female wild-type animals were able to successfully form memories during all stages of their oestrous cycle however, only when learning occurred during proestrus (when oestrogen levels are highest) was the memory retained for a period of 24 h. The acyclic, α-receptor knockout female mice demonstrated impairments in both the formation and the maintenance of social recognition memory, whereas β-receptor female knockouts showed no significant deficits and exhibited the same proestrus-dependent retention of memory at 24 h akin to the wild-type mice. To investigate gender differences, male α- and β-oestrogen receptor knockout mice were also evaluated in the same paradigm demonstrating similar results to the females. The male α-receptor knockouts had normal memory formation and only exhibited the 24 h memory retention deficit, indicating a greater female dependence on α-receptor expression for memory formation during this specific task (Sánchez-Andrade and Kendrick 2011).

Historically, the behavioural phenomenon of social recognition has been extensively studied in males. Insights into the impact of hormonal states on recognition memory (towards a juvenile male) in females have demonstrated that in young adult female rats social recognition memory is negatively affected 3 weeks after ovariectomy and restored with oestrogen replacement, co-incidentally, similar those deficits observed during the NOR task (Fig. 1e, f. Sutcliffe et al., unpublished findings. Complementing these findings it was further observed that 6 weeks after the termination of oestrogen replacement recognition memory was once again impaired (Hlináck 1993).

2.4 Classical and Operant Conditioning

The classical, hippocampal-dependent, trace eyeblink conditioning paradigm is viewed as an associative task in which the animal is presented with a conditioned stimulus (normally white noise) followed closely by an aversive eyelid stimulation, the result of which causes the animal to blink (an unconditioned response). As the animal learns, the unconditioned response becomes a conditioned response (i.e., the animal learns to predict the eyelid stimulation in advance). Prior to training, age-matched male and female Sprague-Dawley rats express similar levels of spontaneous blinking; however, females learn to anticipate the onset of the unconditioned stimulus and thus learn to time the conditioned response (i.e. eyeblink) sooner than males, this response being most evident on the first day of training. At the end of training, there are no gender differences but interestingly, more females than males reach a criterion of 60% conditioned responses (7 out of 8 females compared to 7 out of 10 males reaching the same criterion) (Dalla et al. 2009). Perhaps not surprisingly, re-exposure of trained animals to the conditioned stimulus some weeks later elicits a higher percentage conditioned response in females compared to the males suggesting that the female rats have learnt and retained the rules of this task better than the males (Dalla et al. 2009). Further demonstrating oestrogen sensitivity in this task, removal of the ovaries prevents the sex differences in performance (Wood and Shors 1998) but enhanced conditioned responding is displayed in ovariectomised female rats following two injections of 40 μg estradiol 24 h apart, albeit at supraphysiological doses which produced plasma estradiol levels of greater than 250 pg/mL (Leuner et al. 2004).

Sex differences have also been reported in another type of classical conditioning referred to as fear conditioning. Typically, the animal is trained to associate a cue or a context (e.g., a tone) with an aversive stimulus (i.e. a footshock). Re-exposure to the same cue or context results in the animal "freezing" or to express an enhanced startle reflex in anticipation of the aversive stimulus. A male bias is typically observed during cue fear conditioning since male rats acquire the association between the cue and foot shock quicker than females (Maren et al. 1994). The ability of males to outperform females within this paradigm is also a consistent finding in three strains of laboratory rat, the Wistar, Fischer and Lewis (Pryce et al. 1999). Castration elicits no effect on the conditioned response (Anagnostaras et al. 1998); in contrast, ovariectomy results in female rats displaying comparable levels of fear to males (Gupta et al. 2001).

Although in classical fear conditioning studies males have outperformed the females, during more complex avoidance operant tasks the opposite is true. During an operant task, the animal must make an overt response in order to learn, which is often to escape an aversive stimulus – normally a mild foot shock. During the one-way avoidance task (FR1), the animal must learn to pass through the door way of a shuttle box once to avoid a mild footshock (FR1). This task is often learnt within a day and it has been observed that female rats will learn this task sooner than the males (Dalla et al. 2008; Shors et al. 2007). When the task difficulty is increased to the two-way avoidance task (FR2) where the animals must learn to pass through the doorway twice to terminate the footshock, more striking gender differences have been observed (Dalla et al. 2008; Shors et al. 2007). Females acquire the FR2 task within the first few trials but males require more trials and in some cases never learn the rule (Dalla et al. 2008). It would appear that female rats respond actively to aversive stimuli, whereas males exhibit behavioural inhibition with passive reactions and freezing. Ovary removal does not prevent the gender difference in operant conditioning (Beatty and Beatty 1970; Van Oyen et al. 1981); however, cyclic oestrogen and progesterone influence conditioned avoidance behaviour and escape latencies as evidenced by enhanced performance during pro-oestrous when oestrogen levels are at their highest (Sfikakis et al. 1978).

Another form of operant conditioning which has shown sexual dimorphism is differential-reinforcement-of-low-rate of responding (DRL), a task in which the animal must learn to press a lever for a reward (often food). Irrespective of reward receipt, male animals have shown better performance when compared to females during acquisition of instrumental responding, which is contributed to a higher level of interaction with the lever (van Haaren et al. 1990). On the contrary, when the task difficulty is more complex and the animals have to systematically increase the number of times they press the lever to receive the reward, females outperform the males (van Hest et al. 1987) – although incentive motivation for the reward could be a driving factor. Gender differences in differential reinforcement paradigms are also reliant upon the activational effects of the gonadal hormones, since they are not observed during puberty and are abolished with ovariectomy (Beatty 1973).

Finally, sex differences are also observed in stimulant self-administration paradigms. Female rats are consistently reported to be more sensitive to the reinforcing actions of stimulants and acquire stable, high levels of drug self-administration of low doses of cocaine, methamphetamine, opioids and nicotine at a faster rate than males [for comprehensive reviews see (Fattore et al. 2009; Roth et al. 2004; Lynch et al. 2002)]. Ovariectomy results in slower cannabinoid self-administration acquisition when compared to cycling Long-Evans and Lister-hooded rats. Intriguingly, in these rat strains ovariectomy decreased cannabinoid intake to the same level as those shown by males (Fattore et al. 2007, 2009), suggesting that the ovarian hormones play a crucial role in these responses to cannabinoids.

3 Female Advantage Under Stressful Conditions

The limbic region, critical for processing information, shows plasticity under chronic stress with important gender-related differences (Eichenbaum et al. 2007; Lipton and Eichenbaum 2008). Chronic stress has repeatedly been shown to impair spatial learning and memory in male subjects yet produces different outcomes in females, thus gender is gaining recognition as an important variable acting as a mitigating factor or fundamental aetiology influencing stress-related disorders. Male rats exposed to chronic stress (6 h daily restraint for 21 days) perform poorly on tasks where normally a common male advantage is observed such as the RAM (Luine et al. 1993, 1994; Bowman et al. 2003), Y-maze (Conrad et al. 1996; Wright and Conrad 2005) and the MWM (Markowska 1999). In an almost opposite outcome, stressed female rats perform and complete the maze task with fewer errors and more correct choices (30% enhancement) when compared to unstressed females (Bowman et al. 2003). Not only does this female advantage influence classical tasks for spatial memory, stress has also been demonstrated to enhance female rat performance during the object placement task at an ITI of 2.5 h and 4 h (Beck and Luine 2002), where unstressed females are unable to perform this task at ITIs of greater than 1 h.

Dendritic retraction of the CA3 region of the hippocampus has been implicated in the observed sexual dimorphisms under stressful situations. Chronic stress administered to cycling female rats has been shown to result in mild (Galea et al. 1997) or no (McLaughlin et al. 2010) dendritic retraction in the CA3 region. In ovariectomised rats, chronic stress produces drastic CA3 dendritic retraction (McLaughlin et al. 2010); however, spatial learning and memory remains functional and is even facilitated in ovariectomised and cycling chronically stressed female rats. This disconnection between the CA3 dendritic retractions and spatial memory in females may be attributed to the ovarian hormones. Indeed, within the CA3 region, females express more ERβ immunoreactivity when compared to the CA3 region in male rats (Zhang et al. 2002), implying that oestrogens may be the key neuroprotective agent to prevent stress-induced CA3 dendritic attrition.

4 Mechanisms of Sex Hormone Action for Cognition? The Oestrogen Perspective

Much evidence now supports a role for the interplay of many neurotransmitter systems, neuroplasticity and oestrogen with respect to understanding cognitive abilities. Many of oestrogen's actions in the brain are attributed to the activation of the classical ERs α and β (ERα and ERβ, respectively) and their subsequent impact on synaptic density and morphology, which is concomitant with alterations in learning and memory.

4.1 Hippocampal Architecture

The medial temporal lobe system, including the hippocampal formation (entorhinal cortex, dentate gyrus, areas CA1-4 and subiculum), amygdale, and the parahippocampal cortices are considered to serve as a declarative memory system. The hippocampal formation, along with the frontal cortex, is an extensively explored area of the brain with regard to cognitive competence. This region has been investigated by means of excitotoxic, ablation and radiofrequency lesions, transient neuronal inactivation (lidocaine), lesions, or transections of its connections, by pharmacological N-methyl-D-Aspartate (NMDA) receptor blockade and even via genetic inactivation of the CA1-NMDA receptors (Dere et al. 2007) pre-clinically – evidence of which is supported by human lesion studies (Zola-Morgan et al. 1986). Cytotoxic and radio-frequency lesions to the hippocampus have been shown to result in an object recognition impairment after inter-trial intervals of 5 and 10 min, 1, 4 and 24 h (Ainge et al. 2006; Clark et al. 2000; Mumby et al. 2002). However, the size of the lesion to the hippocampus (>75%) remains important for object recognition memory (Broadbent et al. 2004). Oestrogen-associated changes (direct or indirect) within the hippocampus impact upon neural pathways, acting to alter performance in specific tasks. In the hippocampus of the female rat, Woolley and colleagues (Woolley et al. 1990; Woolley and McEwen 1992, 1994; Woolley et al. 1997; Woolley 1998) discovered that CA1 spine density naturally fluctuates across the female rat oestrous cycle, peaking with an increase of 32% during pro-oestrus when oestrogen levels are at a maximum compared to the di-oestrus phase. Subsequent studies have shown that oestrogen replacement in the ovariectomised female rat increases CA1 spine density (Woolley and McEwen 1992, 1993; Woolley 1998; Birzniece et al. 2006) through NMDA receptor-mediated mechanisms (Woolley and McEwen 1994; Woolley et al. 1997) and ERβ activation (Liu et al. 2008). This research may provide important mechanistic evidence for the female advantage in object and social recognition tasks; furthermore, neuroprotection is observed when oestrogen is administered prior to the NMDA antagonist, PCP (Sutcliffe et al. 2008). This important relationship between glutamate, NMDA and oestrogen is discussed in more detail in Sects. 4.3 and 4.4.

4.2 Oestrogen Receptors

Sex differences exist in the majority of brain regions including many involved in cognitive function such as the hippocampus, amygdala and neocortex (Juraska 1991). In many cases, these gender differences are not evident in overt anatomical structure, but in functional dimensions (e.g., a brain region may differ between the sexes in aspects of neurotransmitter function). The majority of research has investigated the effect of the steroid oestrogen on cognitive function and ERs have been found in several areas of the brain including the amygdala, cerebral cortex, cerebellum and hippocampus (Shughrue and Merchenthaler 2000; Tsutsui et al. 2004). ERs are important not only for the sexual differentiation of the brain but also are known to exert receptor-mediated functions on a number of behavioural functions. While the existence of ERα has been long known (Jensen et al. 2010), the discovery of ERβ has been much more recent (Kuiper et al. 1996). Of particular interest to the present topic is the hippocampus which is known to play a significant role in working and spatial memory (Jarrard 1993) and which expresses both forms (α and β) of the ER (Cahill 2006; Frye 1995; Birzniece et al. 2006); however, their distribution does not completely overlap (Milner et al. 2001, 2005). ERβ is the predominant ER in the cerebral cortex and the hippocampus. Other areas within the brain which are confirmed as supporting mnemonic processes include the frontal cortex and the striatum, and it has become clear that there are strong interconnections (or loops) between the hippocampus

Table 2 Distribution and gender differences in ERβ immunoreactivity in the Wistar rat brain

Region of interest	Female	Male
CA1	+/−	+/−
CA2	+/−	+/−
CA3	++	+
CA4	++	+
Dentate gyrus	++	+
Endopiriform nucleus	+++	+++
Medial septal nucleus	+++	++
Purkinje cells	+++	+++
Lateral and medial amygdaloid	++	+

+++, high; ++, moderate, +, low and +/−, weak.
Source: Shughrue et al. (2000)

and frontal cortex (Vertes 2006). In the frontal cortex of male and female rats, each ER isoform has been shown to display its own unique, selective distribution (Zhang et al. 2002; Kritzer 2002). Expression of ERβ in the medial mammillary nucleus (a limbic area often associated with the hypothalamus) was only detected in the male rat brain. The female rat brain shows a higher predominance of ERβ immunoreactivity (Table 2) in the medial septal nucleus (an area which receives reciprocal connections from the olfactory bulb, hippocampus, amygdala and hypothalamus), pyramidal cells of the CA3 and CA4, the dentate gyrus and the lateral amygdaloid nucleus (Shughrue et al. 2000). Whether such gender differences in anatomical expression occur across species and in man has yet to be shown but the development of novel ER molecules as therapeutic agents remains an exciting prospect, especially considering the growing body of literature suggesting ERβ modulation regulates neuroplasticity and cognition (Liu et al. 2008; Choleris et al. 2008; Walf et al. 2008; Rhodes and Frye 2006).

4.3 Neurotransmission

The ability of oestradiol to induce and increase dendritic spine density is suggested to be due to the reduction of GABA inhibition by oestradiol in the hippocampal area (Murphy et al. 1998; Weaver et al. 1997). Furthermore, this oestrogen-induced increase in spine density is positively correlated with increases in NMDA receptor binding and sensitivity (Daniel and Dohanich 2001) providing one possible mechanism for the ability of oestrogen to improve cognitive function, especially when administered prior to an NMDA receptor antagonist such as PCP (Sutcliffe et al. 2008). Ovariectomy results in significant decreases in choline acetyltransferase (ChAT) activity and high affinity choline uptake (HACU) in the rat basal forebrain, hippocampal formation and cerebral cortex beyond normal ageing (Gibbs et al. 2002), which can be reversed by acute treatment with physiological levels of oestrogen [10 μg/kg oestradiol benzoate (Luine 1985; O'Malley et al. 1987) and

17-β-estradiol silastic capsules (Singh et al. 1994)]. Thus, oestrogen (exogenous or endogenous) may play an important role in the maintenance or regeneration of healthy cholinergic projections to the hippocampus and prefrontal cortex, which, in turn may be correlated with the enhanced cognitive performance observed during some cognitive paradigms when compared to males. One of the mechanisms by which oestrogen may enhance cognitive function is the modulation of the production and release of acetylcholine (ACh), which is shown to enhance the amplitude of synaptic potentials following long-term potentiation in regions such as the dentate gyrus, CA1, piriform cortex, and neocortex. This effect most likely occurs either through an indirect action on NMDA receptors, which are shown to be more abundant in the CA1 region of the female hippocampus compared to the male.

Work in understanding how oestrogen affects cognition has concentrated on the oestrogen-mediated changes in hippocampal spine density and synaptogenesis favouring an NMDA-dependent mechanism of GABAergic disinhibition of pyramidal neurons mediated through non-genomic actions. Recent reports indicate that septal cholinergic inputs are required for this oestrogen-mediated enhancement of working memory and NMDA receptor binding in the CA1 region of the hippocampus (Daniel and Dohanich 2001) and oestrogen-mediated changes in hippocampal spine density in rats (Lam and Leranth 2003), thereby linking these mechanisms.

4.4 Glutamate and Oestrogen

Glutamate is the predominant excitatory amino acid neurotransmitter in the cortical, hippocampal and hypothalamic areas of the brain (Brann 1995) and glutamate receptors are considered to be responsible for the glutamate-mediated post-synaptic excitation of neural cells – important for neural transmission, memory consolidation through synaptic plasticity and accordingly learning. Synergistic interactions between glutamate and the gonadal steroids may underlie multiple limbic roles. The two primary glutamate receptors are NMDA and α-amino-3-hydroxy-5-methyl-4-isoxazolepropionic acid (AMPA), named after their preferred agonists, and it is through the NMDA and AMPA receptors that glutamate may be neurotoxic. Glutamate toxicity may be predominantly mediated by the NMDA subclass of glutamate receptors (Choi et al. 1988). Administration of the non-competitive NMDA receptor antagonist, PCP, has reliably been shown to result in cognitive deficits in the rat (Neill et al. 2010) after both acute and sub-chronic administration, such deficits have been reversed with both the atypical antipsychotics (Grayson et al. 2007) and after pre-treatment with oestradiol-benzoate (Sutcliffe et al. 2008). Protective effects of oestrogens against glutamate toxicity have been described in cultured hippocampal neurons (Goodman et al. 1996). The toxicity experienced by each hippocampal culture was attenuated following a 2 h pre-treatment with 100 nM to 10 mM 17-β Estradiol, estriol or progesterone. This protective influence of oestrogen is further confirmed by the finding that a 24 h pre-treatment with 15–50 nM of 17-β Estradiol significantly decreased the lactate dehydrogenase efflux from primary cortical neurons exposed to glutamate for 5 min (Singer et al. 1996, 1999). Moreover, it was demonstrated that the selective ER modulator, tamoxifen, blocked the protective effects of oestrogen, thus suggesting that classical ER activations are required for oestrogen neuroprotection against glutamate toxicity. Accumulating evidence suggests a sexually dimorphic vulnerability to neurological insults with the ER receptors conveying differential protective capacities (Bryant and Dorsa 2010). Accordingly, a therapeutic opportunity exists for oestrogen modulators, not only in the acute reversal of disease-induced cognitive deficits but also for long-term neuroprotection in neurodegenerative disease.

5 Summary and Conclusions

Gender differences in cognition are evident in both pre-clinical models and in the clinic. Gender differences in learning vary as a function of the demands of the task and this review has primarily focussed on those where a female bias is observed. However, there are paradigms in which males excel such as those which require spatial navigation. In recent years, research in the area surrounding hormonal function in the central nervous system has increased with new discoveries and technology aiding the advance of knowledge regarding the effects of hormones on neurotransmission and consequent interactions within the brain. Despite these advances, research is still only in the early stages and suggests the complex nature of these systems with more questions being raised than answered. Only by increasing our understanding of gender differences with respect to learning and memory we can improve the translational value of our pre-clinical models in order to prevent and more effectively treat cognitive alterations associated with the aetiology and symptomatology of psychiatric and neuropsychological disorders.

References

Ainge JA et al (2006) The role of the hippocampus in object recognition in rats: examination of the influence of task parameters and lesion size. Behav Brain Res 167(1):183–195

Anagnostaras SG et al (1998) Testicular hormones do not regulate sexually dimorphic Pavlovian fear conditioning or perforant-path long-term potentiation in adult male rats. Behav Brain Res 92(1):1–9

Aubele T et al (2008) Effects of gonadectomy and hormone replacement on a spontaneous novel object recognition task in adult male rats. Hormones and Behavior 54(2):244–252

Baker MA (ed) (1987) Sex differences in human performance. Wiley, New York

Beatty WW (1973) Effects of gonadectomy on sex differences in DRL behavior. Physiol Behav 10(1):177–178

Beatty WW, Beatty PA (1970) Hormonal determinants of sex differences in avoidance behavior and reactivity to electric shock in the rat. J Comp Physiol Psychol 73(3):446–455

Beck KD, Luine VN (2002) Sex differences in behavioral and neurochemical profiles after chronic stress: role of housing conditions. Physiol Behav 75(5):661–673

Berry B, McMahan R, Gallagher M (1997) Spatial learning and memory at defined points of the estrous cycle: effects on performance of a hippocampal-dependent task. Behav Neurosci 111(2):267–274
Birzniece V et al (2006) Neuroactive steroid effects on cognitive functions with a focus on the serotonin and GABA systems. Brain Res Rev 51(2):212–239
Bowman RE, Beck KD, Luine VN (2003) Chronic stress effects on memory: sex differences in performance and monoaminergic activity. Horm Behav 43(1):48–59
Brann DW (1995) Glutamate: a major excitatory transmitter in neuroendocrine regulation. Neuroendocrinology 61(3):213–225
Broadbent NJ, Squire LR, Clark RE (2004) Spatial memory, recognition memory, and the hippocampus. Proc Natl Acad Sci USA 101(40):14515–14520
Bryant DN, Dorsa DM (2010) Roles of estrogen receptors alpha and beta in sexually dimorphic neuroprotection against glutamate toxicity. Neuroscience 170(4):1261–1269
Cahill L (2006) Why sex matters for neuroscience. Nat Rev Neurosci 7(6):477–484
Choi DW, Koh JY, Peters S (1988) Pharmacology of glutamate neurotoxicity in cortical cell culture: attenuation by NMDA antagonists. J Neurosci 8(1):185–196
Choleris E et al (2008) Estrogen receptor beta agonists in neurobehavioral investigations. Curr Opin Investig Drugs 9(7):760–773
Clark RE, Zola SM, Squire LR (2000) Impaired recognition memory in rats after damage to the hippocampus. J Neurosci 20(23):8853–8860
Conrad CD et al (1996) Chronic stress impairs rat spatial memory on the Y maze, and this effect is blocked by tianeptine pretreatment. Behav Neurosci 110(6):1321–1334
Corey SM (1930) Sex differences in maze learning by white rats. J Comp Psychol 10:333–338
Dalla C et al (2008) Females do not express learned helplessness like males do. Neuropsychopharmacology 33(7):1559–1569
Dalla C et al (2009) Female rats learn trace memories better than male rats and consequently retain a greater proportion of new neurons in their hippocampi. Proc Natl Acad Sci USA 106(8):2927–2932
Daniel JM, Dohanich GP (2001) Acetylcholine mediates the estrogen-induced increase in NMDA receptor binding in CA1 of the hippocampus and the associated improvement in working memory. J Neurosci 21(17):6949–6956
Dere E, Huston JP, De Souza Silva MA (2007) The pharmacology, neuroanatomy and neurogenetics of one-trial object recognition in rodents. Neurosci Biobehav Rev 31(5):673–704
Eichenbaum H, Yonelinas AP, Ranganath C (2007) The medial temporal lobe and recognition memory. Annu Rev Neurosci 30:123–152
Ennaceur A, Delacour J (1988) A new one-trial test for neurobiological studies of memory in rats 1: behavioral data. Behav Brain Res 31(1):47–59
Faraji J, Metz GA, Sutherland RJ (2010) Characterization of spatial performance in male and female Long-Evans rats by means of the Morris water task and the ziggurat task. Brain Res Bull 81(1):164–172
Fattore L et al (2007) Cannabinoid self-administration in rats: sex differences and the influence of ovarian function. Br J Pharmacol 152(5):795–804
Fattore L, Fadda P, Fratta W (2009) Sex differences in the self-administration of cannabinoids and other drugs of abuse. Psychoneuroendocrinology 34(Supplement 1):S227–S236
Frye CA (1995) Estrus-associated decrements in a water maze task are limited to acquisition. Physiol Behav 57(1):5–14
Galea LA et al (1997) Sex differences in dendritic atrophy of CA3 pyramidal neurons in response to chronic restraint stress. Neuroscience 81(3):689–697
Ghi P et al (1999) Sex differences in memory performance in the object recognition test. Possible role of histamine receptors. Pharmacol Biochem Behav 64(4):761–766
Gibbs RB et al (2002) Effects of long-term hormone replacement and of tibolone on choline acetyltransferase and acetylcholinesterase activities in the brains of ovariectomized, cynomologus monkeys. Neuroscience 113(4):907–914

Goodman Y et al (1996) Estrogens attenuate and corticosterone exacerbates excitotoxicity, oxidative injury, and amyloid beta-peptide toxicity in hippocampal neurons. J Neurochem 66(5):1836–1844
Grayson B, Idris NF, Neill JC (2007) Atypical antipsychotics attenuate a sub-chronic PCP-induced cognitive deficit in the novel object recognition task in the rat. Behav Brain Res 184(1):31–38
Gupta RR et al (2001) Estrogen modulates sexually dimorphic contextual fear conditioning and hippocampal long-term potentiation (LTP) in rats(1). Brain Res 888(2):356–365
Healy SD, Braham SR, Braithwaite VA (1999) Spatial working memory in rats: no differences between the sexes. Proc Biol Sci 266(1435):2303–2308
Hlinack Z (1993) Social recognition in ovariectomized and estradiol-treated female rats. Hormones and Behavior 27(2):159–166
Inagaki T, Gautreaux C, Luine V (2010) Acute estrogen treatment facilitates recognition memory consolidation and alters monoamine levels in memory-related brain areas. Hormones and Behavior 58(3):415–426
Jarrard LE (1993) On the role of the hippocampus in learning and memory in the rat. Behav Neural Biol 60(1):9–26
Jensen EV et al (2010) Estrogen action: a historic perspective on the implications of considering alternative approaches. Physiol Behav 99:151–162
Jonasson Z (2005) Meta-analysis of sex differences in rodent models of learning and memory: a review of behavioral and biological data. Neurosci Biobehav Rev 28(8):811–825
Juraska JM (1991) Sex differences in "cognitive" regions of the rat brain. Psychoneuroendocrinology 16(1–3):105–109
King MV et al (2004) 5-HT6 receptor antagonists reverse delay-dependent deficits in novel object discrimination by enhancing consolidation–an effect sensitive to NMDA receptor antagonism. Neuropharmacology 47(2):195–204
Kritzer MF (2002) Regional, laminar, and cellular distribution of immunoreactivity for ER alpha and ER beta in the cerebral cortex of hormonally intact, adult male and female rats. Cereb Cortex 12(2):116–128
Kuiper GG et al (1996) Cloning of a novel receptor expressed in rat prostate and ovary. Proc Natl Acad Sci USA 93(12):5925–5930
Lam TT, Leranth C (2003) Role of the medial septum diagonal band of Broca cholinergic neurons in oestrogen-induced spine synapse formation on hippocampal CA1 pyramidal cells of female rats. Eur J Neurosci 17(10):1997–2005
Leuner B, Mendolia-Loffredo S, Shors TJ (2004) High levels of estrogen enhance associative memory formation in ovariectomized females. Psychoneuroendocrinology 29(7):883–890
Lipton PA, Eichenbaum H (2008) Complementary roles of hippocampus and medial entorhinal cortex in episodic memory. Neural Plast 2008:258467
Liu F et al (2008) Activation of estrogen receptor-[beta] regulates hippocampal synaptic plasticity and improves memory. Nat Neurosci 11(3):334–343
Luine VN (1985) Estradiol increases choline acetyltransferase activity in specific basal forebrain nuclei and projection areas of female rats. Exp Neurol 89(2):484–490
Luine VN, Spencer RL, McEwen BS (1993) Effects of chronic corticosterone ingestion on spatial memory performance and hippocampal serotonergic function. Brain Res 616(1–2): 65–70
Luine V et al (1994) Repeated stress causes reversible impairments of spatial memory performance. Brain Res 639(1):167–170
Luine VN et al (1998) Estradiol enhances learning and memory in a spatial memory task and effects levels of monoaminergic neurotransmitters. Horm Behav 34(2):149–162
Lynch WJ, Roth ME, Carroll ME (2002) Biological basis of sex differences in drug abuse: preclinical and clinical studies. Psychopharmacology (Berl) 164(2):121–137
Maren S, De Oca B, Fanselow MS (1994) Sex differences in hippocampal long-term potentiation (LTP) and Pavlovian fear conditioning in rats: positive correlation between LTP and contextual learning. Brain Res 661(1–2):25–34

Markham JA, Juraska JM (2007) Social recognition memory: influence of age, sex, and ovarian hormonal status. Physiol Behav 92(5):881–888

Markowska AL (1999) Sex dimorphisms in the rate of age-related decline in spatial memory: relevance to alterations in the estrous cycle. J Neurosci 19(18):8122–8133

McLaughlin KJ et al (2010) Chronic 17beta-estradiol or cholesterol prevents stress-induced hippocampal CA3 dendritic retraction in ovariectomized female rats: possible correspondence between CA1 spine properties and spatial acquisition. Hippocampus 20:768–786

Milner TA et al (2001) Ultrastructural evidence that hippocampal alpha estrogen receptors are located at extranuclear sites. J Comp Neurol 429(3):355–371

Milner TA et al (2005) Ultrastructural localization of estrogen receptor beta immunoreactivity in the rat hippocampal formation. J Comp Neurol 491(2):81–95

Morris RGM (1981) Spatial localization does not require the presence of local ques. Learn Motiv 12:239–260

Mumby DG et al (2002) Hippocampal damage and exploratory preferences in rats: memory for objects, places, and contexts. Learn Mem 9(2):49–57

Murphy DD et al (1998) Estradiol increases dendritic spine density by reducing GABA neurotransmission in hippocampal neurons. J Neurosci 18(7):2550–2559

Neill JC et al (2010) Animal models of cognitive dysfunction and negative symptoms of schizophrenia: Focus on NMDA receptor antagonism. Pharmacol Ther 128:419–432

Nemanic S, Alvarado MC, Bachevalier J (2004) The hippocampal/parahippocampal regions and recognition memory: insights from visual paired comparison versus object-delayed nonmatching in monkeys. J Neurosci 24(8):2013–2026

Olton DS, Samuelson RJ (1976) Remembrance of places past: spatial memory in rats. J Exp Psychol Anim Behav Processes 2:97–116

O'Malley CA et al (1987) Effects of ovariectomy and estradiol benzoate on high affinity choline uptake, ACh synthesis, and release from rat cerebral cortical synaptosomes. Brain Res 403(2):389–392

Pascalis O et al (2004) Visual paired comparison performance is impaired in a patient with selective hippocampal lesions and relatively intact item recognition. Neuropsychologia 42(10):1293–1300

Pilgrim C, Reisert I (1992) Differences between male and female brains – developmental mechanisms and implications. Horm Metab Res 24(8):353–359

Pryce CR, Lehmann J, Feldon J (1999) Effect of sex on fear conditioning is similar for context and discrete CS in Wistar, Lewis and Fischer rat strains. Pharmacol Biochem Behav 64(4): 753–759

Rhodes ME, Frye CA (2006) ERbeta-selective SERMs produce mnemonic enhancing effects in the inhibitory avoidance and water maze tasks. Neurobiol Learn Mem 85(2):183–191

Roth ME, Cosgrove KP, Carroll ME (2004) Sex differences in the vulnerability to drug abuse: a review of preclinical studies. Neurosci Biobehav Rev 28(6):533–546

Sahay A, Hen R (2007) Adult hippocampal neurogenesis in depression. Nat Neurosci 10(9): 1110–1115

Sánchez-Andrade G, Kendrick KM (2011) Roles of [alpha]- and [beta]-estrogen receptors in mouse social recognition memory: effects of gender and the estrous cycle. Hormones and Behavior 59:114–122

Sfikakis A et al (1978) Implication of the estrous cycle on conditioned avoidance behavior in the rat. Physiol Behav 21(3):441–446

Shors TJ et al (2007) Neurogenesis and helplessness are mediated by controllability in males but not in females. Biol Psychiatry 62(5):487–495

Shughrue PJ, Merchenthaler I (2000) Estrogen is more than just a "sex hormone": novel sites for estrogen action in the hippocampus and cerebral cortex. Front Neuroendocrinol 21(1):95–101

Shughrue PJ, Scrimo PJ, Merchenthaler I (2000) Estrogen binding and estrogen receptor characterization (ERalpha and ERbeta) in the cholinergic neurons of the rat basal forebrain. Neuroscience 96(1):41–49

Singer CA et al (1996) Estrogen protects primary cortical neurons from glutamate toxicity. Neurosci Lett 212(1):13–16
Singer CA et al (1999) The mitogen-activated protein kinase pathway mediates estrogen neuroprotection after glutamate toxicity in primary cortical neurons. J Neurosci 19(7):2455–2463
Singh M et al (1994) Ovarian steroid deprivation results in a reversible learning impairment and compromised cholinergic function in female Sprague-Dawley rats. Brain Res 644(2):305–312
Sutcliffe JS, Marshall KM, Neill JC (2007) Influence of gender on working and spatial memory in the novel object recognition task in the rat. Behav Brain Res 177(1):117–125
Tsutsui K et al (2004) Organizing actions of neurosteroids in the Purkinje neuron. Neurosci Res 49(3):273–279
van Haaren F, van Hest A, Heinsbroek RP (1990) Behavioral differences between male and female rats: effects of gonadal hormones on learning and memory. Neurosci Biobehav Rev 14(1):23–33
van Hest A, van Haaren F, van de Poll NE (1987) Behavioral differences between male and female Wistar rats on DRL schedules: effect of stimuli promoting collateral activities. Physiol Behav 39(2):255–261
Van Oyen HG, Walg H, Van De Poll NE (1981) Discriminated lever press avoidance conditioning in male and female rats. Physiol Behav 26(2):313–317
Vertes RP (2006) Interactions among the medial prefrontal cortex, hippocampus and midline thalamus in emotional and cognitive processing in the rat. Neuroscience 142(1):1–20
Walf AA, Rhodes ME, Frye CA (2006) Ovarian steroids enhance object recognition in naturally cycling and ovariectomized, hormone-primed rats. Neurobiol Learn Mem 86(1):35–46
Walf AA, Koonce CJ, Frye CA (2008) Estradiol or diarylpropionitrile administration to wild type, but not estrogen receptor beta knockout, mice enhances performance in the object recognition and object placement tasks. Neurobiol Learn Mem 89(4):513–521
Warren SG, Juraska JM (1997) Spatial and nonspatial learning across the rat estrous cycle. Behav Neurosci 111(2):259–266
Weaver CE Jr et al (1997) 17beta-Estradiol protects against NMDA-induced excitotoxicity by direct inhibition of NMDA receptors. Brain Res 761(2):338–341
Wilson CA, Davies DC (2007) The control of sexual differentiation of the reproductive system and brain. Reproduction 133(2):331–359
Wood GE, Shors TJ (1998) Stress facilitates classical conditioning in males, but impairs classical conditioning in females through activational effects of ovarian hormones. Proc Natl Acad Sci USA 95(7):4066–4071
Woolley CS (1998) Estrogen-mediated structural and functional synaptic plasticity in the female rat hippocampus. Horm Behav 34(2):140–148
Woolley CS, McEwen BS (1992) Estradiol mediates fluctuation in hippocampal synapse density during the estrous cycle in the adult rat. J Neurosci 12(7):2549–2554
Woolley CS, McEwen BS (1993) Roles of estradiol and progesterone in regulation of hippocampal dendritic spine density during the estrous cycle in the rat. J Comp Neurol 336(2):293–306
Woolley CS, McEwen BS (1994) Estradiol regulates hippocampal dendritic spine density via an N-methyl-D-aspartate receptor-dependent mechanism. J Neurosci 14(12):7680–7687
Woolley CS et al (1990) Naturally occurring fluctuation in dendritic spine density on adult hippocampal pyramidal neurons. J Neurosci 10(12):4035–4039
Woolley CS et al (1997) Estradiol increases the sensitivity of hippocampal CA1 pyramidal cells to NMDA receptor-mediated synaptic input: correlation with dendritic spine density. J Neurosci 17(5):1848–1859
Wright RL, Conrad CD (2005) Chronic stress leaves novelty-seeking behavior intact while impairing spatial recognition memory in the Y-maze. Stress 8(2):151–154
Zhang JQ et al (2002) Distribution and differences of estrogen receptor beta immunoreactivity in the brain of adult male and female rats. Brain Res 935(1–2):73–80
Zola-Morgan S, Squire LR, Amaral DG (1986) Human amnesia and the medial temporal region: enduring memory impairment following a bilateral lesion limited to field CA1 of the hippocampus. J Neurosci 6(10):2950–2967

Sex Differences in the Septo-Hippocampal Cholinergic System in Rats: Behavioral Consequences

Dai Mitsushima

Contents

Abstract The hippocampus is processing temporal and spatial information in particular contexts of episodes. Using freely moving rats, we monitored extracellular levels of acetylcholine (ACh), a critical neurotransmitter activating hippocampal circuits. We found that the ACh release in the dorsal hippocampus increases during the period of learning or exploration, exhibiting a sex-specific 24-h release profile. Moreover, neonatal increase in circulating androgen not only androgenizes behavioral and hormonal features, but also produces male-type ACh release profile after the development. The results suggest neonatal sexual differentiation of septo-hippocampal cholinergic system. Environmental conditions (such as stress, housing or food) of animals further affected the ACh release.

Although recent advances of neuroscience successfully revealed molecular/cellular mechanism of learning and memory, most research were performed using

D. Mitsushima
Department of Physiology, Yokohama City University Graduate School of Medicine, 3-9 Fukuura Kanazawaku, Yokohama 236-0004, Japan
e-mail: dm650314@med.yokohama-cu.ac.jp

J.C. Neill and J. Kulkarni (eds.), *Biological Basis of Sex Differences in Psychopharmacology*, Current Topics in Behavioral Neurosciences 8, DOI 10.1007/7854_2010_95,

male animals at specific time period. Sex-specific or time-dependent hippocampal functions are still largely unknown.

Keywords Acetylcholine · Androgen · Diurnal rhythm · Estrogen · Learning and memory · Sex difference

1 General Introduction

The hippocampus is a part of the limbic system and is a brain structure critically involved in learning and memory. The hippocampus processes temporal and spatial information within specific episodes (Komorowski et al. 2009; Gelbard-Sagiv et al. 2008). In freely moving animals, acetylcholine (ACh) release in the hippocampus increases during learning or exploration, showing a temporal 24-h release profile. ACh release changes with spontaneous movement (Day et al. 1991; Mitsushima et al. 1996) that stimulates electrical activity of cholinergic neurons in the basal forebrain (Buzsáki et al. 1988). Moreover, voluntary running enhances learning in mice (van Praag et al. 1999), while a restriction of exploratory behavior impairs ACh release and learning (Mitsushima et al. 1998, 2001). In this review, we focused on in vivo ACh release in the hippocampus to improve our understanding of the role of sexual dimorphism and temporal effects on hippocampal function.

2 Physiological Role of ACh in the Hippocampus

A number of studies suggest that ACh plays an important role in orchestrating major hippocampal functions (Fig. 1). In behavioral studies, ACh release increases during learning (Ragozzino et al. 1996; Stancampiano et al. 1999; Hironaka et al. 2001) and is positively correlated with learning performance (Gold 2003; Parent and Baxter 2004). Bilateral injections of scopolamine into the dorsal hippocampus impair spatial learning ability (Herrera-Morales et al. 2007), suggesting that muscarinic ACh receptors mediate the formation of spatial memory. At the network level, ACh generates a theta rhythm (Lee et al. 1994) that modulates the induction of long-term potentiation (LTP) in hippocampal CA1 neurons (Hyman et al. 2003). Studies exploring a genetic deficiency of muscarinic ACh receptors (M_1 or M_2) further show the impairment of LTP in the CA1 region (Seeger et al. 2004; Shinoe et al. 2005). At the cellular level, both pyramidal and nonpyramidal neurons in the hippocampal CA1 area receive direct cholinergic afferents mediated by muscarinic receptors (Cole and Nicoll 1983; Markram and Segal 1990; Widmer et al. 2006). In vitro studies showed that bath application of carbachol, a cholinergic agonist, induces LTP in CA1 pyramidal neurons without electrical stimulus,

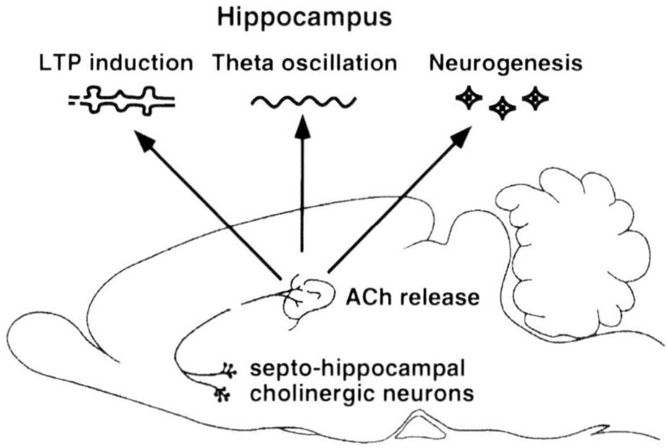

Fig. 1 Schematic illustration of septo-hippocampal cholinergic neurons. The released ACh activates major hippocampal functions. ACh, acetylcholine. LTP, long-term potentiation

suggesting that ACh in the hippocampus plays a principal role in the synaptic plasticity of the CA1 pyramidal neurons (Auerbach and Segal 1996). Furthermore, a recent study revealed an intracellular mechanism of ACh: focal activation of muscarinic ACh receptors in one CA1 pyramidal neuron induces Ca^{2+} release from inositol 1,4,5-trisphosphate-sensitive stores to induce LTP (Fernández de Sevilla et al. 2008). Furthermore, not only is ACh critically involved in synaptic plasticity, but ACh release in the hippocampus is also responsible for neurogenesis in the dentate gyrus. Thus, neurotoxic lesions of forebrain cholinergic neurons or long-term scopolamine treatment significantly decreases the number of newborn cells in the dentate gyrus, approximately 90% of those were also positive for the neuron-specific marker NeuN (Mohapel et al. 2005; Kotani et al. 2006).

3 Monitoring of In Vivo ACh Release

Cholinergic neurons within the basal forebrain provide the major projection to the neocortex and hippocampus (Mesulam et al. 1983). Cortical regions receive cholinergic inputs mainly from the nucleus basalis magnocellularis (NBM) or the diagonal band of Broca, whereas the hippocampus receives cholinergic inputs mostly from the medial septum and horizontal limb of the diagonal band of Broca (Mesulam et al. 1983). Because the cholinergic projections are necessary to maintain learning and memory (Perry et al. 1999, Sarter and Parikh 2005), we hypothesized that in vivo monitoring of ACh release in the hippocampus is necessary to elucidate learning function. To measure ACh release, we have performed in vivo microdialysis studies in freely moving male rats. Briefly, a microdialysis probe with a semipermeable membrane (1.0 mm in length) was inserted into a specific brain area via a surgically pre-implanted guide cannula. We perfused the

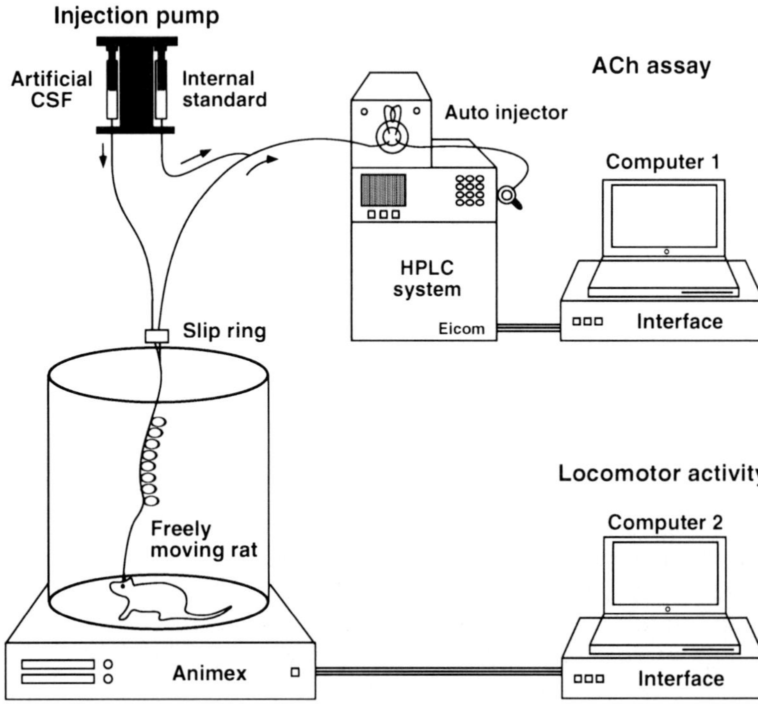

Fig. 2 Experimental setup of in vivo microdialysis system. In order to evaluate the activational effect of sex hormones on ACh release, we simultaneously measured spontaneous locomotor activity in the same subject

inside of the membrane with artificial cerebrospinal fluid, and assayed ACh in dialysates using a high-performance liquid chromatography system. As a result, we were successful in determining an in vivo ACh release profile in selected brain areas in freely moving rats (Fig. 2).

4 ACh Release in the Hippocampus Is Time-Dependent

Using this in vivo measuring system, we showed a temporal 24-h profile of ACh release in the hippocampus. ACh release was episodically observed during the dark phase, but the episodic release was not frequently observed during the light phase (Mitsushima et al. 1998; Masuda et al. 2005). Simultaneous monitoring of spontaneous behavior revealed that the temporal pattern of ACh release is highly correlated with spontaneous movement in freely moving rats (Day et al. 1991; Mizuno et al. 1991; Mitsushima et al. 1996). Since a restriction of exploratory behavior reduces ACh levels and also spatial learning (Mitsushima et al. 1998, 2001), episodic spontaneous behaviors may activate ACh release. In addition, spontaneous behavior stimulates electrical activity of cholinergic neurons in the basal forebrain (Buzsáki et al. 1988). Moreover, voluntary running enhances neurogenesis, spatial learning, and synaptic plasticity in mice (van Praag et al. 1999). Interestingly, this daily change is quite similar to the daily rhythm in hippocampal mitogen activated protein kinase (MAPK) activity and cAMP: phosphorylated extracellular signal-regulated (ERK) protein, GTP-bound Ras protein, and cAMP in the hippocampus show clear daily changes in male mice (Eckel-Mahan et al. 2008). Although the time-resolution of molecular changes may be low at present, it would be of interest to elucidate the intracellular signaling change with spontaneous behavior in future.

5 Sex Differences in ACh Release

We first reported sex-specific ACh release in the hippocampus in 2003 (Mitsushima et al. 2003a). Gonadally intact male rats consistently show a greater ACh release in the hippocampus compared with diestrous or proestrous female rats, suggesting a sexually dimorphic septo-hippocampal cholinergic system. Moreover, we found that sex-dependent ACh release also shows a time-dependent 24-h profile: ACh release in the hippocampus was relatively similar in the light phase, but consistently lower in female compared with male rats in the dark phase (Masuda et al. 2005). Although ACh release clearly showed a daily rhythm in female rats, females exhibited smaller amplitude of daily change than males. However, it is necessary to rule out the possibility that the sex difference in ACh release reflects the differences in spontaneous locomotor activity levels. By simultaneous monitoring of ACh levels and spontaneous locomotor activity, we revealed a real sex difference in the "ACh release property" (Fig. 3, Mitsushima et al. 2009): males showed higher ACh release than females while displaying similar levels of behavioral activity. Although female rats showed slightly higher overall spontaneous activity than intact male rats, male rats showed higher ACh release than female rats. Simple linear regression analysis was used to evaluate the relationship between ACh levels and spontaneous locomotor activity (Fig. 3). Pearson's correlation coefficient (r) or slope of the best fit line was calculated for each rat, and sex difference was evaluated using ANOVA. We found that the data from intact males had a steep slope of fit line, while the data from females had a gentle slope. These results suggest that sex-specific ACh release is not due to the change in spontaneous behavior, but due to actual differences in the ACh release property in gonadally intact rats (Mitsushima et al. 2009).

To analyze the sex difference in the septo-hippocampal cholinergic neurons, we performed immunocytochemistry. Stereological analysis showed that no sex difference was observed in the number of choline acetyltransferase immunoreactive (ChAT-ir) cells in the medial septum or horizontal limb of diagonal band (Takase et al. 2009). Since the number of septo-hippocampal cholinergic neurons does not appear to be involved in the sex difference in ACh release in the hippocampus, we

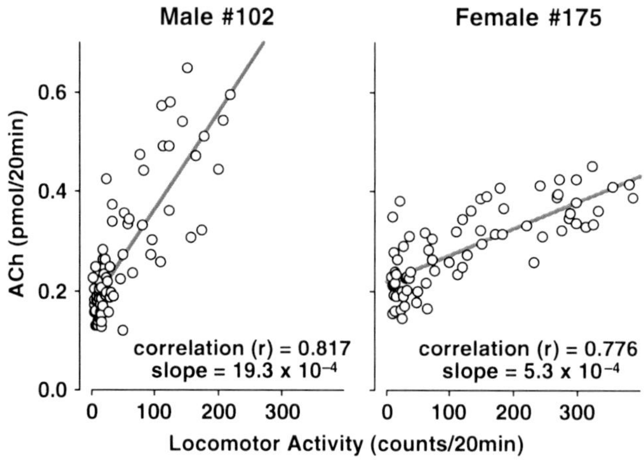

Fig. 3 Sex specific ACh release property in behaving rats. Representative data from a male (#102) and a female (#175) rat were shown. Simple linear regression analysis revealed a sex-specific "ACh release property." Male rats showed higher ACh release than females undergoing similar behavioral activity levels. Although both sexes showed a high correlation, male rats showed a steeper slope than female rats. (see Mitsushima et al. 2009)

hypothesized that sex-specific neural circuits or substance(s) may control the endogenous release.

6 Neural Control of Septo-Hippocampal Cholinergic Neurons

Neurotransmitters may be involved in expression of the sex difference in ACh release. For instance, dopaminergic neurons in the ventral tegmental area (A10) have been shown to control septo-hippocampal cholinergic neurons through the A10-septal dopaminergic pathway in male rats (Swanson 1982; Nilsson et al. 1992; Yanai et al. 1993). A neuroanatomical study suggested that dopamine D_2 receptors rather than D_1 receptors mediate the dopaminergic control of septo-hippocampal cholinergic neurons (Weiner et al. 1991). It has been shown that opiatergic neurons also control septo-hippocampal cholinergic neurons in male rats (Mizuno and Kimura 1996); the injection of naloxone, a μ opioid receptor antagonist, into the medial septum markedly increased ACh release in the hippocampus, while a μ opioid receptor agonist decreased its release (Mizuno and Kimura 1996). In contrast, GABA seems to inhibit septo-hippocampal cholinergic neurons; the injection of muscimol, a GABA receptor agonist, into the medial septum decreased ACh release in the hippocampus, while the injection of bicuculline, a GABA receptor antagonist, increased it (Moor et al. 1998). Although the neural systems are still unknown for female rats, it seems likely that neural control of septo-hippocampal cholinergic neurons is involved in the expression of sex differences in ACh release. It will be important to investigate these neural systems in female rats in future studies.

7 Circulating Sex Steroids Activate ACh Release

Not only neurotransmitters, but also circulating sex steroids, may regulate cholinergic neurons. In fact, neuroanatomical studies have demonstrated that, in intact male and female rats, a number of dopaminergic neurons in the A10 region have androgen receptor immunoreactivity (Kritzer 1997) and 45–60% of cholinergic neurons in the medial septum have estrogen receptor α immunoreactivity (Miettinen et al. 2002; Mufson et al. 1999). Taken together with the fact that female rats show a greater circulating estrogen concentration than male rats (Shors et al. 2001; Mitsushima et al. 2003b) and male rats show a greater circulating androgen concentration than female rats (Falvo et al. 1974; Rush and Blake 1982), it is possible that cholinergic neurons are affected by sex steroids differently in male and female rats.

The activational effects of sex steroids on cholinergic neurons have been suggested by previous neuroanatomical and neurochemical findings. For example, male gonadectomy decreases the density of cholinergic fibers in the dorsal hippocampus, while testosterone replacement in gonadectomized male rats maintains fiber density (Nakamura et al. 2002). Also, estradiol increases the induction of choline acetyltransferase in the basal forebrain in gonadectomized female rats (Luine et al. 1986; McEwen and Alves 1999). A previous in vitro study demonstrated that estradiol treatment increases both high affinity choline uptake and ACh synthesis in basal forebrain neurons (Pongrac et al. 2004). Furthermore, we recently reported an activational effect of sex steroids on the maintenance of stress-induced ACh release in the dorsal hippocampus in immobilized rats (Mitsushima et al. 2008). These findings suggest the activational effect of sex steroids on ACh release in the dorsal hippocampus, and we presented conclusive evidence of activational effects on dynamic ACh changes in behaving animals. To analyze the precise effects of sex steroids on ACh release, we simultaneously analyzed ACh release and spontaneous locomotor activity to determine the precise effect of sex steroids. Simultaneous analysis revealed that gonadectomy severely impaired ACh release without affecting spontaneous locomotor activity levels. Moreover, the activational effect on ACh release was apparent, especially during the active period, i.e., the dark phase, but not during the rest period, the light phase (Fig. 4 and Mitsushima et al. 2009). Our results provide the first evidence that the sex-specific 24-h profile of ACh release is highly dependent on the presence of sex steroids.

Moreover, we found that after gonadectomy, the positive correlation between ACh release and locomotor activity levels was severely impaired, suggesting that hippocampal function may not always be activated at low sex steroid levels (Mitsushima et al. 2009). This therefore suggests that learning impairment in gonadectomized rats (Gibbs and Pfaff 1992; Daniel et al. 1997; Kritzer et al. 2001;

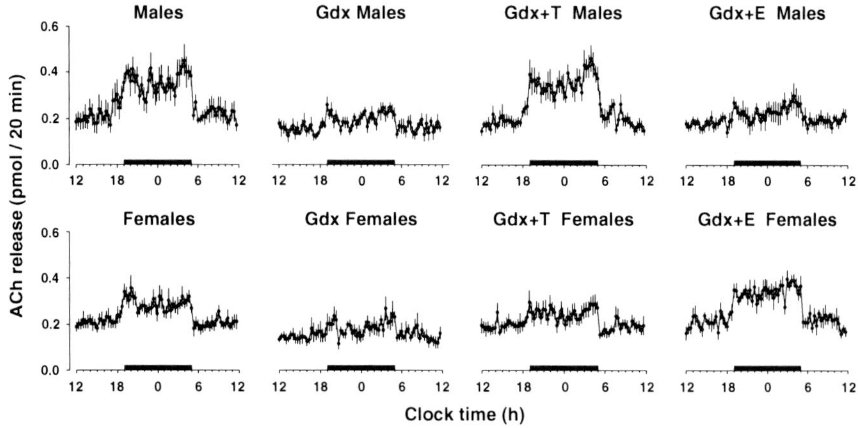

Fig. 4 ACh release in the hippocampus is time-dependent, sex-specific, and hormone-dependent. Experiments were performed 2 weeks after gonadectomy or steroid replacement. Gdx, gonadectomized. +T, testosterone-priming. +E, estradiol-priming. The number of animals was 6–8 in each group. 19–5 h is the dark phase, shown as *black bars* on the x axes. (see Mitsushima et al. 2009)

Markowska and Savonenko 2002; Luine et al. 2003) may be due to insufficient activation of hippocampus at the appropriate time. Because the replacement of sex-specific steroids restored the high positive correlation between ACh release and activity levels, the correlation appears to depend on the presence of sex steroids. These results suggest that circulating sex steroids strengthen the coupling between spontaneous behavior and ACh release (Mitsushima et al. 2009).

8 Sexual Differentiation Produces the Sex-Specific Activational Effect

The activational effect of sex steroids was sex-specific (Fig. 4). Testosterone replacement in gonadectomized female rats failed to increase ACh release to levels seen in gonadectomized testosterone-primed male rats. Similarly, estradiol replacement was unable to restore ACh release in gonadectomized male rats. Moreover, estradiol consistently increases N-methyl-D-aspartate receptor binding and spine density in the CA1 area of gonadectomized female rats, although the treatment fails to increase these same parameters in gonadectomized male rats (Romeo et al. 2005; Parducz et al. 2006). These results suggest that sex-specific steroids are important for maintaining hippocampal function. Based on our data, we hypothesized that the action of sex-specific steroids is due to neonatal sexual differentiation rather than the activational effects of sex steroids in adult rats. Moreover, in the latest study, we found that neonatal androgenization in females increased ACh release to resemble that of normal males without affecting spontaneous activity levels (Mitsushima et al. 2009). These results indicate an organizational effect on sex-specific ACh release in behaving rats, and support currently accepted theories of sexual differentiation.

Because testosterone can be aromatized to estradiol in the forebrain, neonatal sex steroids activate both estrogen and androgen receptors (McEwen 1981). In our study, both testosterone and estradiol treatment in neonatal female pups masculinized ACh release profile in adults, suggesting an estrogen receptor-mediated masculinization of septo-hippocampal cholinergic systems (Mitsushima et al. 2009). These results are consistent with the previous finding that testosterone or estradiol treatment in neonatal female pups improves their adult spatial performance, whereas neonatal gonadectomy in male pups impairs the performance (Williams and Meck 1991). In contrast, dihydrotestosterone treatment failed to masculinize the ACh release profile. Although dihydrotestosterone has been classically considered as a prototypical androgen receptor agonist, a metabolite of dihydrotestosterone, 3β-diol, has a higher affinity for estrogen receptor β (Lund et al. 2006). Therefore, dihydrotestosterone and its metabolites may stimulate both androgen receptor and estrogen receptor β, whereas estradiol stimulates estrogen receptors α and β. Considering the action of sex steroids and their metabolites, estrogen receptor α may mediate the organizational effect on the septo-hippocampal cholinergic system.

9 Interaction with Environmental Conditions

Various environmental conditions may interact with the activational effects of sex steroids. First, we reported an interaction between stress and sex steroids. Although sex steroids did not show activational effects on baseline levels of ACh release, sex steroids clearly activated the immobility stress-induced ACh release response. In addition, we found that the contributing sex hormone effect to maintain the ACh release response was sex-specific: testosterone enhanced the ACh release response in male rats, while estradiol maintained the response in females (Mitsushima et al. 2008). Second, we reported an interaction between the light/dark cycle and sex steroids. Although sex steroids slightly enhanced ACh release during the light phase, the activational effects were much stronger during the dark phase (Fig. 4). Considering the fact that the time-dependent activational effect was also sex-specific and hormone-dependent, environmental conditions seem to have complicated interactions with sex steroids (Mitsushima et al. 2009).

Some other environmental effects may affect the basal forebrain cholinergic system. Environmental conditions, such as complex or restricted (Brown 1968; Smith 1972), enriched or impoverished (Greenough et al. 1972), social or isolated conditions (Hymovitch 1952; Juraska et al. 1984; Seymoure et al. 1996), seem to affect spatial learning ability in a sex-specific manner. For example, male rats exhibited superior performance in learning maze tests compared with female rats if they were housed socially (Einon 1980). But if they were housed in isolation, female rats exhibited a performance superior to that of male rats (Einon 1980). Although few studies were performed on the relationship between the sex-specific environmental effects and ACh release in the brain, we have reported that 4-day housing in a small cage attenuates the ACh release in the hippocampus in male rats (Mitsushima et al. 1998), but not in female rats (Masuda et al. 2005). Taken together, these results suggest that housing conditions contribute to the sex difference in ACh release and spatial learning ability.

Feeding conditions after weaning also affect spatial learning ability. If fed pelleted diet (i.e., standard laboratory diet), male rats show performance superior to that of female rats (Beatty 1984; Williams and Meck 1991). But when fed powdered diet, female rats, but not male rats, showed improved performance (Endo et al. 1994; Takase et al. 2005a). In our study, it was found that feeding with powdered diet after weaning increased ACh release in the hippocampus in female rats, but not in male rats (Takase et al. 2005b). A 24-h ACh release in female rats fed powdered diet was as high as that in male rats fed either powdered or pelleted diet, showing no sex difference. Since feeding with powdered diet improved spatial learning ability in female rats (Endo et al. 1994), the increase in the ACh release in the hippocampus in female rats fed powdered diet may partly contribute to this effect. Our findings provide evidence that environmental conditions such as housing or feeding may play a role in sex-specific hippocampal function.

10 A Possible Treatment Strategy for Alzheimer's Disease

Activational effects of sex steroids are very important in humans, since circulating sex steroid levels decline with age. A reduction in ACh synthesis is known as a common feature of Alzheimer's disease (Coyle et al. 1983), afflicting more than 18 million people worldwide (Ferri et al. 2005; Mount and Downtown 2006). The disease is the most common form of dementia (Cummings 2004) and is frequently accompanied by insomnia, poor concentration, and day/night confusion (McCurry et al. 2004; Starkstein et al. 2005). The centrally active acetylcholinesterase inhibitor (donepezil) is effective in not only mild, but also moderate to severe cases (Petersen et al. 2005; Winblad et al. 2006), proving the importance of endogenous ACh in humans. In addition, women are twice as likely to develop the disease (Swaab and Hofman 1995), and estradiol seems to play a protective role (Zandi et al. 2002; Norbury et al. 2007). A recent study using single photon emission tomography showed that estrogen replacement therapy in healthy postmenopausal women increases muscarinic M_1/M_4 receptor binding in the hippocampus (Norbury et al. 2007). Conversely in men, testosterone but not estradiol seems to play a protective role (Moffat et al. 2004; Rosario et al. 2004) and testosterone supplementation clearly improved hippocampal-dependent learning deficits in men with Alzheimer's disease (Cherrier et al. 2005). These results suggest a sex-specific activational effect of gonadal steroids on the cholinergic system in humans. Thus, there are many similarities between the rat model and the human studies, supporting the idea that gonadal steroid replacement therapy or an increase in bioavailability is beneficial when there is a subthreshold level of the hormone. Based on the neonatal sexual differentiation of the septo-hippocampal cholinergic system, we may have to search for sex-specific clinical strategies for Alzheimer's disease.

11 Conclusions

Gonadally intact male rats consistently show a greater ACh release in the hippocampus compared with diestrous or proestrous female rats. The activational effects of sex steroids are important for sex-specific ACh release in the hippocampus, since impaired ACh release in gonadectomized rats does not show sex-specific effects. Neonatal treatment with either testosterone or estradiol clearly increased ACh release in female rats, suggesting neonatal sex differentiation of septo-hippocampal cholinergic systems. Moreover, environmental effects on the basal forebrain cholinergic system seem to be sex-specific; housing in a small cage attenuated ACh release in male rats only, while feeding with powdered diet after sexual maturation increases ACh release in female rats only. These results indicate that: (1) sex-specific circulating sex steroids are necessary for sex-specific ACh release, (2) neonatal activation of estrogen receptors is sufficient to mediate masculinization of the septo-hippocampal cholinergic system, and (3) sex-specific effects of environmental conditions may suggest an interaction with the effect of sex hormones.

Understanding the importance of gonadal steroids and the sex-specific effects in cognitive disorders such as Alzheimer's disease is essential for real improvements in therapy.

Acknowledgements This work was supported by Grant-in-Aid 18590219 from the Ministry of Education, Culture, Sports, Science and Technology of Japan (to D.M.).

References

Auerbach JM, Segal M (1996) Muscarinic receptors mediating depression and long-term potentiation in rat hippocampus. J Physiol 492:479–493

Beatty WW (1984) Hormonal organization of sex differences in play fighting and spatial behavior. Prog Brain Res 61:315–330

Brown RT (1968) Early experience and problem-solving ability. J Comp Physiol Psychol 65:433–440

Buzsáki G, Bickford RG, Ponomareff G, Thal LJ, Mandel R, Gage FH (1988) Nucleus basalis and thalamic control of neocortical activity in the freely moving rat. J Neurosci 8:4007–4026

Cherrier MM, Matsumoto AM, Amory JK, Asthana S, Bremner W, Peskind ER, Raskind MA, Craft S (2005) Testosterone improves spatial memory in men with Alzheimer disease and mild cognitive impairment. Neurology 64:2063–2068

Cole AE, Nicoll RA (1983) Acetylcholine mediates a slow synaptic potential in hippocampal pyramidal cells. Science 221:1299–1301

Coyle JT, Price DL, DeLong MR (1983) Alzheimer's disease: a disorder of cortical cholinergic innervation. Science 219:1184–1190

Cummings JL (2004) Alzheimer's disease. N Engl J Med 351:56–67

Daniel JM, Fader AJ, Spencer AL, Dohanich GP (1997) Estrogen enhances performance of female rats during acquisition of a radial arm maze. Horm Behav 32:217–225

Day J, Damsma G, Fibiger HC (1991) Cholinergic activity in the rat hippocampus, cortex and striatum correlates with locomotor activity: an in vivo microdialysis study. Pharmacol Biochem Behav 38:723–729

Eckel-Mahan KL, Phan T, Han S, Wang H, Chan GC, Scheiner ZS, Storm DR (2008) Circadian oscillation of hippocampal MAPK activity and cAMP: implications for memory persistence. Nat Neurosci 11:1074–1082

Einon D (1980) Spatial memory and response strategies in rats: age, sex and rearing differences in performance. Q J Exp Psychol 32:473–489

Endo Y, Mizuno T, Fujita K, Funabashi T, Kimura F (1994) Soft-diet feeding during development enhances later learning abilities in female rats. Physiol Behav 56:629–633

Falvo RE, Buhl A, Nalbandov AV (1974) Testosterone concentrations in the peripheral plasma of androgenized female rats and in the estrous cycle of normal female rats. Endocrinology 95:26–29

Fernández de Sevilla D, Núñez A, Borde M, Malinow R, Buño W (2008) Cholinergic-mediated IP3-receptor activation induces long-lasting synaptic enhancement in CA1 pyramidal neurons. J Neurosci 28:1469–1478

Ferri CP, Prince M, Brayne C, Brodaty H, Fratiglioni L, Ganguli M, Hall K, Hasegawa K, Hendrie H, Huang Y, Jorm A, Mathers C, Menezes PR, Rimmer E, Scazufca M (2005) Global prevalence of dementia: a Delphi consensus study. Lancet 366:2112–2117

Gelbard-Sagiv H, Mukamel R, Harel M, Malach R, Fried I (2008) Internally generated reactivation of single neurons in human hippocampus during free recall. Science 322:96–101

Gibbs RB, Pfaff DW (1992) Effects of estrogen and fimbria/fornix transaction on p75NGFR and ChAT expression in the medial septum and diagonal band of Broca. Exp Neurol 116:23–39

Gold PE (2003) Acetylcholine modulation of neural systems involved in learning and memory. Neurobiol Learn Mem 80:194–210

Greenough WT, Madden TC, Fleischmann TB (1972) Effects of isolation, daily handling, and enriched rearing on maze learning. Psychon Sci 27:279–280

Herrera-Morales W, Mar I, Serrano B, Bermúdez-Rattoni F (2007) Activation of hippocampal postsynaptic muscarinic receptors is involved in long-term spatial memory formation. Eur J Neurosci 25:1581–1588

Hironaka N, Tanaka K, Izaki Y, Hori K, Nomura M (2001) Memory-related acetylcholine efflux from the rat prefrontal cortex and hippocampus: a microdialysis study. Brain Res 901:143–150

Hyman JM, Wyble BP, Goyal V, Rossi CA, Hasselmo ME (2003) Stimulation in hippocampal region CA1 in behaving rats yields long-term potentiation when delivered to the peak of theta and long-term depression when delivered to the trough. J Neurosci 23:11725–11731

Hymovitch B (1952) The effects of experimental variations on problem solving in the rat. J Comp Physiol Psychol 45:313–321

Juraska JM, Henderson C, Muller J (1984) Differential rearing experience, gender, and radial maze performance. Dev Psychobiol 17:209–215

Komorowski RW, Manns JR, Eichenbaum H (2009) Robust conjunctive item-place coding by hippocampal neurons parallels learning what happens where. J Neurosci 29:9918–9929

Kotani S, Yamauchi T, Teramoto T, Ogura H (2006) Pharmacological evidence of cholinergic involvement in adult hippocampal neurogenesis in rats. Neuroscience 142:505–514

Kritzer MF (1997) Selective colocalization of immunoreactivity for intracellular gonadal hormone receptors and tyrosine hydroxylase in the ventral tegmental area, substantia nigra, and retrorubral fields in the rat. J Comp Neurol 379:247–260

Kritzer MF, McLaughlin PJ, Smirlis T, Robinson JK (2001) Gonadectomy impairs T-maze acquisition in adult male rats. Horm Behav 39:167–174

Lee MG, Chrobak JJ, Sik A, Wiley RG, Buzsáki G (1994) Hippocampal theta activity following selective lesion of the septal cholinergic system. Neuroscience 62:1033–1047

Luine VN, Renner KJ, McEwen BS (1986) Sex-dependent differences in estrogen regulation of choline acetyltransferase are altered by neonatal treatments. Endocrinology 119:874–878

Luine V, Jacome LF, MacLusky NJ (2003) Rapid enhancement of visual and place memory by estrogens in rats. Endocrinology 144:2836–2844

Lund TD, Hinds LR, Handa RJ (2006) The androgen 5α-dihydrotestosterone and its metabolite 5α-androstan-3β, 17β-diol inhibit the hypothalamo-pituitary-adrenal response to stress by acting through estrogen receptor β-expressing neurons in the hypothalamus. J Neurosci 26:1448–1456

Markowska AJ, Savonenko AV (2002) Effectiveness of estrogen replacement in restoration of cognitive function after long-term estrogen withdrawal in aging rats. J Neurosci 22: 10985–10995

Markram H, Segal M (1990) Long-lasting facilitation of excitatory postsynaptic potentials in the rat hippocampus by acetylcholine. J Physiol 427:381–393

Masuda J, Mitsushima D, Funabashi T, Kimura F (2005) Sex and housing conditions affect the 24-h acetylcholine release profile in the hippocampus in rats. Neuroscience 132:537–542

McCurry SM, Logsdon RG, Vitiello MV, Teri L (2004) Treatment of sleep and nighttime disturbances in Alzheimer's disease: a behavior management approach. Sleep Med 5:373–377

McEwen BS (1981) Neural gonadal steroid actions. Science 211:1303–1311

McEwen BS, Alves SE (1999) Estrogen actions in the central nervous system. Endocr Rev 20:279–307

Mesulam MM, Mufson EJ, Wainer BH, Levey AI (1983) Central cholinergic pathways in the rat: an overview based on an alternative nomenclature (Ch1-Ch6). Neuroscience 10:1185–1201

Miettinen RA, Kalesnykas G, Koivisto EH (2002) Estimation of the total number of cholinergic neurons containing estrogen receptor-a in the rat basal forebrain. J Histochem Cytochem 50:891–902

Mitsushima D, Mizuno T, Kimura F (1996) Age-related changes in diurnal acetylcholine release in the prefrontal cortex of male rats as measured by microdialysis. Neuroscience 72:429–434

Mitsushima D, Yamanoi C, Kimura F (1998) Restriction of environmental space attenuates locomotor activity and hippocampal acetylcholine release in male rats. Brain Res 805:207–212

Mitsushima D, Funabashi T, Shinohara K, Kimura F (2001) Impairment of maze learning in rats by restricting environmental space. Neurosci Lett 297:73–76

Mitsushima D, Masuda J, Kimura F (2003a) Sex differences in the stress-induced release of acetylcholine in the hippocampus and corticosterone from the adrenal cortex in rats. Neuroendocrinology 78:234–240

Mitsushima D, Tin-Tin-Win-Shwe, Kimura F (2003b) Sexual dimorphism in the GABAergic control of gonadotropin release in intact rats. Neurosci Res 46:399–405

Mitsushima D, Takase K, Funabashi T, Kimura F (2008) Gonadal steroid hormones maintain the stress-induced acetylcholine release in the hippocampus: simultaneous measurements of the extracellular acetylcholine and serum corticosterone levels in the same subjects. Endocrinology 149:802–811

Mitsushima D, Takase K, Funabashi T, Kimura F (2009) Gonadal steroids maintain 24-h acetylcholine release in the hippocampus: organizational and activational effects in behaving rats. J Neurosci 29:3808–3815

Mizuno T, Kimura F (1996) Medial septal injection of naloxone elevates acetylcholine release in the hippocampus and induces behavioral seizures in rats. Brain Res 713:1–7

Mizuno T, Endo Y, Arita J, Kimura F (1991) Acetylcholine release in the rat hippocampus as measured by the microdialysis method correlates with motor activity and exhibits a diurnal variation. Neuroscience 44:607–612

Moffat SD, Zonderman AB, Metter EJ, Kawas C, Blackman MR, Harman SM, Resnick SM (2004) Free testosterone and risk for Alzheimer disease in older men. Neurology 62:188–193

Mohapel P, Leanza G, Kokaia M, Lindvall O (2005) Forebrain acetylcholine regulates adult hippocampal neurogenesis and learning. Neurobiol Aging 26:939–946
Moor E, DeBoer P, Westerink BHC (1998) GABA receptors and benzodiazepine binding sites modulate hippocampal acetylcholine release in vivo. Eur J Pharmacol 359:119–126
Mount C, Downtown D (2006) Alzheimer disease: progress or profit? Nat Med 12:780–784
Mufson EJ, Cai WJ, Jaffar S, Chen E, Stebbins G, Sendera T, Kordower JH (1999) Estrogen receptor immunoreactivity within subregions of the rat forebrain: neuronal distribution and association with perikarya containing choline acetyltransferase. Brain Res 849:253–274
Nakamura N, Fujita H, Kawata M (2002) Effects of gonadectomy on immunoreactivity for choline acetyltransferase in the cortex, hippocampus, and basal forebrain of adult male rats. Neuroscience 109:473–485
Nilsson OG, Leanza G, Bjorklund A (1992) Acetylcholine release in the hippocampus: regulation by monoaminergic afferents as assessed by in vivo microdialysis. Brain Res 584:132–140
Norbury R, Travis MJ, Erlandsson K, Waddington W, Ell PJ, Murphy DGM (2007) Estrogen therapy and brain muscarinic receptor density in healthy females: a SPET study. Horm Behav 51:249–257
Parducz A, Hajszan T, Maclusky NJ, Hoyk Z, Csakvari E, Kurunczi A, Prange-Kiel J, Leranth C (2006) Synaptic remodeling induced by gonadal hormones: neuronal plasticity as a mediator of neuroendocrine and behavioral responses to steroids. Neuroscience 138:977–985
Parent MB, Baxter MG (2004) Septohippocampal acetylcholine: involved in but not necessary for learning and memory? Learn Mem 11:9–20
Perry E, Walker M, Grace J, Perry R (1999) Acetylcholine in mind: a neurotransmitter correlate of consciousness? Trend Neurosci 22:273–280
Petersen RC, Thomas RG, Grundman M, Bennett D, Doody R, Ferris S, Galasko D, Jin S, Kaye J, Levey A, Pfeiffer E, Sano M, van Dyck CH, Thal LJ (2005) Vitamin E and donepezil for the treatment of mild cognitive impairment. N Engl J Med 352:2379–2388
Pongrac JL, Gibbs RB, Defranco DB (2004) Estrogen-mediated regulation of cholinergic expression in basal forebrain neurons requires extracellular signal-regulated kinase activity. Neuroscience 124:809–816
Ragozzino ME, Unick KE, Gold PE (1996) Hippocampal acetylcholine release during memory testing in rats: augmentation by glucose. Proc Natl Acad Sci USA 93:4693–4698
Romeo RD, McCarthy JB, Wang A, Milner TA, McEwen BS (2005) Sex differences in hippocampal estradiol-induced N-methyl-D-aspartic acid binding and ultrastructural localization of estrogen receptor-α. Neuroendocrinology 81:391–399
Rosario ER, Chang L, Stanczyk FZ, Pike CJ (2004) Age-related testosterone deplation and the development of Alzheimer disease. JAMA 292:1431–1432
Rush ME, Blake CA (1982) Serum testosterone concentrations during the 4-day estrous cycle in normal and adrenalectomized rats. Proc Soc Exp Biol Med 169:216–221
Sarter M, Parikh V (2005) Choline transporters, cholinergic transmission and cognition. Nat Neurosci 6:48–56
Seeger T, Fedorova I, Zheng F, Miyakawa T, Koustova E, Gomeza J, Basile AS, Alzheimer C, Wess J (2004) M_2 muscarinic acetylcholine receptor knock-out mice show deficits in behavioral flexibility, working memory, and hippocampal plasticity. J Neurosci 24: 10117–10127
Seymoure P, Dou H, Juraska JM (1996) Sex differences in radial maze performance: influence of rearing environment and room cues. Psychobiology 24:33–37
Shinoe T, Matsui M, Taketo MM, Manabe T (2005) Modulation of synaptic plasticity by physiological activation of M_1 muscarinic acetylcholine receptors in the mouse hippocampus. J Neurosci 25:11194–11200
Shors TJ, Chua C, Falduto J (2001) Sex differences and opposite effects of stress on dendritic spine density in the male versus female hippocampus. J Neurosci 21:6292–6297
Smith HV (1972) Effects of environmental enrichment on open-field activity and Hebb–Williams problem solving in rats. J Comp Physiol Psychol 80:163–168

Stancampiano R, Cocco S, Cugusi C, Sarais L, Fadda F (1999) Serotonin and acetylcholine release response in the rat hippocampus during a spatial memory task. Neuroscience 89:1135–1143
Starkstein SE, Jorge R, Mizrahi R, Robinson RG (2005) The construct of minor and major depression in Alzheimer's disease. Am J Psychiatry 162:2086–2093
Swaab DF, Hofman MA (1995) Sexual differentiation of the human hypothalamus in relation to gender and sexual orientation. Trend Neurosci 18:264–270
Swanson LW (1982) The projections of the ventral tegmental area and adjacent regions: a combined fluorescent retrograde tracer and immunofluorescence study in the rat. Brain Res Bull 9:321–353
Takase K, Funabashi T, Mogi K, Mitsushima D, Kimura F (2005a) Feeding with powdered diet after weaning increases visuospatial ability in association with increases in the expression of N-methyl-D-aspartate receptors in the hippocampus of female rats. Neurosci Res 53:169–175
Takase K, Mitsushima D, Masuda J, Mogi K, Funabashi T, Endo Y, Kimura F (2005b) Feeding with powdered diet after weaning affects sex difference in acetylcholine release in the hippocampus in rats. Neuroscience 136:593–599
Takase K, Kimura F, Yagami T, Mitsushima D (2009) Sex-specific 24-h acetylcholine release profile in the medial prefrontal cortex: simultaneous measurement of spontaneous locomotor activity in behaving rats. Neuroscience 159:7–15
van Praag H, Christie BR, Sejnowski TJ, Gage FH (1999) Running enhances neurogenesis, learning and long-term potentiation in mice. Proc Natl Acad Sci USA 96:13427–13431
Weiner DM, Levey AI, Sunahara RK, Niznik HB, O'Dowd BF, Seeman P, Brann MR (1991) D_1 and D_2 dopamine receptor mRNA in rat brain. Proc Natl Acad Sci USA 88:1859–1863
Widmer H, Ferrigan L, Davies CH, Cobb SR (2006) Evoked slow muscarinic acetylcholinergic synaptic potentials in rat hippocampal interneurons. Hippocampus 16:617–628
Williams CL, Meck WH (1991) The organizational effects of gonadal steroids on sexually dimorphic spatial ability. Psychoneuroendocrinology 16:155–176
Winblad B, Kilander L, Eriksson S, Minthon L, Båtsman S, Wetterholm AL, Jansson-Blixt C, Haglund A (2006) Donepezil in patients with severe Alzheimer's disease: double-blind, parallel-group, placebo-controlled study. Lancet 367:1057–1065
Yanai J, Rogel-Fuchs Y, Pick CG, Slotkin T, Seidler FJ, Zahalka EA, Newman ME (1993) Septohippocampal cholinergic changes after destruction of the A10-septal dopaminergic pathways. Neuropharmacology 32:113–117
Zandi PP, Carlson MC, Plassman BL, Welsh-Bohmer KA, Mayer LS, Steffens DC, Breitner JCS (2002) Hormone replacement therapy and incidence of Alzheimer disease in older women. JAMA 288:2123–2129

Females Are More Vulnerable to Drug Abuse than Males: Evidence from Preclinical Studies and the Role of Ovarian Hormones

Justin J. Anker and Marilyn E. Carroll

Contents

Abstract Human and animal research indicates the presence of sex differences in drug abuse. These data suggest that females, compared to males, are more vulnerable to key phases of the addiction process that mark transitions in drug use such as initiation, drug bingeing, and relapse. Recent data indicate that the female gonadal hormone estrogen may facilitate drug abuse in women. For example, phases of the menstrual cycle when estrogen levels are high are associated with enhanced positive subjective measures following cocaine and amphetamine administration in women. Furthermore, in animal research, the administration of estrogen increases drug taking and facilitates the acquisition, escalation, and reinstatement of cocaine-seeking behavior. Neurobiological data suggest that estrogen may

J.J. Anker (✉) and M.E. Carroll
Department of Psychiatry, University of Minnesota, MMC 392, Minneapolis, MN 55455, USA
e-mail: anke0022@umn.edu

J.C. Neill and J. Kulkarni (eds.), *Biological Basis of Sex Differences in Psychopharmacology*, Current Topics in Behavioral Neurosciences 8, DOI 10.1007/7854_2010_93,

facilitate drug taking by interacting with reward- and stress-related systems. This chapter discusses sex differences in and hormonal effects on drug-seeking behaviors in animal models of drug abuse. The neurobiological basis of these differences and effects are also discussed.

Keywords Drug abuse · Estrogen · Progesterone · Rats · Sex differences · Sex hormones

1 Introduction

Historically, drug abuse has been considered a male disease. Research from several areas including epidemiology, behavioral pharmacology, and neuroscience has taken a male-centric approach when studying factors and/or treatments that influence drug abuse. This approach has led to neglect of factors underlying drug abuse in women such as ovarian hormones. In fact, over the centuries, the predominant sex that abused drugs has varied from female to male, depending on cultural conditions (Kornetsky 2007). However, in recent years, epidemiological research has shown that females are catching up and exceeding males in their drug use, particularly among younger populations. Thus, an important direction for current research is to acknowledge that sex is a vulnerability factor in drug abuse and to study the neurobiological basis for this trend and its implications for drug abuse treatment (Ashley et al. 2003; Marsh et al. 2004).

Clinical and preclinical research indicates that females, compared to males, exhibit greater vulnerability toward drug abuse at stages of the addiction process that mark transitions in drug use. These stages include drug initiation, bingeing, withdrawal, and relapse and may be modeled in animals using acquisition, escalation, withdrawal/extinction, and reinstatement procedures, (Carroll et al. 2009a). Clinical reports indicate that women are more likely than men to initiate drug use at an earlier age (Chen and Kandel 2002), engage in binge-like patterns of drug intake (Becker and Hu 2008; Brady and Randall 1999; Lynch et al. 2002; Mann et al. 2005; Randall et al. 1999), report greater difficulty in quitting (Becker and Hu 2008; Carpenter et al. 2006; Lynch et al. 2002), exhibit greater drug craving (Robbins et al. 1999), relapse (Ignjatova and Raleva 2009), and resume higher levels of drug use following relapse (Gallop et al. 2007). One area in which males exceed females in the phases of drug abuse is during withdrawal where males experience more severe withdrawal effects than females (Carroll et al. 2009b; Perry et al. 2008). Thus, elevated drug use in females may be due not only to their greater sensitivity to rewarding effects but also to their resilience to the negative effects of drugs. This differential sensitivity to the rewarding and aversive aspects of drug use has parallels in other addiction-prone and -resistant phenotypes (Carroll et al. 2009a), and it is an emerging area of research that should yield interesting developments.

Animal models of drug abuse add further support to enhanced female vulnerability across most stages of the addiction process. Female rats acquire drug self-administration at a faster rate, they exhibit greater binge-like patterns of drug intake, and they are more vulnerable to relapse of drug-seeking behavior (for reviews, see Carroll and Anker 2009; Carroll et al. 2009a; Lynch et al. 2002). The animal data suggest that females may be more vulnerable than males due to an underlying biological predisposition related to ovarian hormones or developmental/organizational differences in male and female neurobiology (Becker and Hu 2008). As a consequence, it is important to identify biological vulnerability factors that contribute to the onset and progression of drug addiction in women.

A growing number of findings suggest that the biological basis for increased drug abuse vulnerability in women may be attributed to female gonadal hormones (Becker and Hu 2008; Carroll et al. 2004; Festa and Quinones-Jenab 2004; Lynch et al. 2002; Terner and de Wit 2006). More specifically, enhanced drug-seeking and subjective effects in women are associated with higher levels of endogenous estrogen (EST) (Evans 2007; Terner and de Wit 2006). Further work supports this and has demonstrated that endogenous or exogenous EST (e.g., estradiol benzoate, 17β-estradiol) facilitates the acquisition, escalation, and reinstatement of cocaine-seeking behavior in female rats (for a review, see Carroll and Anker 2010). Preclinical work further suggests that EST's potentiating effects on drug-related responses may involve activation of the EST-receptor subtype β (ER-β), whereas ER-α had little influence on drug seeking (Larson and Carroll 2007). In contrast to the potentiating effects of EST, progesterone (PROG), another female gonadal hormone, attenuates responses to drugs of abuse in both humans and animals.

This chapter emphasizes sex and hormonal influences across several phases of the human drug abuse process that are represented by animal models. Particular attention is given to the effects of EST on drug-related responses; however, as PROG often opposes EST's behavioral effects, its influence on drug-seeking behavior is also important to consider. The final section of this chapter discusses possible neurobiological mechanisms underlying sex differences and EST's effects on addiction-related behaviors. EST's interactions within the mesolimbic dopamine (DA) system, the hypothalamic-pituitary-adrenal (HPA) axis, and the involvement of ER-β receptors are considered. Since a majority of research on sex differences and hormonal influences on drug abuse involves psychomotor stimulants, this drug class is the primary focus.

2 Menstrual Cycle and Hormonal Effects on Responses to Drugs of Abuse: Clinical Evidence

Natural fluctuations of hormones during the menstrual cycle correspond to differences in the physiological and subjective effects of some stimulant drugs

(Evans 2007; Terner and de Wit 2006). Cardiovascular and/or positive subjective responses to cocaine (Evans and Foltin 2006; Evans et al. 2002; Sofuoglu et al. 1999) and amphetamine (Justice and de Wit 1999, 2000; White et al. 2002), but not nicotine (Terner and de Wit 2006), were enhanced during the EST-dominant follicular phase of the menstrual cycle compared with the luteal phase (when EST levels are low) in women. These results were also extended to measures of stress reactivity and craving elicited by cocaine-related stimuli in cocaine-dependent women (Sinha et al. 2007). In this study, women in the follicular phase showed higher systolic/diastolic blood pressure measures, and they scored higher on self-reported measures of anxiety and drug craving following presentations of stressful or drug-related stimuli than women during the midluteal phase (Sinha et al. 2007). The finding with stress-induced craving is especially important given that clinical and preclinical reports implicate stress as a primary factor in drug bingeing and relapse (Covington et al. 2005; Sinha 2008), and stress is associated with heightened drug abuse vulnerability in women (Fox and Sinha 2009). Taken together, these results indicated that the phase of the menstrual cycle in which EST levels were highest was associated with high positive affective responses to cocaine and enhanced cue- and stress-induced cocaine craving. Thus, EST may operate as a vulnerability factor that facilitates the positive and aversive aspects of cocaine abuse in women. Others have found no differences in the physiological and subjective responses to cocaine during the female menstrual cycle (Lukas et al. 1996; Mendelson et al. 1999). However, methodological differences related to the dose and route of cocaine administration may account for the discrepancy in results.

Compared to results with stimulants, behavioral/subjective responses to other drugs of abuse do not consistently vary with the phase of the menstrual cycle in humans. Subjective measures were insensitive to menstrual cycle effects following alcohol (Freitag and Adesso 1993; Hay et al. 1984; Holdstock and de Wit 2000; Nyberg et al. 2004; Sutker et al. 1987), nicotine (Allen et al. 1999, 2004; Pomerleau et al. 1992, 2000; Snively et al. 2000), marijuana (Lex et al. 1984), and opioid (Gear et al. 1996) administration in women.

In contrast to the results with EST, PROG has an opposite effect on subjective measures following drug administration (for review, see Evans 2007). Women treated with PROG showed a decrease in the positive-subjective effects of smoked (Evans and Foltin 2006; Sofuoglu et al. 2002) and iv (Sofuoglu et al. 2004) cocaine compared with placebo-treated controls. In addition, high circulating plasma levels of PROG were associated with decreased craving following a stress- or drug-related cue in cocaine-dependent women (Sinha et al. 2007).

Overall, results from clinical studies indicate that the female gonadal hormone, EST, may be associated with sex differences in cocaine abuse, as phases of the menstrual cycle associated with heightened EST corresponded to increases in positive-subjective measures following cocaine. In contrast, PROG had an attenuating effect on these measures and suppressed stress- and cue-induced drug craving.

3 Sex and Ovarian Hormones Influence Drug Seeking and Drug Taking: Preclinical Evidence

Preclinical studies corroborate clinical findings and confirm the importance of sex differences and hormonal influences during key phases of the addiction process as they are modeled in animals. In animal research, the self-administration paradigm is considered a valid model of human drug addiction (Panlilio et al. 2007), as subjects have control over their self-administration of the drug. Animal models of drug self-administration allow a controlled longitudinal approach to the study of factors that predict drug abuse in addition to potential treatments during critical transition phases of the drug abuse process (Carroll et al. 2009a). Phases to be discussed are acquisition (initiation) of drug use, maintenance of steady drug intake, bingeing or escalation of drug intake, extinction (withdrawal), and reinstatement (relapse). The following section discusses sex differences and the effects of EST with regard to these important transition phases of drug addiction. Results from these studies are summarized in Table 1.

3.1 Acquisition

In animal research, acquisition or the initiation of drug self-administration is measured using several techniques that primarily involve automatic (e.g., autoshaping) or experimenter-administered priming infusions of drug prior to, or at the beginning of, each self-administration session. Animals achieve acquisition criteria once they earn a predefined number of self-administered drug infusions.

3.1.1 Sex Differences

Similar to clinical findings (Chen and Kandel 2002), females acquired drug self-administration faster than male rats across a wide range of drugs including cannabinoids (Fattore et al. 2007), cocaine (Jackson et al. 2006; Lynch 2008; Lynch and Carroll 1999b), methamphetamine (Roth and Carroll 2004a), nicotine (Chaudhri et al. 2005), and heroin (Carroll et al. 2002; Lynch and Carroll 1999b). Research with monkeys also confirmed that females acquired PCP self-administration more successfully than males, with 100% of the females acquiring compared to only 36.4% of the males (Carroll et al. 2000). Together, these results indicated that females had increased vulnerability to initiate drug use compared with males.

3.1.2 Hormonal Influences

Preclinical work implicates EST in sex differences in the acquisition of drug abuse. Two methods are involved in investigating the contributions of EST and PROG in

Table 1 Summary of the sex differences and estrogen effects on behavioral responses to drugs of abuse across phases of the drug abuse process: animal models

Independent variable	Dependent measure	Drug	Finding	Reference
Sex differences (monkey)	Acquisition	PCP	F > M	Carroll et al. (2000)
	Escalation	PCP	F > M	Carroll et al. (2005)
Sex differences (rat)	Acquisition	Cannabinoids	F > M	Fattore et al. (2007)
		Cocaine	F > M	Lynch and Carroll (1999b), Lynch (2008), Jackson et al. (2006)
		Methamphetamine	F > M	Roth and Carroll (2004a)
		Nicotine	F > M	Chaudhri et al. (2005)
		Heroin	F > M	Lynch and Carroll (1999b), Carroll et al. (2002)
	Escalation	Cocaine	F > M	Lynch et al. (2000), Lynch and Taylor (2004, 2005), Roth and Carroll (2004b)
	Extinction	Cocaine	F > M	Lynch and Carroll (2000), Kippin et al. (2005), Lynch et al. (2005), Kerstetter et al. (2008), Perry et al. (2008)
	Cocaine-primed reinstatement	Cocaine	F > M	Lynch and Carroll (2000), Kerstetter et al. (2008), Anker et al. (2009)
	Stress-induced reinstatement	Cocaine	F > M	Anker and Carroll (2010)
	Cue-induced reinstatement	Cocaine	M < F	Fuchs et al. (2005)
Systemic estrogen administration	Acquisition	Cocaine	OVX-E > OVX-V	Lynch et al. (2001), Jackson et al. (2006)
			OVX-E > OVX-E+P	Jackson et al. (2006)
		Heroin	OVX-E > OVX-V	Roth et al. (2002)
	Escalation	Cocaine	OVX-E > OVX-V	Lynch and Taylor (2005), Larson et al. (2007)
			OVX-E > OVX-E+P	Larson et al. (2007)
	Cocaine-primed reinstatement	Cocaine	OVX-E > OVX-V	Larson et al. (2005), Anker et al. (2007), Larson and Carroll (2007)
			OVX-E > OVX-E+P	Anker et al. (2007)
			OVX-ERbeta > OVX-ERalpha, OVX-V	Larson and Carroll (2007)

F Female, *M* Male, *E* Estrogen, *V* vehicle, *P* PROG, OVX ovariectomy

animal models of drug abuse. The first involves comparing drug-seeking behavior across different phases of the ovarian hormone cycle, and the second involves depleting naturally occurring levels of hormones through ovariectomy (OVX), administering EST and/or PROG, measuring the addiction-related response, and comparing this response to gonadally intact sham (SH)-operated controls.

The acquisition rates for iv cocaine self-administration were reduced in OVX female rats compared with SH-operated females (Jackson et al. 2006), whereas the administration of EST in OVX females facilitated the acquisition of cocaine (Jackson et al. 2006; Lynch et al. 2001) and heroin (Roth et al. 2002) self-administration. The injection of the EST receptor antagonist, tamoxifen, blocked EST's facilitation of cocaine acquisition in OVX female rats (Lynch et al. 2001). In contrast to EST, PROG had an opposite effect on the acquisition of drug self-administration. For example, Jackson et al. (2006) showed that PROG treatment blocked the effects of EST on the acquisition of cocaine self-administration in OVX female rats. Together, the results suggest that EST and PROG have opposite roles during the initiation of drug self-administration in females.

3.2 *Escalation*

The transition from steady to dysregulated drug consumption characterizes the escalation phase of the drug abuse process (Ahmed and Koob 1998, 1999; Lynch and Carroll 1999a). In humans, escalation represents out-of-control drug bingeing that is linked to overdose and death (Kalivas and Volkow 2005). Females, compared to males, are more susceptible to binge-like drug intake (Brady and Randall 1999; Mann et al. 2005; Randall et al. 1999). Thus, it is important to use animal models to identify factors that contribute to the development of this critical aspect of drug addiction in women. Animal models offer a means to examine sex differences and hormonal influences on binge-like drug intake.

3.2.1 Sex Differences

Drug bingeing is modeled in animals using an extended-access procedure. In these studies, long access (LgA) to a self-administered drug (e.g., 6 h) results in increased drug intake over subsequent days (Ahmed and Koob 1999). In a study by Roth and Carroll (2004b), female rats escalated cocaine intake to a greater extent than males during LgA. Furthermore, females responded significantly more for iv infusions of cocaine at lower doses (Roth and Carroll 2004b). Similar results were reported with rhesus monkeys self-administering PCP (Carroll et al. 2005) providing cross-species evidence for sex differences. Females and males did not differ in PCP intake under a short-access condition (ShA) (3 h); however, females exceeded males in mg/kg PCP intake when access was extended to 6 h. Dose–response functions under a ShA progressive-ratio (PR) schedule before and after the LgA condition indicated both groups experienced a rightward shift in their dose–response curves following LgA and that females (vs. males) consumed more PCP across several concentrations under the post-LgA PR condition (Carroll et al. 2005).

Escalation of drug intake has also been modeled in animals using a dose self-selection procedure in which animals achieve a preferred dose of drug by responding on two levers that respectively increase or decrease the infusion duration (Lynch and Carroll 2001; Lynch et al. 1998). Female rats exhibited greater dysregulation of drug intake compared with males as determined by a lower correlation between the interdose interval and the preceding dose size, and females responded on the dose-increasing lever more than males (Lynch et al. 2000).

Another method for assessing excessive drug intake in animal models is the discrete-trials procedure that allows two to four 10-min trials/h during self-administration. In a study by Lynch and Taylor (2004), male and female rats self-administered similar amounts of cocaine under ShA FR 1 and PR schedules. However, when rats were subsequently placed on the discrete-trials procedure for 7 days, females self-administered significantly more cocaine than males and showed greater disruption in diurnal self-administration patterns (Lynch and Taylor 2005). When performance under the PR schedule was reassessed 10 days following the discrete-trials procedure, females surpassed males in cocaine infusions.

3.2.2 Hormonal Influences

In a study by Larson et al. (2007), the effects of EST and PROG were examined on cocaine self-administration under ShA and LgA conditions. Five groups were compared: OVX-VEH, OVX-EST, OVX-EST+PROG, SH-VEH, and SH-PROG. Prior to LgA, all groups exhibited similar levels of ShA cocaine intake, a finding consistent with previous studies (Cain et al. 2004; Larson et al. 2005; Lynch and Carroll 2000; Roth and Carroll 2004b). However, when access was extended to 6 h/day (LgA), groups SH-VEH, OVX-EST, and OVX-VEH escalated cocaine intake, whereas the PROG-treated groups (SH-PROG, OVX-EST+PROG) did not. Furthermore, OVX EST-treated rats escalated their drug intake more rapidly and self-administered more cocaine during LgA than OVX-VEH rats. Thus, exogenously administered EST facilitated the escalation of cocaine intake, whereas PROG attenuated it.

Similar results have been reported with the 24 h/day discrete-trials procedure. Lynch and Taylor (2005) demonstrated that OVX rats treated with VEH earned fewer cocaine infusions under the discrete-trials procedure when compared with SH-operated controls, and EST enhanced drug intake in OVX female rats relative to VEH-treated controls to levels of the SH-operated rats (Lynch and Taylor 2005). Taken together, these results suggest that EST is involved in enhanced escalation of drug taking in females relative to males.

3.3 Extinction/Reinstatement (Relapse)

Relapse is one of the most difficult aspects of drug abuse to treat due to the craving and other withdrawal effects that result in its high rate of recurrence (McKay and Weiss 2001). As previously mentioned, women are especially vulnerable to drug abuse relapse following a period of abstinence (Ignjatova and Raleva 2009), and once relapse occurs, women are prone to consume excessive amounts of drug (Gallop et al. 2007). Relapse is modeled in animals using the reinstatement procedure. Typically, animals are allowed to self-administer a drug for several days (usually 10–14 days). Drug solutions are then removed or replaced with saline, and animals subsequently extinguish their responding for the drug over the next 2–3 weeks. Subsequently, a priming stimulus consisting of the drug, a drug-associated cue, or a physical (e.g., shock) or chemical (e.g., yohimbine) stressor is introduced. Subsequent responding on the device previously associated with the drug delivery following one or more of the priming stimuli is considered a measure of reinstatement and a predictor of relapse in humans (Katz and Higgins 2003; Shaham et al. 2003). Results from animal reinstatement studies confirm female vulnerability during this critical stage of the drug abuse process and implicate EST in the facilitation of these measures. The following section discusses these findings.

3.3.1 Sex Differences

Several studies have demonstrated sex differences in the extinction and reinstatement of drug-seeking behavior. During extinction, female (vs. male) rats show greater resistance to extinguishing responding that was previously maintained by iv cocaine infusions (Anker and Carroll 2010; Kerstetter et al. 2008; Kippin et al. 2005; Lynch and Carroll 2000; Lynch et al. 2005; Perry et al. 2008). Elevated responding in female rats relative to males extends to the reinstatement phase as well; however, this depends on the type of reinstatement stimulus used. Female (vs male) rats responded more on a lever previously associated with cocaine self-administration than male rats following a priming injection of cocaine (Anker et al. 2009; Kerstetter et al. 2008; Lynch and Carroll 2000) and after 1, 14, 60, and 180 days of cocaine withdrawal (Kerstetter et al. 2008). In another study, Anker and Carroll (2010) showed that females reinstated significantly more than males following an injection of the pharmacological stressor, yohimbine. This is especially pertinent, as previous clinical studies indicate that female cocaine addicts are more vulnerable to stress-induced relapse than male cocaine addicts (Fox and Sinha 2009).

3.3.2 Hormonal Influences

During the estrus phase, female rats were more resistant to extinction of lever pressing previously reinforced with cocaine than during any other phases of the rat estrous cycle (Feltenstein and See 2007; Kerstetter et al. 2008). Furthermore, systemic injection of EST enhanced the cocaine-primed reinstatement responding in OVX female rats relative to OVX rats treated with VEH (Anker et al. 2007; Larson and Carroll 2007; Larson et al. 2005). Larson and Carroll (2007) examined the effects of ER-α and ER-β on cocaine-seeking behavior under a reinstatement procedure. Following extinction of lever pressing, OVX rats received acute systemic injections of EST (ER-α and ER-β agonist), the ER-α agonist propylpyrazole-triol (PPT), or the ER-β agonist diarylpropionitrile (DPN). They were then tested on cocaine-primed reinstatement of cocaine seeking. The results indicated that EST- and DPN-treated OVX rats reinstated significantly more than OVX rats treated with VEH, while there were no differences in reinstatement responding in OVX rats treated with PPT compared with those treated with VEH. Thus, EST may facilitate reinstatement responding via activation of ER-β. In contrast to the results with EST, PROG and its metabolite, allopregnanolone (ALLO), have opposite effects on animal models of relapse. Increases in plasma PROG levels in freely cycling female rats were associated with decreased reinstatement responding following a cocaine priming injection (Feltenstein and See 2007). In addition, systemic injections of PROG in SH- and EST-treated OVX female rats attenuated reinstatement responding relative to SH rats treated with VEH and OVX rats treated with EST alone (Anker et al. 2007). The suppression of reinstatement responding by PROG may be attributed to its metabolism into ALLO. In a follow-up study, coadministering finasteride, a 5-α reductase inhibitor that prevents the conversion of PROG into ALLO, blocked PROG's attenuating effects on cocaine-primed reinstatement (Anker et al. 2009). Taken together, these results indicate that female gonadal hormones are involved in susceptibility toward (EST) and protection against (PROG) relapse of cocaine seeking in females.

4 Neurobiological Basis of Sex Differences and EST Effects in Drug Seeking

Drugs interact with motivational systems that regulate survival behaviors, and drug and nondrug stimuli activate common neurobiological systems (Spanagel and Weiss 1999; Wise 1996). General reward-mediated responding involves the interaction of several neuronal systems within the ventral and midbrain areas of the brain that contain the nucleus accumbens (NA) and the ventral tegmental area, collectively referred to as the mesolimbic reward pathway. This section discusses male/female brain dimorphism and the interaction between female sex hormones, and neurotransmitter systems within this motivational pathway, with respect to drug abuse. Most of this research has centered on stimulants, and consequently this drug class will be the focus.

4.1 *Brain Dimorphism and Sex Differences in Drug Addiction*

Sexual dimorphism in areas of the brain involved in motivation and/or hormone-DA system interactions may play a key role in sex differences in drug abuse. These topics have been reviewed elsewhere (Becker 2009) and are briefly covered here. Masculinization of the brain occurs early during maturation (perinatal period) and is largely attributed to the gonadal hormone testosterone (Becker 2009; McCarthy et al. 1997), while feminization occurs in the absence of testosterone. Gonadectomy during this early period of brain sexualization decreased amphetamine-induced DA increases that occurred during adulthood in female rats, while it had no effect on adult males (Becker and Ramirez 1981a). Thus, it is hypothesized that sexual differentiation of key components of the DA system during periadolescence may later sensitize rats to the facilitating effects of EST and contribute to the sex differences in the reinforcing effects of stimulants (Becker 2009).

Morphological differences in areas of the brain that regulate cocaine craving are also observed between sexes in adult humans. For example, men and women differ in the relative size of mesolimbic and mesocortical structures that are implicated in responses to drugs of abuse such as the cerebral cortex (Rabinowicz et al. 1999), medial amygdala (Mizukami et al. 1983), and the hippocampus (Fattore et al. 2008; Filipek et al. 1994).

4.2 *Role of DA*

4.2.1 Sex Differences

Several lines of research indicate that sex differences and the influence of EST affect neurotransmitter systems that operate in the mesolimbic reward pathway to regulate the abuse-related effects of stimulants. In the striatum, there are clear sex differences in baseline DA tone and activation following exposure to drugs of abuse. Striatal D_1 DA receptors decrease while D_2 receptors increase cocaine-seeking behavior (Becker and Hu 2008; Self et al. 1996). Interestingly, there are approximately 10% more striatal D_1 DA receptors in male rats compared to females (Andersen et al. 1997), which may explain why females outperform males on several measures of cocaine seeking. However, there are reportedly no sex differences in D_2 receptor densities in striatal regions in humans (Farde et al. 1995; Munro et al. 2006). There are also sex differences in extracellular striatal DA concentrations. For example, basal (Castner et al. 1993) and K^+-stimulated (Walker et al. 2000) DA concentrations are greater in female rats compared to males, a finding that may be due to differential affinity for the DA transporter in presynaptic terminals (Walker et al. 2000, 2006). Protein kinase A (PKA) signaling has been shown to alter DA transmission within the mesolimbic reward pathway and is implicated in drug abuse (Nairn et al. 2004; Nestler 2005). In a study by Lynch et al. (2007), females exhibited higher levels of PKA-mediated phosphorylization of DARPP-32 (DA- and cyclic AMP-regulated phosphoprotein) in the striatum and NA (Nazarian et al. 2009; Zhou et al. 2009), while Nazarian et al. (2009) reported similar results with PKA protein levels in the NA.

Administration of stimulants enhances sex differences in dopaminergic activation in the mesolimbic pathway, and this may also lead to subsequent sex differences in drug reinforcement. For example, females exhibited increased striatal DA following amphetamine than males (Becker and Cha 1989; Becker and Ramirez 1981b), and they were more sensitive to the facilitating effects of cocaine on electrically stimulated DA release than males (Walker et al. 2006). Males and females also differed in activation of the DARPP-32 pathway in the NA following cocaine administration (females < males) (Lynch et al. 2007; Zhou et al. 2009). Striatal DA levels were greater in females than males following the administration of other drugs of abuse. For example, using in vivo microdialysis, Blanchard and Glick (1995) demonstrated that mesolimbic DA levels in the NA were greater in female rats following administration of low-to-moderate doses of alcohol, and female rats consumed more alcohol at these doses than male rats (Blanchard and Glick 1995; Blanchard et al. 1993). Females also exhibited an increased number of DA transporters in the NA following repeated injections of intravenous nicotine (Harrod et al. 2004).

4.2.2 Estrous Cycle and EST

Sex differences in the activation of the mesolimbic DA system have been attributed to circulating hormones. Several studies have demonstrated that EST treatment enhances striatal DA release (Becker 1990a,b, 1999; Becker and Ramirez 1981a; Dazzi et al. 2007; McEwen and Alves 1999; Zhang et al. 2008) and induces conditioned place preference when injected in large doses (Frye and Rhodes 2006) in OVX rats relative to VEH-treated controls. Striatal DA levels are also significantly higher in gonadally intact females compared to OVX female rats (Becker and Beer 1986; Becker et al. 1984; Becker and Ramirez 1981a), suggesting that the absence of EST may decrease DA levels. This may explain why a lack of EST, due to natural fluctuations or pharmacological and/or surgical manipulation (Anker et al. 2007; Larson and Carroll 2007; Larson et al. 2005, 2007; Lynch et al. 2001), leads to attenuated cocaine seeking.

Ligand-bound EST receptors regulate the transcription of proteins involved in the DA system (Jones and Miller 2008). Indeed, D_2 receptor densities in the striatum and other areas of the brain implicated in addiction vary across the estrous cycle in rats. They are greater following natural elevations of EST or following systemic EST administration (Bazzett and Becker 1994; Czoty et al. 2009; Di Paolo et al. 1988; Pazos et al. 1985; Zhou et al. 2002) and decrease significantly within 2 weeks following OVX (Le Saux et al. 2006). In contrast, in one study using positron emission tomography, D2 receptor concentrations were significantly lower during the EST-dominant follicular compared to the luteal phase of the menstrual cycle.

There is also evidence indicating that intracellular DA activity changes across the estrous cycle in female rats, and this may also contribute to cycle-dependent alterations in responses to drugs of abuse. Weiner et al. (2009) demonstrated that phosphorylated DARPP-32 levels in female rats were significantly lower during the estrus phase compared to all other phases of the estrous cycle. This result was explained as a consequence of heightened DA levels in the NA during the estrus phase, and lower DARPP-32 levels reflected a compensatory mechanism to stabilize excessive DA concentrations (Weiner et al. 2009).

Research also indicates hormone cycle mediation of dopaminergic responses to stimulants. For example, amphetamine-stimulated DA release in striatal tissue was increased during the estrus phase of the estrous cycle as determined using in vitro infusion (Becker and Ramirez 1981b), microdialysis (Becker and Cha 1989), and voltammetry (Becker 1990b). Several studies also implicate EST in modulating stimulant-induced dopaminergic activity. For example, EST treatment in OVX rats promoted DA neuronal sensitivity to cocaine, while DA neurons in OVX rats treated with VEH produced no change (Zhang et al. 2008). In another study, the induction of DA by cocaine- and amphetamine-regulated transcript (CART), a protein the regulates mesolimbic function in response to stimulants (Kuhar et al. 2005), was enhanced by EST administration in OVX rats relative to OVX rats treated with VEH (Shieh and Yang 2008). Furthermore, this effect was attributed to an intracellular mechanism as only administration of EST that was permeable to cellular membranes facilitated CART-induced DA turnover (Shieh and Yang 2008). EST treatment also facilitated nicotine-evoked DA release in the striatum in female, but not male rats (Dluzen and Anderson 1997), suggesting that the effects of EST on stimulant-induced DA are sex specific.

4.2.3 Estrogen Receptor Subtype β

Facilitation by EST on the reinforcing effects of cocaine may be attributed to the interaction between ER-β and DA neurotransmission in the mesolimbic pathway. ER-β is found in DA neurons (Laflamme et al. 1998) and has been shown to influence DA receptor expression and neurotransmission (Morissette et al. 2008; Schultz et al. 2009) in the mesolimbic DA pathway. Furthermore, administration of the ER-β agonist DPN, but not the ER-α agonist PPT, reversed OVX-induced decreases in D_2 receptors and DA turnover within the striatum and NA core (Le Saux and Di Paolo 2006). ER-β also regulates cocaine-seeking behaviors. As previously noted, administration of the ER-β agonist DPN, but not the ER-α agonist PPT, enhanced cocaine-primed reinstatement (Larson and Carroll 2007) and amphetamine-induced CPP (Silverman and Koenig 2007) in OVX female rats. Also, administration of tamoxifen, a partial antagonist at ER-α, but pure antagonist at ER-β, reduced EST's-facilitating effects on the acquisition of cocaine self-administration in EST-treated OVX female rats (Lynch et al. 2001). The contribution of ER-β to the rewarding effects of cocaine may be attributed to an intracellular mechanism involving downregulation of the regulator of G-protein signaling (RGS) 9-2. RGS9-2 regulates intracellular D_2 receptor activity and is highly localized in the NA following chronic cocaine exposure (Rahman et al. 2003; Wood 2007). Mice lacking this gene show enhanced responsiveness to cocaine (Rahman et al. 2003). Silverman and Koenig (2007) demonstrated that administration with an ER-β agonist, but not an ER-α agonist, reduced RGS9-2 expression in the core of the NA, a structure also implicated in cocaine-induced reinstatement (Ping et al. 2008), and enhanced amphetamine-induced CPP in OVX female rats. Together, these results suggest that EST enhancement of stimulant-related behaviors involves an interaction between ER-β and the mesolimbic DA pathway via intracellular transcription-related mechanisms that influence D2 receptor activity.

4.3 Progestins' Influence on the DA System

Far less work has been conducted on the effects of progestins on the mesolimbic DA pathway. However, the PROG metabolite, ALLO, attenuated stress-induced increases in DA (Dazzi et al. 2002), and altered DA release in the striatum and NA (Barrot et al. 1999; Jaworska-Feil et al. 1998; Laconi et al. 2007; Rouge-Pont et al. 2002). PROG also modulate DA levels in the striatum; however, results were equivocal and dependent on the time of testing, manner of administration, and presence or absence of EST (Dluzen and Ramirez 1984; 1987a,b; Fernandez-Ruiz et al. 1989). Thus, PROG and ALLO may interact with DA systems to influence drug-seeking behavior, but further work is needed to substantiate this.

4.4 Gamma-Aminobutyric Acid

An additional mechanism that may underlie EST and PROG's influence on cocaine seeking may be related to gamma-aminobutyric acid (GABA) neurotransmission. The facilitation and inhibition of GABA receptor neurotransmission resulted in the suppression and enhancement of mesolimbic DA, respectively (Tam and Roth 1985, 1990). Decreased GABA release was also associated with increased DA mediated behavior such as drug seeking in rats (Caille and Parsons 2004; Tang et al. 2005), while the administration of GABA receptor agonists decreased cocaine seeking (Campbell et al. 1999, 2002). This suggests that decreased GABA is associated with heightened vulnerability to cocaine-seeking behavior, whereas increased GABA may attenuate this behavior. Previous work indicated that EST inhibited activation of medium spiny GABAergic neurons in the striatum (Mermelstein et al. 1996), increased striatal GABA release (Hu et al.

2006), enhanced DA metabolism and turnover (Di Paolo et al. 1985; Shimizu and Bray 1993), and enhanced stimulant-elicited DA release in the striatum (Becker 1990a,b; Becker and Beer 1986; Castner et al. 1993). Conversely, PROG and its metabolites promoted striatal-GABA activity (Schumacher et al. 1989a,b) and decreased DA in the striatum (Dazzi et al. 2002; Dluzen and Ramirez 1987b; Jaworska-Feil et al. 1998; Laconi et al. 2007; Shimizu and Bray 1993). Thus, the different effects of EST and ALLO on cocaine seeking may be explained by their opposite effects on GABA and/or DA neurotransmission in areas of the brain that regulate drug seeking.

4.5 *HPA*

Stress is a major contributor to drug abuse (Fox and Sinha 2009), and activation of the stress system, the hypothalamic-pituitary-adrenal (HPA) axis, is associated with enhanced drug reward (Goeders 2002a,b). Preclinical work indicates that EST potentiated the release of CRF (Patchev et al. 1995; Swanson and Simmons 1989) and increased adrenocorticotropic hormone and corticosterone (Burgess and Handa 1992), which led to increased HPA activity (Dallman et al. 2004). These effects also extended to cocaine-induced HPA activation. In a study by Niyomchai and colleagues (2005), EST increased cocaine-induced corticosterone levels relative to VEH-treated controls. Behavioral studies provided further support for the role of EST on stress-related responses. EST facilitated fear-potentiated startle in OVX female rats relative to OVX rats treated with VEH, while PROG attenuated this facilitation (Hiroi and Neumaier 2006; Toufexis et al. 2004). Interestingly, ALLO attenuated HPA activation (Drugan et al. 1993; Frye et al. 2006; Owens et al. 1992; Patchev et al. 1994; Purdy et al. 1991), and it blocked stress-induced reinstatement in rats (Anker and Carroll, 2010). Further work is needed to examine the interaction between EST, the effects of stress, and behavior associated with drug abuse.

Overall, the neurobiological findings indicate that there are sex differences in DA receptor densities, intracellular DA neuronal activity, and extracellular DA levels in areas of the brain implicated in drug abuse vulnerability. Results presented in this chapter implicate EST in these differences possibly through its interaction with the ER-β. EST also decreases striatal GABA activation, and that may facilitate DA activation leading to increased sensitivity to the rewarding effects of drugs of abuse. Further, EST facilitates neurobiological and behavioral substrates of stress, which is a primary vulnerability factor in drug abuse. In contrast to EST, PROG and its metabolite ALLO exert an opposite effect on DA neurotransmission and potentiate GABA receptors and HPA activation that may lead to an attenuation of drug reinforcement.

In conclusion, the results indicate that females are more vulnerable to drug abuse than males during almost all of the critical phases of drug abuse: initiation, maintenance of rewarding effects, escalation of intake, extinction/craving, and reinstatement/relapse. Females are also more responsive than males to a wide range of behavioral and pharmacological interventions that reduce drug taking and drug seeking. In contrast, males exhibited greater withdrawal effects than females, suggesting they are more sensitive to the aversive effects of drugs, while females are more responsive to the rewarding effects. These differences in responsiveness to the positive and negative effects of drugs are an emerging area of interest for medication development and other treatment approaches for drug abuse. The enhanced vulnerability to drug seeking in females may be attributed to an interaction between EST and the mesolimbic reward pathway, specifically the ER-β, and its interaction with DA neuronal activity. In contrast to EST, PROG and its metabolite ALLO exert an opposite effect on DA neurotransmission and potentiate GABA receptors that may attenuate drug reinforcement. Thus, EST and PROG differentially interact with the HPA axis, GABA, DA to influence sensitivity to the rewarding effects of drugs, and they are the basis of sex differences in drug abuse. The role of EST and PROG should be considered in the development of sex-specific prevention and treatment approaches for drug abuse.

References

Ahmed SH, Koob GF (1998) Transition from moderate to excessive drug intake: change in hedonic set point. Science 282:298–300

Ahmed SH, Koob GF (1999) Long-lasting increase in the set point for cocaine self-administration after escalation in rats. Psychopharmacology (Berl) 146:303–312

Allen SS, Hatsukami DK, Christianson D, Nelson D (1999) Withdrawal and pre-menstrual symptomatology during the menstrual cycle in short-term smoking abstinence: effects of menstrual cycle on smoking abstinence. Nicotine Tob Res 1:129–142

Allen SS, Hatsukami DK, Bade T, Center B (2004) Transdermal nicotine use in postmenopausal women: does the treatment efficacy differ in women using and not using hormone replacement therapy? Nicotine Tob Res 6:777–788

Andersen SL, Rutstein M, Benzo JM, Hostetter JC, Teicher MH (1997) Sex differences in dopamine receptor overproduction and elimination. Neuroreport 8:1495–1498

Anker JJ, Carroll ME (2010) Sex differences in the effects of allopregnanolone on yohimbine-induced reinstatement of cocaine seeking in rats. Drug Alcohol Depend. 107:264–267

Anker JJ, Larson EB, Gliddon LA, Carroll ME (2007) Effects of progesterone on the reinstatement of cocaine-seeking behavior in female rats. Exp Clin Psychopharmacol 15:472–480

Anker JJ, Holtz NA, Zlebnik N, Carroll ME (2009) Effects of allopregnanolone on the reinstatement of cocaine-seeking behavior in male and female rats. Psychopharmacology (Berl) 203:63–72

Ashley OS, Marsden ME, Brady TM (2003) Effectiveness of substance abuse treatment programming for women: a review. Am J Drug Alcohol Abuse 29:19–53

Barrot M, Vallee M, Gingras MA, Le Moal M, Mayo W, Piazza PV (1999) The neurosteroid pregnenolone sulphate increases dopamine release and the dopaminergic response to morphine in the rat nucleus accumbens. Eur J Neurosci 11:3757–3760

Bazzett TJ, Becker JB (1994) Sex differences in the rapid and acute effects of estrogen on striatal D2 dopamine receptor binding. Brain Res 637:163–172

Becker JB (1990a) Direct effect of 17 beta-estradiol on striatum: sex differences in dopamine release. Synapse 5:157–164

Becker JB (1990b) Estrogen rapidly potentiates amphetamine-induced striatal dopamine release and rotational behavior during microdialysis. Neurosci Lett 118:169–171

Becker JB (1999) Gender differences in dopaminergic function in striatum and nucleus accumbens. Pharmacol Biochem Behav 64:803–812

Becker JB (2009) Sexual differentiation of motivation: a novel mechanism? Horm Behav 55: 646–654

Becker JB, Beer ME (1986) The influence of estrogen on nigrostriatal dopamine activity: behavioral and neurochemical evidence for both pre- and postsynaptic components. Behav Brain Res 19:27–33

Becker JB, Cha JH (1989) Estrous cycle-dependent variation in amphetamine-induced behaviors and striatal dopamine release assessed with microdialysis. Behav Brain Res 35:117–125

Becker JB, Hu M (2008) Sex differences in drug abuse. Front Neuroendocrinol 29:36–47

Becker JB, Ramirez VD (1981a) Experimental studies on the development of sex differences in the release of dopamine from striatal tissue fragments in vitro. Neuroendocrinology 32:168–173

Becker JB, Ramirez VD (1981b) Sex differences in the amphetamine stimulated release of catecholamines from rat striatal tissue in vitro. Brain Res 204:361–372

Becker JB, Beer ME, Robinson TE (1984) Striatal dopamine release stimulated by amphetamine or potassium: influence of ovarian hormones and the light-dark cycle. Brain Res 311:157–160

Blanchard BA, Glick SD (1995) Sex differences in mesolimbic dopamine responses to ethanol and relationship to ethanol intake in rats. Recent Dev Alcohol 12:231–241

Blanchard BA, Steindorf S, Wang S, Glick SD (1993) Sex differences in ethanol-induced dopamine release in nucleus accumbens and in ethanol consumption in rats. Alcohol Clin Exp Res 17:968–973

Brady KT, Randall CL (1999) Gender differences in substance use disorders. Psychiatr Clin North Am 22:241–252

Burgess LH, Handa RJ (1992) Chronic estrogen-induced alterations in adrenocorticotropin and corticosterone secretion, and glucocorticoid receptor-mediated functions in female rats. Endocrinology 131:1261–1269

Caille S, Parsons LH (2004) Intravenous heroin self-administration decreases GABA efflux in the ventral pallidum: an in vivo microdialysis study in rats. Eur J Neurosci 20:593–596

Cain ME, Smith CM, Bardo MT (2004) The effect of novelty on amphetamine self-administration in rats classified as high and low responders. Psychopharmacology (Berl) 176:129–138

Campbell UC, Lac ST, Carroll ME (1999) Effects of baclofen on maintenance and reinstatement of intravenous cocaine self-administration in rats. Psychopharmacology (Berl) 143:209–214

Campbell UC, Morgan AD, Carroll ME (2002) Sex differences in the effects of baclofen on the acquisition of intravenous cocaine self-administration in rats. Drug Alcohol Depend 66:61–69

Carpenter MJ, Upadhyaya HP, LaRowe SD, Saladin ME, Brady KT (2006) Menstrual cycle phase effects on nicotine withdrawal and cigarette craving: a review. Nicotine Tob Res 8:627–638

Carroll ME, Anker JJ (2010) Sex differences and ovarian steroid hormones in animal models of drug dependence and withdrawal. Horm Behav. 58:44–56

Carroll ME, Roth ME, Voeller RK, Nguyen PD (2000) Acquisition of oral phencyclidine self-administration in rhesus monkeys: effect of sex. Psychopharmacology (Berl) 149:401–408

Carroll ME, Morgan AD, Lynch WJ, Campbell UC, Dess NK (2002) Intravenous cocaine and heroin self-administration in rats selectively bred for differential saccharin intake: phenotype and sex differences. Psychopharmacology (Berl) 161:304–313

Carroll ME, Lynch WJ, Roth ME, Morgan AD, Cosgrove KP (2004) Sex and estrogen influence drug abuse. Trends Pharmacol Sci 25:273–279

Carroll ME, Batulis DK, Landry KL, Morgan AD (2005) Sex differences in the escalation of oral phencyclidine (PCP) self-administration under FR and PR schedules in rhesus monkeys. Psychopharmacology (Berl) 180:414–426

Carroll ME, Anker JJ, Perry JL (2009a) Modeling risk factors for nicotine and other drug abuse in the preclinical laboratory. Drug Alcohol Depend 104(Suppl 1):S70–S78

Carroll ME, Mach JL, La Nasa RM, Newman JL (2009b) Impulsivity as a behavioral measure of withdrawal of orally delivered PCP and nondrug rewards in male and female monkeys. Psychopharmacology (Berl) 207:85–98

Castner SA, Xiao L, Becker JB (1993) Sex differences in striatal dopamine: in vivo microdialysis and behavioral studies. Brain Res 610:127–134

Chaudhri N, Caggiula AR, Donny EC, Booth S, Gharib MA, Craven LA, Allen SS, Sved AF, Perkins KA (2005) Sex differences in the contribution of nicotine and nonpharmacological stimuli to nicotine self-administration in rats. Psychopharmacology (Berl) 180:258–266

Chen K, Kandel D (2002) Relationship between extent of cocaine use and dependence among adolescents and adults in the United States. Drug Alcohol Depend 68:65–85

Covington HE III, Kikusui T, Goodhue J, Nikulina EM, Hammer RP Jr, Miczek KA (2005) Brief social defeat stress: long lasting effects on cocaine taking during a binge and zif268 mRNA expression in the amygdala and prefrontal cortex. Neuropsychopharmacology 30:310–321

Czoty PW, Riddick NV, Gage HD, Sandridge M, Nader SH, Garg S, Bounds M, Garg PK, Nader MA (2009) Effect of menstrual cycle phase on dopamine D2 receptor availability in female cynomolgus monkeys. Neuropsychopharmacology 34:548–554

Dallman MF, Akana SF, Strack AM, Scribner KS, Pecoraro N, La Fleur SE, Houshyar H, Gomez F (2004) Chronic stress-induced effects of corticosterone on brain: direct and indirect. Ann N Y Acad Sci 1018:141–150

Dazzi L, Serra M, Vacca G, Ladu S, Latrofa A, Trapani G, Biggio G (2002) Depletion of cortical allopregnanolone potentiates stress-induced increase in cortical dopamine output. Brain Res 932:135–139

Dazzi L, Seu E, Cherchi G, Barbieri PP, Matzeu A, Biggio G (2007) Estrous cycle-dependent changes in basal and ethanol-induced activity of cortical dopaminergic neurons in the rat. Neuropsychopharmacology 32:892–901

Di Paolo T, Rouillard C, Bedard P (1985) 17 beta-Estradiol at a physiological dose acutely increases dopamine turnover in rat brain. Eur J Pharmacol 117:197–203

Di Paolo T, Falardeau P, Morissette M (1988) Striatal D-2 dopamine agonist binding sites fluctuate during the rat estrous cycle. Life Sci 43:665–672

Dluzen DE, Anderson LI (1997) Estrogen differentially modulates nicotine-evoked dopamine release from the striatum of male and female rats. Neurosci Lett 230:140–142

Dluzen DE, Ramirez VD (1984) Bimodal effect of progesterone on in vitro dopamine function of the rat corpus striatum. Neuroendocrinology 39:149–155

Dluzen DE, Ramirez VD (1987a) In vivo release of dopamine and its metabolites following a direct infusion of L-dopa into the caudate nucleus of awake, freely behaving rats using a push-pull cannula. Neurosci Lett 82:29–34

Dluzen DE, Ramirez VD (1987b) Intermittent infusion of progesterone potentiates whereas continuous infusion reduces amphetamine-stimulated dopamine release from ovariectomized estrogen-primed rat striatal fragments superfused in vitro. Brain Res 406:1–9

Drugan RC, Park R, Kaufman L, Holmes PV (1993) Etiology of the sexual dimorphism in renal peripheral benzodiazepine receptor response to stress in rats. Horm Behav 27:348–365

Evans SM (2007) The role of estradiol and progesterone in modulating the subjective effects of stimulants in humans. Exp Clin Psychopharmacol 15:418–426

Evans SM, Foltin RW (2006) Exogenous progesterone attenuates the subjective effects of smoked cocaine in women, but not in men. Neuropsychopharmacology 31:659–674

Evans SM, Haney M, Foltin RW (2002) The effects of smoked cocaine during the follicular and luteal phases of the menstrual cycle in women. Psychopharmacology (Berl) 159:397–406

Farde L, Hall H, Pauli S, Halldin C (1995) Variability in D2-dopamine receptor density and affinity: a PET study with [11C]raclopride in man. Synapse 20:200–208

Fattore L, Spano MS, Altea S, Angius F, Fadda P, Fratta W (2007) Cannabinoid self-administration in rats: sex differences and the influence of ovarian function. Br J Pharmacol 152:795–804

Fattore L, Altea S, Fratta W (2008) Sex differences in drug addiction: a review of animal and human studies. Womens Health (Lond Engl) 4:51–65

Feltenstein MW, See RE (2007) Plasma progesterone levels and cocaine-seeking in freely cycling female rats across the estrous cycle. Drug Alcohol Depend 89:183–189
Fernandez-Ruiz JJ, Amor JC, Ramos JA (1989) Time-dependent effects of estradiol and progesterone on the number of striatal dopaminergic D2-receptors. Brain Res 476:388–395
Festa ED, Quinones-Jenab V (2004) Gonadal hormones provide the biological basis for sex differences in behavioral responses to cocaine. Horm Behav 46:509–519
Filipek PA, Richelme C, Kennedy DN, Caviness VS Jr (1994) The young adult human brain: an MRI-based morphometric analysis. Cereb Cortex 4:344–360
Fox HC, Sinha R (2009) Sex differences in drug-related stress-system changes: implications for treatment in substance-abusing women. Harv Rev Psychiatry 17:103–119
Freitag WJ, Adesso VJ (1993) Mood effects of alcohol and expectancies across the menstrual cycle. Alcohol 10:291–298
Frye CA, Rhodes ME (2006) Administration of estrogen to ovariectomized rats promotes conditioned place preference and produces moderate levels of estrogen in the nucleus accumbens. Brain Res 1067:209–215
Frye CA, Rhodes ME, Petralia SM, Walf AA, Sumida K, Edinger KL (2006) 3alpha-hydroxy-5alpha-pregnan-20-one in the midbrain ventral tegmental area mediates social, sexual, and affective behaviors. Neuroscience 138:1007–1014
Fuchs RA, Evans KA, Mehta RH, Case JM, See RE (2005) Influence of sex and estrous cyclicity on conditioned cue-induced reinstatement of cocaine-seeking behavior in rats. Psychopharmacology (Berl) 179:662–672
Gallop RJ, Crits-Christoph P, Ten Have TR, Barber JP, Frank A, Griffin ML, Thase ME (2007) Differential transitions between cocaine use and abstinence for men and women. J Consult Clin Psychol 75:95–103
Gear RW, Miaskowski C, Gordon NC, Paul SM, Heller PH, Levine JD (1996) Kappa-opioids produce significantly greater analgesia in women than in men. Nat Med 2:1248–1250
Goeders NE (2002a) Stress and cocaine addiction. J Pharmacol Exp Ther 301:785–789
Goeders NE (2002b) The HPA axis and cocaine reinforcement. Psychoneuroendocrinology 27:13–33
Harrod SB, Mactutus CF, Bennett K, Hasselrot U, Wu G, Welch M, Booze RM (2004) Sex differences and repeated intravenous nicotine: behavioral sensitization and dopamine receptors. Pharmacol Biochem Behav 78:581–592
Hay WM, Nathan PE, Heermans HW, Frankenstein W (1984) Menstrual cycle, tolerance and blood alcohol level discrimination ability. Addict Behav 9:67–77
Hiroi R, Neumaier JF (2006) Differential effects of ovarian steroids on anxiety versus fear as measured by open field test and fear-potentiated startle. Behav Brain Res 166:93–100
Holdstock L, de Wit H (2000) Effects of ethanol at four phases of the menstrual cycle. Psychopharmacology (Berl) 150:374–382
Hu M, Watson CJ, Kennedy RT, Becker JB (2006) Estradiol attenuates the K+-induced increase in extracellular GABA in rat striatum. Synapse 59:122–124
Ignjatova L, Raleva M (2009) Gender difference in the treatment outcome of patients served in the mixed-gender program. Bratisl Lek Listy 110:285–289
Jackson LR, Robinson TE, Becker JB (2006) Sex differences and hormonal influences on acquisition of cocaine self-administration in rats. Neuropsychopharmacology 31:129–138
Jaworska-Feil L, Budziszewska B, Leskiewicz M, Lason W (1998) Opposite effects of inhibitory and excitatory neurosteroids on [3H]dopamine release from rat nucleus accumbens. Pol J Pharmacol 50:449–452
Jones DC, Miller GW (2008) The effects of environmental neurotoxicants on the dopaminergic system: a possible role in drug addiction. Biochem Pharmacol 76:569–581
Justice AJ, de Wit H (1999) Acute effects of d-amphetamine during the follicular and luteal phases of the menstrual cycle in women. Psychopharmacology (Berl) 145:67–75
Justice AJ, de Wit H (2000) Acute effects of d-amphetamine during the early and late follicular phases of the menstrual cycle in women. Pharmacol Biochem Behav 66:509–515

Kalivas PW, Volkow ND (2005) The neural basis of addiction: a pathology of motivation and choice. Am J Psychiatry 162:1403–1413

Katz JL, Higgins ST (2003) The validity of the reinstatement model of craving and relapse to drug use. Psychopharmacology (Berl) 168:21–30

Kerstetter KA, Aguilar VR, Parrish AB, Kippin TE (2008) Protracted time-dependent increases in cocaine-seeking behavior during cocaine withdrawal in female relative to male rats. Psychopharmacology (Berl) 198:63–75

Kippin TE, Fuchs RA, Mehta RH, Case JM, Parker MP, Bimonte-Nelson HA, See RE (2005) Potentiation of cocaine-primed reinstatement of drug seeking in female rats during estrus. Psychopharmacology (Berl) 182:245–252

Kornetsky C (2007) A walk through the history of research in drug abuse trends and fads. The Nathan B. Eddy Award address given at CPDD 2005: part II. Drug Alcohol Depend 90:312–316

Kuhar MJ, Jaworski JN, Hubert GW, Philpot KB, Dominguez G (2005) Cocaine- and amphetamine-regulated transcript peptides play a role in drug abuse and are potential therapeutic targets. AAPS J 7:E259–E265

Laconi MR, Reggiani PC, Penissi A, Yunes R, Cabrera RJ (2007) Allopregnanolone modulates striatal dopamingergic activity of rats under different gonadal hormones conditions. Neurol Res 29:622–627

Laflamme N, Nappi RE, Drolet G, Labrie C, Rivest S (1998) Expression and neuropeptidergic characterization of estrogen receptors (ERalpha and ERbeta) throughout the rat brain: anatomical evidence of distinct roles of each subtype. J Neurobiol 36:357–378

Larson EB, Carroll ME (2007) Estrogen receptor beta, but not alpha, mediates estrogen's effect on cocaine-induced reinstatement of extinguished cocaine-seeking behavior in ovariectomized female rats. Neuropsychopharmacology 32:1334–1345

Larson EB, Roth ME, Anker JJ, Carroll ME (2005) Effect of short- vs. long-term estrogen on reinstatement of cocaine-seeking behavior in female rats. Pharmacol Biochem Behav 82: 98–108

Larson EB, Anker JJ, Gliddon LA, Fons KS, Carroll ME (2007) Effects of estrogen and progesterone on the escalation of cocaine self-administration in female rats during extended access. Exp Clin Psychopharmacol 15:461–471

Le Saux M, Di Paolo T (2006) Influence of oestrogenic compounds on monoamine transporters in rat striatum. J Neuroendocrinol 18:25–32

Le Saux M, Morissette M, Di Paolo T (2006) ERbeta mediates the estradiol increase of D2 receptors in rat striatum and nucleus accumbens. Neuropharmacology 50:451–457

Lex BW, Mendelson JH, Bavli S, Harvey K, Mello NK (1984) Effects of acute marijuana smoking on pulse rate and mood states in women. Psychopharmacology (Berl) 84:178–187

Lukas SE, Sholar M, Lundahl LH, Lamas X, Kouri E, Wines JD, Kragie L, Mendelson JH (1996) Sex differences in plasma cocaine levels and subjective effects after acute cocaine administration in human volunteers. Psychopharmacology (Berl) 125:346–354

Lynch WJ (2008) Acquisition and maintenance of cocaine self-administration in adolescent rats: effects of sex and gonadal hormones. Psychopharmacology (Berl) 197:237–246

Lynch WJ, Carroll ME (1999a) Regulation of intravenously self-administered nicotine in rats. Exp Clin Psychopharmacol 7:198–207

Lynch WJ, Carroll ME (1999b) Sex differences in the acquisition of intravenously self-administered cocaine and heroin in rats. Psychopharmacology (Berl) 144:77–82

Lynch WJ, Carroll ME (2000) Reinstatement of cocaine self-administration in rats: sex differences. Psychopharmacology (Berl) 148:196–200

Lynch WJ, Carroll ME (2001) Regulation of drug intake. Exp Clin Psychopharmacol 9:131–143

Lynch WJ, Taylor JR (2004) Sex differences in the behavioral effects of 24-h access to cocaine under a discrete trial procedure. Neuropsychopharmacology 29:943–951

Lynch WJ, Taylor JR (2005) Decreased motivation following cocaine self-administration under extended access conditions: effects of sex and ovarian hormones. Neuropsychopharmacology 30:927–935

Lynch WJ, LaBounty LP, Carroll ME (1998) A novel paradigm to investigate regulation of drug intake in rats self-administering cocaine or heroin intravenously. Exp Clin Psychopharmacol 6:22–31

Lynch WJ, Arizzi MN, Carroll ME (2000) Effects of sex and the estrous cycle on regulation of intravenously self-administered cocaine in rats. Psychopharmacology (Berl) 152:132–139

Lynch WJ, Roth ME, Mickelberg JL, Carroll ME (2001) Role of estrogen in the acquisition of intravenously self-administered cocaine in female rats. Pharmacol Biochem Behav 68:641–646

Lynch WJ, Roth ME, Carroll ME (2002) Biological basis of sex differences in drug abuse: preclinical and clinical studies. Psychopharmacology (Berl) 164:121–137

Lynch WJ, Mangini LD, Taylor JR (2005) Neonatal isolation stress potentiates cocaine seeking behavior in adult male and female rats. Neuropsychopharmacology 30:322–329

Lynch WJ, Kiraly DD, Caldarone BJ, Picciotto MR, Taylor JR (2007) Effect of cocaine self-administration on striatal PKA-regulated signaling in male and female rats. Psychopharmacology (Berl) 191:263–271

Mann K, Ackermann K, Croissant B, Mundle G, Nakovics H, Diehl A (2005) Neuroimaging of gender differences in alcohol dependence: are women more vulnerable? Alcohol Clin Exp Res 29:896–901

Marsh JC, Cao D, D'Aunno T (2004) Gender differences in the impact of comprehensive services in substance abuse treatment. J Subst Abuse Treat 27:289–300

McCarthy MM, Davis AM, Mong JA (1997) Excitatory neurotransmission and sexual differentiation of the brain. Brain Res Bull 44:487–495

McEwen BS, Alves SE (1999) Estrogen actions in the central nervous system. Endocr Rev 20:279–307

McKay JR, Weiss RV (2001) A review of temporal effects and outcome predictors in substance abuse treatment studies with long-term follow-ups. Preliminary results and methodological issues. Eval Rev 25:113–161

Mendelson JH, Mello NK, Sholar MB, Siegel AJ, Kaufman MJ, Levin JM, Renshaw PF, Cohen BM (1999) Cocaine pharmacokinetics in men and in women during the follicular and luteal phases of the menstrual cycle. Neuropsychopharmacology 21:294–303

Mermelstein PG, Becker JB, Surmeier DJ (1996) Estradiol reduces calcium currents in rat neostriatal neurons via a membrane receptor. J Neurosci 16:595–604

Mizukami S, Nishizuka M, Arai Y (1983) Sexual difference in nuclear volume and its ontogeny in the rat amygdala. Exp Neurol 79:569–575

Morissette M, Le Saux M, D'Astous M, Jourdain S, Al Sweidi S, Morin N, Estrada-Camarena E, Mendez P, Garcia-Segura LM, Di Paolo T (2008) Contribution of estrogen receptors alpha and beta to the effects of estradiol in the brain. J Steroid Biochem Mol Biol 108:327–338

Munro CA, McCaul ME, Wong DF, Oswald LM, Zhou Y, Brasic J, Kuwabara H, Kumar A, Alexander M, Ye W, Wand GS (2006) Sex differences in striatal dopamine release in healthy adults. Biol Psychiatry 59:966–974

Nairn AC, Svenningsson P, Nishi A, Fisone G, Girault JA, Greengard P (2004) The role of DARPP-32 in the actions of drugs of abuse. Neuropharmacology 47(Suppl 1):14–23

Nazarian A, Sun WL, Zhou L, Kemen LM, Jenab S, Quinones-Jenab V (2009) Sex differences in basal and cocaine-induced alterations in PKA and CREB proteins in the nucleus accumbens. Psychopharmacology (Berl) 203:641–650

Nestler EJ (2005) Is there a common molecular pathway for addiction? Nat Neurosci 8:1445–1449

Niyomchai T, Russo SJ, Festa ED, Akhavan A, Jenab S, Quinones-Jenab V (2005) Progesterone inhibits behavioral responses and estrogen increases corticosterone levels after acute cocaine administration. Pharmacol Biochem Behav 80:603–610

Nyberg S, Wahlstrom G, Backstrom T, Sundstrom Poromaa I (2004) Altered sensitivity to alcohol in the late luteal phase among patients with premenstrual dysphoric disorder. Psychoneuroendocrinology 29:767–777

Owens MJ, Ritchie JC, Nemeroff CB (1992) 5 alpha-pregnane-3 alpha, 21-diol-20-one (THDOC) attenuates mild stress-induced increases in plasma corticosterone via a non-glucocorticoid mechanism: comparison with alprazolam. Brain Res 573:353–355

Panlilio LV, Thorndike EB, Schindler CW (2007) Blocking of conditioning to a cocaine-paired stimulus: testing the hypothesis that cocaine perpetually produces a signal of larger-than-expected reward. Pharmacol Biochem Behav 86:774–777

Patchev VK, Shoaib M, Holsboer F, Almeida OF (1994) The neurosteroid tetrahydroprogesterone counteracts corticotropin-releasing hormone-induced anxiety and alters the release and gene expression of corticotropin-releasing hormone in the rat hypothalamus. Neuroscience 62:265–271

Patchev VK, Hayashi S, Orikasa C, Almeida OF (1995) Implications of estrogen-dependent brain organization for gender differences in hypothalamo-pituitary-adrenal regulation. FASEB J 9:419–423

Pazos A, Stoeckel ME, Hindelang C, Palacios JM (1985) Autoradiographic studies on dopamine D2 receptors in rat pituitary: influence of hormonal states. Neurosci Lett 59:1–7

Perry JL, Nelson SE, Carroll ME (2008) Impulsive choice as a predictor of acquisition of IV cocaine self- administration and reinstatement of cocaine-seeking behavior in male and female rats. Exp Clin Psychopharmacol 16:165–177

Ping A, Xi J, Prasad BM, Wang MH, Kruzich PJ (2008) Contributions of nucleus accumbens core and shell GluR1 containing AMPA receptors in AMPA- and cocaine-primed reinstatement of cocaine-seeking behavior. Brain Res 1215:173–182

Pomerleau CS, Garcia AW, Pomerleau OF, Cameron OG (1992) The effects of menstrual phase and nicotine abstinence on nicotine intake and on biochemical and subjective measures in women smokers: a preliminary report. Psychoneuroendocrinology 17:627–638

Pomerleau CS, Mehringer AM, Marks JL, Downey KK, Pomerleau OF (2000) Effects of menstrual phase and smoking abstinence in smokers with and without a history of major depressive disorder. Addict Behav 25:483–497

Purdy RH, Morrow AL, Moore PH Jr, Paul SM (1991) Stress-induced elevations of gamma-aminobutyric acid type A receptor-active steroids in the rat brain. Proc Natl Acad Sci U S A 88:4553–4557

Rabinowicz T, Dean DE, Petetot JM, de Courten-Myers GM (1999) Gender differences in the human cerebral cortex: more neurons in males; more processes in females. J Child Neurol 14:98–107

Rahman Z, Schwarz J, Gold SJ, Zachariou V, Wein MN, Choi KH, Kovoor A, Chen CK, DiLeone RJ, Schwarz SC, Selley DE, Sim-Selley LJ, Barrot M, Luedtke RR, Self D, Neve RL, Lester HA, Simon MI, Nestler EJ (2003) RGS9 modulates dopamine signaling in the basal ganglia. Neuron 38:941–952

Randall CL, Roberts JS, Del Boca FK, Carroll KM, Connors GJ, Mattson ME (1999) Telescoping of landmark events associated with drinking: a gender comparison. J Stud Alcohol 60:252–260

Robbins SJ, Ehrman RN, Childress AR, O'Brien CP (1999) Comparing levels of cocaine cue reactivity in male and female outpatients. Drug Alcohol Depend 53:223–230

Roth ME, Carroll ME (2004a) Sex differences in the acquisition of IV methamphetamine self-administration and subsequent maintenance under a progressive ratio schedule in rats. Psychopharmacology (Berl) 172:443–449

Roth ME, Carroll ME (2004b) Sex differences in the escalation of intravenous cocaine intake following long- or short-access to cocaine self-administration. Pharmacol Biochem Behav 78:199–207

Roth ME, Casimir AG, Carroll ME (2002) Influence of estrogen in the acquisition of intravenously self-administered heroin in female rats. Pharmacol Biochem Behav 72:313–318

Rouge-Pont F, Mayo W, Marinelli M, Gingras M, Le Moal M, Piazza PV (2002) The neurosteroid allopregnanolone increases dopamine release and dopaminergic response to morphine in the rat nucleus accumbens. Eur J Neurosci 16:169–173

Schultz KN, von Esenwein SA, Hu M, Bennett AL, Kennedy RT, Musatov S, Toran-Allerand CD, Kaplitt MG, Young LJ, Becker JB (2009) Viral vector-mediated overexpression of estrogen receptor-alpha in striatum enhances the estradiol-induced motor activity in female rats and estradiol-modulated GABA release. J Neurosci 29:1897–1903

Schumacher M, Coirini H, McEwen BS (1989a) Regulation of high-affinity GABAa receptors in specific brain regions by ovarian hormones. Neuroendocrinology 50:315–320

Schumacher M, Coirini H, McEwen BS (1989b) Regulation of high-affinity GABAA receptors in the dorsal hippocampus by estradiol and progesterone. Brain Res 487:178–183

Self DW, Barnhart WJ, Lehman DA, Nestler EJ (1996) Opposite modulation of cocaine-seeking behavior by D1- and D2-like dopamine receptor agonists. Science 271:1586–1589

Shaham Y, Shalev U, Lu L, de Wit H, Stewart J (2003) The reinstatement model of drug relapse: history, methodology and major findings. Psychopharmacology (Berl) 168:3–20

Shieh KR, Yang SC (2008) Effects of estradiol on the stimulation of dopamine turnover in mesolimbic and nigrostriatal systems by cocaine- and amphetamine-regulated transcript peptide in female rats. Neuroscience 154:1589–1597

Shimizu H, Bray GA (1993) Effects of castration, estrogen replacement and estrus cycle on monoamine metabolism in the nucleus accumbens, measured by microdialysis. Brain Res 621:200–206

Silverman JL, Koenig JI (2007) Evidence for the involvement of ERbeta and RGS9-2 in 17-beta estradiol enhancement of amphetamine-induced place preference behavior. Horm Behav 52:146–155

Sinha R (2008) Chronic stress, drug use, and vulnerability to addiction. Ann N Y Acad Sci 1141:105–130

Sinha R, Fox H, Hong KI, Sofuoglu M, Morgan PT, Bergquist KT (2007) Sex steroid hormones, stress response, and drug craving in cocaine-dependent women: implications for relapse susceptibility. Exp Clin Psychopharmacol 15:445–452

Snively TA, Ahijevych KL, Bernhard LA, Wewers ME (2000) Smoking behavior, dysphoric states and the menstrual cycle: results from single smoking sessions and the natural environment. Psychoneuroendocrinology 25:677–691

Sofuoglu M, Dudish-Poulsen S, Nelson D, Pentel PR, Hatsukami DK (1999) Sex and menstrual cycle differences in the subjective effects from smoked cocaine in humans. Exp Clin Psychopharmacol 7:274–283

Sofuoglu M, Babb DA, Hatsukami DK (2002) Effects of progesterone treatment on smoked cocaine response in women. Pharmacol Biochem Behav 72:431–435

Sofuoglu M, Mitchell E, Kosten TR (2004) Effects of progesterone treatment on cocaine responses in male and female cocaine users. Pharmacol Biochem Behav 78:699–705

Spanagel R, Weiss F (1999) The dopamine hypothesis of reward: past and current status. Trends Neurosci 22:521–527

Sutker PB, Goist KC Jr, King AR (1987) Acute alcohol intoxication in women: relationship to dose and menstrual cycle phase. Alcohol Clin Exp Res 11:74–79

Swanson LW, Simmons DM (1989) Differential steroid hormone and neural influences on peptide mRNA levels in CRH cells of the paraventricular nucleus: a hybridization histochemical study in the rat. J Comp Neurol 285:413–435

Tam SY, Roth RH (1985) Selective increase in dopamine metabolism in the prefrontal cortex by the anxiogenic beta-carboline FG 7142. Biochem Pharmacol 34:1595–1598

Tam SY, Roth RH (1990) Modulation of mesoprefrontal dopamine neurons by central benzodiazepine receptors. I. Pharmacological characterization. J Pharmacol Exp Ther 252:989–996

Tang AC, Nakazawa M, Romeo RD, Reeb BC, Sisti H, McEwen BS (2005) Effects of long-term estrogen replacement on social investigation and social memory in ovariectomized C57BL/6 mice. Horm Behav 47:350–357

Terner JM, de Wit H (2006) Menstrual cycle phase and responses to drugs of abuse in humans. Drug Alcohol Depend 84:1–13

Toufexis DJ, Davis C, Hammond A, Davis M (2004) Progesterone attenuates corticotropin-releasing factor-enhanced but not fear-potentiated startle via the activity of its neuroactive metabolite, allopregnanolone. J Neurosci 24:10280–10287

Walker QD, Rooney MB, Wightman RM, Kuhn CM (2000) Dopamine release and uptake are greater in female than male rat striatum as measured by fast cyclic voltammetry. Neuroscience 95:1061–1070

Walker QD, Ray R, Kuhn CM (2006) Sex differences in neurochemical effects of dopaminergic drugs in rat striatum. Neuropsychopharmacology 31:1193–1202

Weiner J, Sun WL, Zhou L, Kreiter CM, Jenab S, Quinones-Jenab V (2009) PKA-mediated responses in females' estrous cycle affect cocaine-induced responses in dopamine-mediated intracellular cascades. Neuroscience 161:865–876

White TL, Justice AJ, de Wit H (2002) Differential subjective effects of D-amphetamine by gender, hormone levels and menstrual cycle phase. Pharmacol Biochem Behav 73:729–741

Wise RA (1996) Neurobiology of addiction. Curr Opin Neurobiol 6:243–251

Wood RI (2007) Sex and drugs: comment on "Evidence for involvement of erbeta and rgs9-2 in 17-beta estradiol enhancement of amphetamine-induced place preference behavior" by Silverman and Koenig. Horm Behav 52:143–145

Zhang D, Yang S, Yang C, Jin G, Zhen X (2008) Estrogen regulates responses of dopamine neurons in the ventral tegmental area to cocaine. Psychopharmacology (Berl) 199:625–635

Zhou W, Cunningham KA, Thomas ML (2002) Estrogen regulation of gene expression in the brain: a possible mechanism altering the response to psychostimulants in female rats. Brain Res Mol Brain Res 100:75–83

Zhou L, Nazarian A, Sun WL, Jenab S, Quinones-Jenab V (2009) Basal and cocaine-induced sex differences in the DARPP-32-mediated signaling pathway. Psychopharmacology (Berl) 203:175–183

Sex Differences in Response to Stress and Expression of Depressive-Like Behaviours in the Rat

Christina Dalla, Pothitos M. Pitychoutis, Nikolaos Kokras, and Zeta Papadopoulou-Daifoti

Contents

Abstract Women are more susceptible than men to certain stress-related psychiatric disorders, such as depression. Preclinical studies aim to understand these sex differences by studying male and female rats in stress models. In this chapter, we review sex differences in behavioural aspects, as well as neurochemical and neurobiological findings derived from acute, repeated and chronic stress models. In particular, we focus on sex differences in depressive-like symptomatology expressed in the forced swim test, the chronic mild stress (CMS) and the learned helplessness models, the Flinders Sensitive Line rats (FSL), which is a genetic model of depression and in the lipopolysaccharide (LPS)-induced *sickness behaviour*, a putative inflammatory model of depression. Also, sex differences in stress effects

C. Dalla, P.M. Pitychoutis, N. Kokras, and Z. Papadopoulou-Daifoti (✉)
Department of Pharmacology, Medical School, University of Athens, Mikras Asias 75, Goudi, 11527 Athens, Greece
e-mail: zdaifoti@med.uoa.gr

J.C. Neill and J. Kulkarni (eds.), *Biological Basis of Sex Differences in Psychopharmacology*, 97
Current Topics in Behavioral Neurosciences 8, DOI 10.1007/7854_2010_94,

on learning and memory parameters are discussed, because cognitive alterations are often seen in sex-differentiated psychiatric disorders. The observed behavioural alterations are often linked with abnormalities in the endophenotype, such as in hormonal, neurochemical, immune and neuroplasticity indices. From these data, it is clear that all stress models have strengths and limitations that need to be recognized in order to use them effectively in the investigation of sex differences in affective disorders.

Keywords Chronic mild stress · Depression · Flinders sensitive line · Forced swim test · Gender · Immunity · Learned helplessness · Learning · Serotonin · Sex differences · Sickness behaviour · Stress

Abbreviations

8-OHDPAT	(+)8-Hydroxy-N,N-dipropyl-2-aminotetralin hydrobromide
ACTH	Adrenocorticotropin hormone
AVP	Arginine vasopressin
CMS	Chronic mild stress
FSL	Flinders sensitive line
FST	Forced swim test
HPA	Hypothalamic-pituitary-adrenal
HPG	Hypothalamus-pituitary-gonadal
LPS	Lipopolysaccharide
LC	Locus coeruleus

1 Introduction

The organism's stress response is an evolutionary mechanism of great adaptive value; however, when stress exposure is prolonged and uncontrollable the response may be inadequate and maladaptive for the organism, presumably leading to physical and mental disorders, such as major depression, and anxiety disorders, which are major causes of disability in contemporary western societies (Lupien et al. 2009). Women are more susceptible than men to stress-related psychiatric disorders, such as major depression, generalized anxiety disorder, acute and post-traumatic stress disorder (Holden 2005). It is suggested that because of genetic, hormonal, biochemical and social factors, different responses to stressful life events and different coping strategies used by men *versus* women ultimately contribute to sex differences in the occurrence of stress-related disorders (Kendler et al. 2001; Maciejewski et al. 2001; Nemeroff et al. 2006; Sherrill et al. 1997). However, the basic neuroendocrine stress response, especially the one associated with acute stress, does not differ substantially between the two sexes (Klein and Corwin 2002).

Understanding sex differences in the stress response may contribute to the improvement of psychiatric diagnosis and treatment. Animal models have been widely used to study the effects of stress exposure on multiple aspects, such as hormonal, neurochemical, neurobiological and behavioural. Factors that determine the stress response include the type, the duration and the characteristics of the stressors, as well as the strain, the age of the animal and its genetic predisposition (Anisman and Matheson 2005). During the last decades, we have acknowledged the sex of the animal as an important factor that influences the stress response (Palanza 2001; Dalla et al. 2010). Although comparative studies on males and females are still limited, they have been conducted throughout the lifespan of the animal and include different protocols of stressor exposure, as well as genetic manipulations. Additionally, because of the role of stress as a predisposing and precipitating factor in the onset of depression (Anisman and Zacharko 1990; Holsboer 2001) and the marked sex differences in the prevalence of major depression in humans (Marcus et al. 2005), there has been a considerable interest in sex differences in depressive-like symptomatology expressed in several animal stress models. Other behavioural parameters that are studied and are often differentiated between the two sexes are cognitive, anxiety and activity responses. In particular, learning and memory parameters are widely studied in stress paradigms, because cognitive alterations are often seen in sex-differentiated psychiatric disorders, such as depression and post-traumatic stress disorder (Moore 2009; Sotiropoulos et al. 2008). In this chapter, we will review sex differences in the stress response, focusing on behavioural and neurochemical changes.

2 Sex Differences in Response to Acute and Repeated Stressful Experiences

2.1 *Sex Differences in Neuroendocrine Responses*

The organism's acute stress response mainly includes the activation of the locus coeruleus–norepinephrine (LC-NE) system and the hypothalamic-pituitary-adrenal (HPA) axis. The morphology of the LC is sexually dimorphic with females having a greater volume and number of neurons than males (Pinos et al. 2001). Also, its neuronal activity is more responsive to certain stressors in females than in males and this is probably regulated by corticotropin-releasing hormone (CRH) activity (Curtis et al. 2006).

On the other hand, HPA axis activation seems to be responsible for the slower response to stress. It is activated within minutes after the stressful stimulus and is initiated by the release of CRH and arginine vasopressin (AVP) by the hypothalamus (DeBold et al. 1984) and subsequently the release of adrenocorticotropin (ACTH) from the hypophysis. Such a cascade of events results in the release of glucocorticoids, such as corticosterone in rodents, from the adrenal cortex.

Under normal conditions, the secreted glucocorticoids inhibit further release of ACTH by feedback inhibition, thus enabling termination of the stress response and a return to homeostasis (Heinrichs and Koob 2004). This feedback is accomplished by the action of glucocorticoids on corticosteroid (GR and MR) receptors that are abundant in the limbic system, especially in the hypothalamus and hippocampus and their activation is sex-dependent (Karandrea et al. 2000, 2002). Sex differences are also evident in other aspects of HPA axis activity; female rats have higher resting levels of corticosterone and display greater diurnal changes in both ACTH and corticosterone than males (Handa et al. 1994; Kitay 1961). Also, in comparison to males, female rats have higher glucocorticoid levels following acute and repeated stress exposure and this seems to be dependent on circulating gonadal hormones (Galea et al. 1997; Seale et al. 2004a,b).

Prolonged activation of the HPA axis, as in the case of repeated or chronic stress exposure, can lead to "exhaustion" of the stress response system and halt the return to homeostasis. In this state, stress often has detrimental effects on the integrity and function of the brain. These effects include alterations in the activity of monoaminergic systems, in dendritic and synaptic remodelling in the hippocampus and frontal cortex, as well as in levels of adult hippocampal neurogenesis (Galea et al. 1997; McEwen 2002; Shors et al. 2007). In many cases, these changes have been linked to behavioural disturbances, such as depressive-like symptomatology and learning/memory deficits (Bekris et al. 2005; Bowman 2005; Dalla et al. 2005, 2008a; Kitraki et al. 2004; Shors 2006).

2.2 *Sex Differences in Depressive Symptomatology*

2.2.1 Swim Stress Exposure: Forced Swim Test

As mentioned, the HPA axis is activated in both sexes after acute or short-term stress exposure. For example, when male and female rats are exposed to two sessions of swim stress in two consecutive days (forced swim test; FST), corticosterone levels are enhanced in both sexes (Drossopoulou et al. 2004). At a behavioural level, both sexes increase their immobility levels during the second FST session, but females exhibit higher levels than males, which is indicative of increased despair and depressive-like symptomatology (Dalla et al. 2008a; Drossopoulou et al. 2004; Pitychoutis et al. 2009b) (Fig. 1). Accordingly, in the open space swimming test, which is a similar animal model to the FST, female rats appear more vulnerable to swim stress than males (Sun and Alkon 2006). During the FST, female rats also exhibit lower levels of head swinging behaviour than males and this has been linked to sex differences in postsynaptic serotonergic $5\text{-}HT_{2A}$ receptor activity (Darmani 1996; Drossopoulou et al. 2004; Matuszewich and Yamamoto 2003). At the neurochemical level, FST induces a decrease in serotonergic activity in the hippocampus and the hypothalamus of female rats, while serotonergic activity is increased in the hypothalamus of males. Moreover,

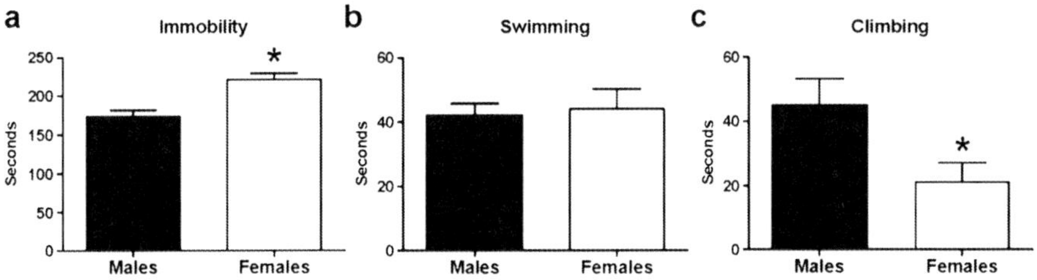

Fig. 1 *Behaviour of male and female rats during the Forced swim test.* Female rats spend more time immobile than male rats during the second session of the Forced swim test (FST), indicating enhanced levels of despair. Also, the duration of climbing, which represents an effort to escape from the cylinder, is higher in males than in females (* = $p < 0.05$) (Dalla et al. 2008a; Drossopoulou et al. 2004)

hypothalamic serotonin 5-HT_{1A} mRNA levels are decreased in female rats; while hippocampal 5-HT_{1A} mRNA levels are increased in males (Drossopoulou et al. 2004). Dopaminergic activity is also increased in the hippocampus and prefrontal cortex of male rats exposed to FST, while there is no effect in females (Dalla et al. 2008a). In both sexes, there is a tendency for enhanced GR mRNA levels 24h after the second swim session (Drossopoulou et al. 2004). However, following longer periods of repeated swim exposure, no alterations in GR or MR mRNA levels are detected in the female hippocampus and hypothalamus, while in males there is a down-regulation of GR mRNA in the hippocampus (Karandrea et al. 2002).

Thus, it seems that male rats have an adaptive response to swim stress by exhibiting enhanced corticosterone levels and increases in monoaminergic activity. On the other hand, female rats seem to be more vulnerable to swim stress than males and this is expressed by enhanced levels of despair and decreased serotonergic activity in the hippocampus and hypothalamus.

2.2.2 Shock Exposure. Learned Helplessness Model

In response to another kind of stressor (i.e. footshock or tailshock), males and females also exhibit elevated corticosterone levels (Heinsbroek et al. 1991). Numerous studies have shown that most males exposed to this kind of stress that they cannot control, develop helplessness behaviour, because when they are tested on a new task (e.g. avoid the footshock by going through a door-way at a shuttle box twice), they do not escape the stress even if they have the opportunity to do so (Maier 1984). This behaviour has been equated with a sense of "giving up", experienced by humans with major depression (Miller and Seligman 1975). In male rats, the expression of learned helplessness behaviour in operant-conditioning tasks, is accompanied by a wide range of physiological changes in response to shock, such as enhanced c-fos activity in the hippocampus, amygdala and prefrontal cortex, as well as monoaminergic activation in several brain regions and reduced rates of neurogenesis in the dentate gyrus of the hippocampus (Heinsbroek et al. 1990, 1991; Maier and Watkins 2005; Malberg and Duman 2003; Shors et al. 2007;

Trentani et al. 2003). When these stressors are repeated for longer periods and half of the rats can control the stressor by avoiding the shock exposure, corticosterone levels are still enhanced in all groups (i.e. in male rats exposed to controllable or uncontrollable stress) (Shors et al. 1989). However, only male rats, which are exposed to uncontrollable stress, exhibit learned helplessness behaviour, serotonergic activation and decreases in hippocampal neurogenesis (Amat et al. 2005; Bland et al. 2006; Heinsbroek et al. 1991; Shors et al. 2007).

Regarding females, although corticosterone levels and monoaminergic activity is also enhanced in female rats in response to footshock (Heinsbroek et al. 1991), most female rats do not express helplessness behaviour after acute or repeated uncontrollable footshock stress exposure (i.e. they learn to escape on the new task), while there is no effect of stress on neurogenesis levels in their hippocampus (Dalla et al. 2008b; Shors et al. 2007). Also, controllability over the stressor influences the degree of monoaminergic activation in a sex-dependent manner. In particular, the activation of noradrenaline and dopamine in the frontal cortex is larger after uncontrollable shock than after controllable shock and these differences are most evident in females. Moreover, uncontrollable shock induces higher serotonin levels in the frontal cortex of both sexes than controllable shock (Heinsbroek et al. 1991).

Notably, sex differences in learned helplessness behaviour cannot be attributed to gonadal hormonal differences, because they are not reversed by gonadectomy in adulthood of males or females (Dalla et al. 2008b) (Fig. 2). Sex differences in the learned helplessness model are probably mediated by sex differences in baseline escape behaviour in the operant-conditioning task, as well as in general activity levels. Indeed, unstressed female rats are in general more active and express less freezing behaviour than males (Dalla and Shors 2009; Padilla et al. 2009).

Overall, it seems that in this model, sex differences in the phenotype (i.e. behavioural sex differences in learned helplessness behaviour) do not always reflect sex differences in the endophenotype (e.g. corticosterone levels are equally affected in males and females by uncontrollable stress). However, levels of adult neurogenesis in the hippocampus seem to correlate with helplessness behaviour (Malberg and Duman 2003; Shors et al. 2007). Finally, controllability over the stressor seems to be an important factor, which determines the stress response in both sexes.

2.3 *Sex Differences in Stress Effects on Learning*

2.3.1 Associative Learning

Stress effects on different types of learning also differ between male and female rats. Regarding associative learning, exposure to an acute or repeated painful stressful event (i.e. tailshock, footshock or swim stress, but not restraint) greatly enhances learning during eyeblink classical conditioning in male rats, whereas the same stressor impairs learning in females (Shors 2006). Interestingly, animals that are exposed to the same amount of controllable shock do not express any change in this type of learning (Leuner et al. 2004). Corticosterone levels are

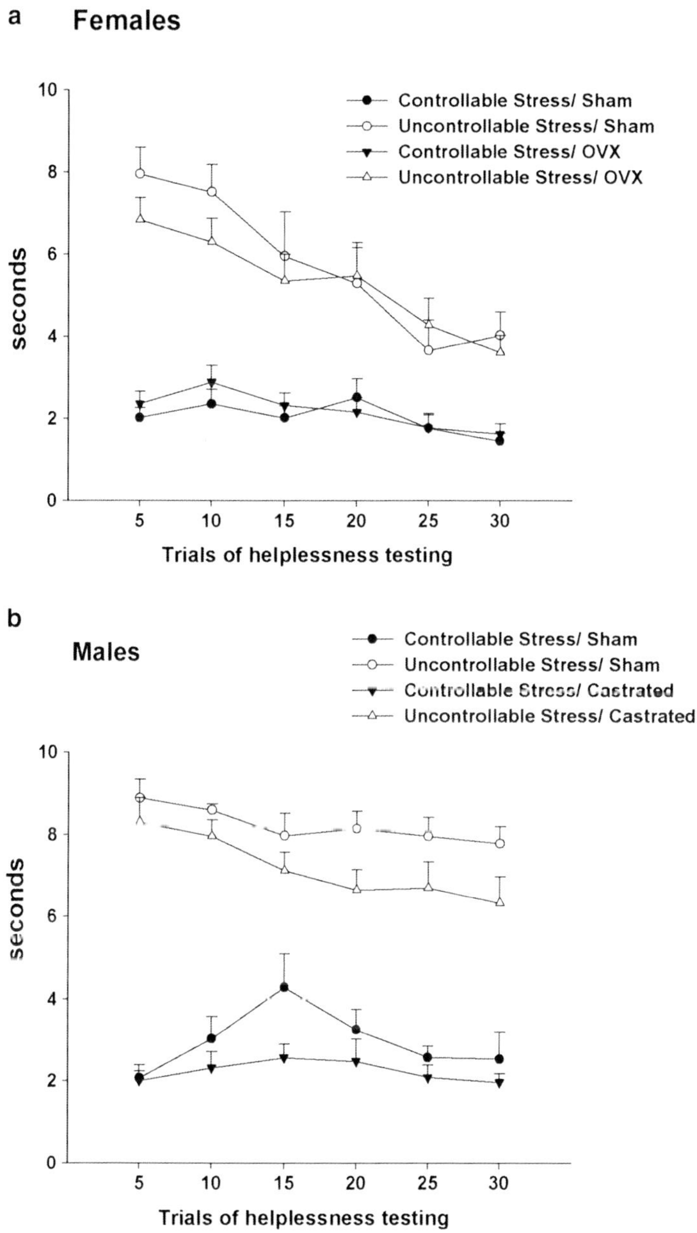

Fig. 2 *Males express learned helplessness behaviour whereas females do not.* During testing for learned helplessness behaviour, the rats have to cross a shuttle-box twice to escape the footshock. The graph depicts the escape latencies of male and female sham-operated and gonadectomized rats that were previously exposed to 7 sessions of controllable or uncontrollable footshock stress. Male sham- operated and castrated rats from the uncontrollable stress groups, did not learn to escape during testing and thus exhibited learned helplessness behaviour ($p > 0.05$). On the other hand, female sham-operated and ovariectomized (OVX) rats from the uncontrollable stress groups learned to escape during testing and did not exhibit learned helplessness behaviour ($p < 0.001$) (Dalla et al. 2008b)

enhanced in both sexes in response to the stressor, but adrenalectomy and gonadectomy studies have shown that males require the presence of glucocorticoids to enhance learning after stress, whereas females require the presence of ovarian hormones in order to impair learning after stress (Beylin and Shors 2003; Wood and Shors 1998). In particular, the impairment of associative learning after stress exposure emerges only in adult females with a mature oestrous cycle and is evident when females start training with eyeblink conditioning in proestrus when oestrogen levels are high (Shors et al. 1998).

Interestingly, the emergence of the stress effects on this type of associative learning in both males and females require an intact hippocampus (Bangasser and Shors 2007). Also, the same acute stressful experience increases the density of dendritic spines in the CA1 area of the male hippocampus and decreases it in the females, suggesting a link between density of spines in the hippocampus and associative learning (Shors et al. 2001, 2004). Notably, masculinized females injected with testosterone on the day of birth, exhibit enhanced learning during eyeblink conditioning and possess more spines in the CA1 area of the hippocampus after acute tailshock exposure (Bangasser and Shors 2008; Dalla et al. 2009).

Thus, females appear to be more vulnerable to shock exposure when associative learning and spine density in the hippocampus is evaluated. This is in contrast to the expression of helplessness behaviour that is mainly evident in male rats. The discrepancy between the two types of learning is probably due to inherent differences between the operant conditioning and the classical conditioning tasks. During operant conditioning that is used for assessment of learned helplessness behaviour, the animal must emit a voluntary motor response in order to change the outcome and learn, whereas during classical conditioning, which is indicative of associative learning, the animal emits an obligatory unconditioned response to the unconditioned stimulus, irrespective of volition (Shors 1998). Thus, it is not surprising that males and females respond differently in the two tasks and that sex effects are differently mediated by gonadal hormones.

2.3.2 Spatial Learning

In contrast to classical conditioning tasks, opposite effects of stress on performance of male and female rats in spatial learning and memory tasks have been repeatedly reported (Bowman 2005; Bowman et al. 2003). Following acute restraint stress, spatial memory in a Y-maze is impaired in males and is facilitated in females, irrespective of the oestrous cycle, while corticosterone levels are enhanced in both sexes (Conrad et al. 2004). Acute restraint stress has also been shown to enhance serotonergic activity in the basolateral amygdala of both sexes, but female rats show a greater response than males. Moreover, dopaminergic activity is increased in female rats, but not in males (Mitsushima et al. 2006). In the hippocampus, acute restraint stress significantly decreases brain-derived neurotrophic factor (BDNF) in both males and females in the CA3 area. In the dentate gyrus of the hippocampus, stress increases BDNF levels only in estradiol-treated ovariectomized rats, while it decreases its levels in oestrogen-deprived rats (Franklin and Perrot-Sinal 2006). This finding can probably be linked with reduced levels of adult neurogenesis in the female hippocampus when oestrogen levels are low (Tanapat et al. 1999).

After one week of repeated restraint stress exposure, male rats are impaired in the object recognition test, while females exhibit enhanced performance during spatial memory tasks. These behavioural changes are accompanied by increased noradrenergic activity in the hippocampus and amygdala of females and decreased activity in males (Bowman et al. 2009). In the medial prefrontal cortex, the same protocol decreases the number and length of the apical neuronal dendritic branches of males, whereas in females, the same stress exposure increases apical dendritic length and this is prevented by ovariectomy (Garrett and Wellman 2009).

The detrimental effect of stress on the Morris water maze task is most evident in male rats exposed to longer periods of repeated restraint stress (21 days), while females exhibit improved memory scores (Bowman et al. 2003; Kitraki et al. 2004). Also, behavioural changes are accompanied by a decrease in GR receptor immunoreactivity in the male and an increase in the female hippocampus (Kitraki et al. 2004). The same stress protocol also decreases dopaminergic activity in the frontal cortex and amygdala of males but not females; whereas, in the hippocampus, stress increases levels of serotonin and noradrenaline in females, but not in males. These effects seem to depend on both organizational and activational effects of gonadal hormones (Luine 2002). Twenty one days of repeated restraint stress also induces apical dendritic atrophy of the CA3 pyramidal neurons in the male hippocampus, while this effect is not evident in females (Galea et al. 1997). However, spinophilin levels, which are indicative of new spine formation, are elevated in the male, but not female hippocampus following repeated restraint stress exposure (Khurana and Devaud 2007)

Thus, it seems that, in contrast to swim or shock stress exposure, males are more vulnerable to restraint stress than females, because they exhibit learning and memory deficits in spatial tasks, as well as decreases in monoamine levels and activity and decreases in certain neuroplasticity indices. On the other hand, females seem to adapt better to this type of stress exposure.

3 Sex Differences in the Effects of Chronic Mild Stress Exposure

Chronic mild stress (CMS) was developed in the late 1980s and is one of the most extensively investigated animal models of depression to-date. CMS's advantage over other models of depression lies on the fact that it employs relatively realistic inducing conditions (construct validity), simulates *anhedonia* which is a core symptom of major depression (face validity), and responds appropriately to antidepressant drugs (predictive validity) (Willner 1997, 2005; Willner et al. 1987). The CMS procedure involves continuous exposure to a variety of low-grade stressors, such as periods of food and water deprivation, small temperature

Table 1 Example of a weekly CMS protocol

Monday 10:00 h	Cage cleaning followed by no stress
Monday 20:00 h	Food and water deprivation for 14 h
Tuesday 10:00 h	Sucrose test, followed by food or water deprivation for 10 h
Tuesday 20:00 h	Paired housing for 14 h
Wednesday 10:00 h	Lights switched on and off every 2 h for 10 h
Wednesday 20:00 h	Soiled cage (250 ml of water was poured into the sawdust bedding) for 14 h
Thursday 10:00 h	Cage cleaning, followed by water deprivation for 10 h
Thursday 20:00 h	Paired housing for 14 h
Friday 10:00 h	Stroboscopic illumination in darkness for 10 h
Friday 20:00 h	Food deprivation for 14 h
Saturday 10:00 h	Tilting of the cages backwards (45 degrees) for 10 h
Saturday 20:00 h	Cages were put back in straight position/followed by no stress
Sunday 10:00 h	Stroboscopic illumination in darkness for 10 h
Sunday 20:00 h	Soiled cage (250 ml of water was poured into the sawdust bedding) for 14 h

reductions, changes of cage mates, and other similar mild manipulations (Table 1). Animals subjected to a battery of mild stressors for a period of several weeks develop a wide spectrum of behavioural, biochemical and physiological alterations, which can be effectively reversed upon chronic antidepressant treatment (Willner 1997, 2005). Exposure to a CMS regime ultimately leads to the induction of *anhedonia* (i.e. loss of pleasure derived from normally rewarding activities), which accounts for the impairment of rodents' preference for a palatable sucrose (or saccharin) solution.

Beyond *anhedonia*, this model simulates other symptoms of depression as well, such as decreased sexual behaviour and self-care, changes in sleep architecture and locomotor activity, as well as cardiovascular and immune alterations (Willner 1997, 2005). However, there have been some replication problems in CMS among laboratories (Willner 2005), mainly due to the failure of eliciting a significant decrease in sucrose consumption. This has been partly attributed to individual differences regarding rats' hedonic *status quo* (Duncko et al. 2003), as well as to differences between different strains of rats (Bekris et al. 2005; Konkle et al. 2003). Thus, it can be suggested that, besides sucrose consumption, we may be in need of more robust and extensive outputs to evaluate how CMS exerts its "*depressogenic*" influences on rats of both sexes. It is worth noting that the induction of *anhedonia* after CMS exposure has also been validated with the intracranial self stimulation paradigm, but inter-individual differences have also been reported (Nielsen et al. 2000). Moreover, the duration of stress application (e.g. 3 *versus* 7 weeks) and the stress regime itself (i.e. the cyclicity of the stressors) may also explain some of the discrepancies between different studies.

3.1 Sex Differences in Behavioural Responses

Sex differences in reward sensitivity are revealed when the CMS model of depression is applied to male and female rats. Rewarding reactivity to sucrose is impaired in both sexes (Duncko et al. 2001), but this phenomenon appears to be more robust

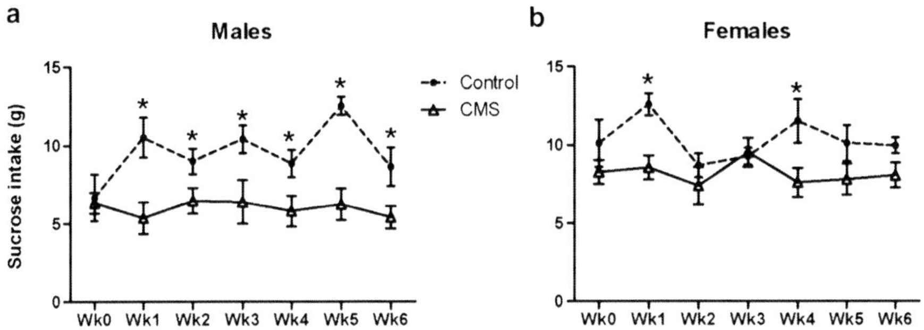

Fig. 3 *Sucrose intake in male and female rats during Chronic mild stress.* Male rats exposed to Chronic mild stress (CMS) consume a lower amount of a sucrose solution (1%) than control rats, during all weeks of CMS exposure. This effect is evident in female rats only during the first and fourth week of CMS exposure (* = $p < 0.05$) (Dalla et al. 2005)

in male compared to female CMS-exposed rats (Dalla et al. 2005, 2008a; Grippo et al. 2005; Kamper et al. 2009) (Fig. 3). As a matter of fact, it has been suggested that sucrose intake might not be an appropriate behavioural index for female rats, because unstressed females tend to drink more sucrose than males and show a more erratic increase in their consumption (Dalla et al. 2005). Interestingly, it has been shown that pair- or group- housing conditions may diminish the effects of CMS on body weight and sucrose consumption, in comparison to single-housed female rats (Baker and Bielajew 2007). On the other hand, another study that measured sucrose intake during 24 h periods has revealed a gradual reduction of sucrose consumption in female, but not in male CMS treated rats (Konkle et al. 2003). However, no changes in thresholds for brain stimulation reward were observed in two strains of female rats (Baker et al. 2006).

As far as other behavioural measurements are concerned, female rats display less exploratory behaviour in a novel open field environment following three weeks of CMS application (Dalla et al. 2005). In another study, total activity in the open field test did not differ between stressed male and female rats, but CMS-exposed male rats exhibited enhanced locomotor activity during the first minute of the session, suggesting increased anxiety (Duncko et al. 2001).

3.2 *Sex Differences in Neuroendocrine Responses*

CMS application has been reported to dysregulate both the HPA and the hypothalamus-pituitary-gonadal (HPG) axes. For example, CMS induces severe disruptions of normal oestrous cyclicity. These alterations typically involve desynchronization of oestrous cycling in CMS-treated female rats that is either expressed by staying in one phase of the cycle or lengthening of the oestrous cycle (Baker et al. 2006; Dalla et al. 2005; Grippo et al. 2005; Konkle et al. 2003). However, in most cases when the stressful manipulations cease, the normal cyclicity is restored (Baker et al. 2006).

Impaired function of the HPA axis has repeatedly been noticed in patients with major depression (Holsboer 2000). As such, following, 3–4 weeks of CMS exposure both male and female rats displayed a tendency towards higher basal corticosterone levels (Duncko et al. 2001; Grippo et al. 2005). However, following 6 weeks of CMS exposure, a sex-specific elevation of corticosterone concentrations was observed only in female rats, suggesting that females are more vulnerable in the CMS model of depression than males (Dalla et al. 2005). Overall, it seems that there is an important time factor that differentiates corticosteroid responses between the two sexes upon chronic stress application.

3.3 Sex Differences in Neurochemical and Neurobiological Alterations

CMS exposure in male rats induces alterations in noradrenergic, dopaminergic and serotonergic *status*. In our initial studies in male rats, we observed an increase in dopaminergic and a decrease in serotonergic activity in the prefrontal cortex in two strains of male rats (*Sprague-Dawley* and *Wistar* rats) exposed to 7 weeks of CMS (Bekris et al. 2005). In the same study, we found a decrease in striatal dopaminergic activity and an increase in serotonergic activity in the hippocampus of male rats. All these effects were reversed by chronic imipramine treatment (Bekris et al. 2005). In further comparative studies between male and female rats, we used a milder CMS protocol, which did not induce neurochemical alterations in male rats, but resulted in a decrease in hippocampal serotonergic activity and prefrontocortical dopaminergic activity in females (Dalla et al. 2005, 2008a).

Accordingly, exposure to CMS induces a wide spectrum of relevant neurobiological alterations in specific brain regions implicated in the pathophysiology of major depression. For instance, it has been reported that hippocampal $5HT_{1A}$ receptors are increased by CMS in male rats (Papp et al. 1994). Four weeks of CMS exposure attenuates ACTH responses following systemic administration of a selective $5\text{-}HT_{1A}$ receptor agonist [(+)8-hydroxy-N,N-dipropyl-2-aminotetralin hydrobromide; 8-OH-DPAT] in both male and female rats (Grippo et al. 2005). These results were extrapolated as being indicative of CMS-driven alterations in $5\text{-}HT_{1A}$ receptor function in specific subpopulations of neurons in the central nervous system (Grippo et al. 2005). Also, it has been shown that CMS exposure is associated with a sex-specific enhancement of CRH mRNA levels in the hypothalamus of male rats (Duncko et al. 2001). Additionally, tyrosine hydroxylase mRNA levels in the LC were significantly decreased in response to CMS in both sexes, indicating an impairment of the central noradrenergic function (Duncko et al. 2001).

Thus, male and female rats are differentially affected by CMS application depending on the neurobiological indices that are measured. In some studies, females appear to be more vulnerable (e.g. serotonergic activity) while in other studies the two sexes are equally affected.

3.4 Sex Differences in Immune Parameters

CMS has long been associated with alterations in central and peripheral monoaminergic systems, as well as immunoreactivity (Willner 2005). CMS application has been shown to induce robust changes in thyroid hormone levels and reduction of the thymus weight in male rats (Kioukia-Fougia et al. 2002; Kioukia et al. 2000). In a recent study, we reported for the first time that both chronic clomipramine treatment and CMS application, exerted sexually dimorphic effects on cellular immunoreactivity [natural killer (NK) and lymphokine-activated killer (LAK) cell cytotoxicity and interleukin-2-induced T-cell proliferation], with female rats presenting a relatively immunosuppressed phenotype compared to males (Pitychoutis et al. 2009a). In addition, CMS and chronic clomipramine treatment induced sex-dependent alterations in the monoamine profile of the thymus. Thymic DA levels were augmented only in CMS-treated female rats with this increase being partially reversed by chronic clomipramine treatment. Further, while CMS application did not alter 5-HT and NA concentrations in either sex, chronic antidepressant treatment elevated thymic NA levels only in male rats, irrespective of stress application (Pitychoutis et al. 2009a). Intriguingly, clomipramine treatment rendered thymic 5-HT a positive modulator of LAK cytotoxicity; thymic monoamine alterations being associated with functional measures of cellular immunity are suggestive of a thymus-dependent route by which antidepressants could affect cell-mediated immunity (Pitychoutis et al. 2009a; Pitychoutis et al. 2010).

4 Sex Differences in Sickness Behaviour, a Putative Inflammatory Model of Depression

Data accumulated largely during the last two decades advocate the innate immune response as a mechanism that may be implicated in the pathophysiology of major depression, mainly due to the documented alterations in the ability of immune cells to secrete proinflammatory cytokines (Dantzer et al. 2008). Challenging the innate immune machinery with the proinflammatory agent lipopolysaccharide (LPS) induces a mild state of *nosothymia*, termed as *sickness behaviour*. *Sickness behaviour* is conceptualized through numerous depressive-like behavioural and physiological manifestations, most of which overlap with the clinical symptoms of depression (i.e. *anhedonia*, depression in motor/exploratory activity and a reduction in food intake, activation of the HPA axis, as well as alterations in central and peripheral monoamine utilization) (Dantzer et al. 2008; Zampeli et al. 2009).

Behavioural responses in LPS/cytokine-induced sickness have been studied in both male and female rats in only a few studies. Female rats exhibit greater sensitivity than males to LPS and/or cytokine administration in several aspects of behaviour, including sexual activity and sucrose reward (Avitsur and Yirmiya 1999; Merali et al. 2003). However, *in vitro* experiments have shown that

LPS-challenged macrophages derived from male mice produced higher levels of inflammatory cytokines than similarly treated female-derived cells, suggesting that males may be more susceptible to bacterial sepsis than females (Marriott et al. 2006). Furthermore, female rats develop behavioural tolerance to repeated LPS administration more quickly than males and this phenomenon was found to be oestrous cycle-dependent (Engeland et al. 2006).

Traditional measures of sickness (i.e. impairment of social exploration of a juvenile conspecific and weight loss) have been shown to be equally affected in male and female rats, except for *anorexia* (lack of appetite for food) establishment where males appeared to be more vulnerable (Pitychoutis et al. 2009b). Sickness establishment also alters the way the two sexes react upon stress exposure. In particular, LPS administration (100 μg/kg) induced a beneficial female-specific enhancement of coping ability in the stressful FST, as evidenced by the increased swimming durations achieved during the second swim session induced (Pitychoutis et al. 2009b). However, it has been shown that higher doses of LPS (e.g. 2 mg/kg) reliably induced an increase in floating time in rats of both sexes, at 24 h post-injection (Tonelli et al. 2008). Finally, *anhedonia*, which is a core symptom of depression, was equally established in both sexes at 24 h post-LPS administration, with this effect being reversed within the following day (Pitychoutis et al. 2009b) (Fig. 4).

Ex vivo neurochemical analysis at 2 h post-LPS administration, when many symptoms of sickness reach a *plateau*, indicated that central serotonergic activity in female rats is enhanced in all limbic sites examined (i.e. hypothalamus, hippocampus, prefrontal cortex, hippocampus, amygdala and striatum). On the other hand, serotonergic *status* in male rats was only modestly altered (Pitychoutis et al. 2009b). Dopaminergic indices were primarily affected in female rats, especially in the striatum, while there were no apparent alterations in males following LPS administration. In addition, neuroendocrine corticosteroid responses further confirm that females are more vulnerable to LPS sensitization compared to males

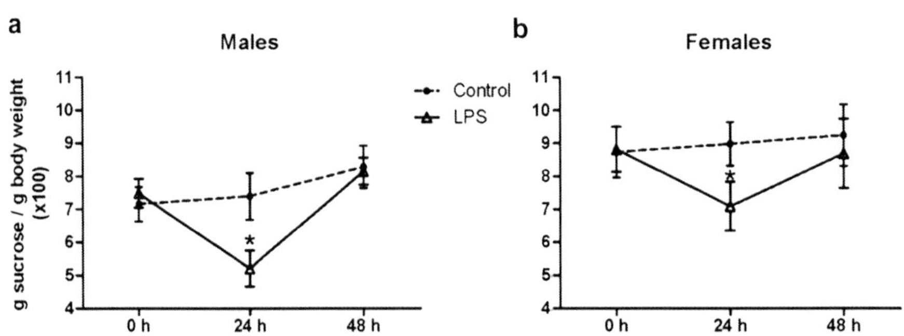

Fig. 4 *Sucrose intake after Lipopolysaccharide (LPS) administration*. Anhedonia was established equally in rats of both sexes at 24 h post-LPS (100 μg/kg, ip) administration, as evidenced by the relative consumption of a sucrose solution (1%). This effect was reversed within the following day (* = $p < 0.05$) (Pitychoutis et al. 2009b)

(Frederic et al. 1993; Pitychoutis et al. 2009b; Spinedi et al. 1994; Tonelli et al. 2008).

The apparent sex differences underlying the neurochemical and behavioural profile of *sickness behaviour* point to the important role of the immune system activation in the pathophysiology of depression. Taking into account the higher prevalence of affective disorders in females, a focus on the basic science of sex differences that underlie *sickness behaviour* is useful in delineating the neuroimmunological substrate for the appearance, course and outcome of these conditions (Pitychoutis and Papadopoulou-Daifoti 2010).

5 Sex Differences in a Genetic Model of Depression: Flinders Sensitive Line

Current evidence points towards a gene-environment interaction in the ability to cope with stress and with the predisposition to stress-related disorders, such as depression (Cryan and Slattery 2007). In agreement with this, many laboratories study genetic models of depression, such as the *Flinders Sensitive Line* (FSL) of rats. These rats have been created by selectively breeding *Sprague-Dawley* rats for their hyper-sensitivity to cholinergic agonists; a characteristic that has also been observed in depressed humans (Overstreet et al. 2005). FSL rats present a number of behavioural similarities to depressed individuals, such as reduced general activity, disturbed appetite, REM-sleep abnormalities and *anhedonia* (Yadid et al. 2000). Also, FSL rats display a clear depressive-like behavioural profile, particularly showing increased immobility, already observed from the first swim session of the FST paradigm. Chronic treatment with different antidepressant agents successfully reverses the depressive-like behaviour of male FSL rats (Kokras et al. 2009a; Overstreet et al. 2005). Regarding females, we have found that they do not exhibit enhanced immobility levels, but they show decreased latency to become immobile, in comparison to *Sprague-Dawley* controls. Interestingly, sex differences in FST performance are alleviated following treatment with the tri-cyclic antidepressant, clomipramine (Kokras et al. 2009a) (Fig. 5). Sex differences are also present in the frequency of head swinging behaviour, with female FSL rats exhibiting lower levels than males, as do *Wistar* rats (Drossopoulou et al. 2004). However, antidepressant treatment does not affect head swinging behaviour and sex differences are still apparent after treatment (Kokras et al. 2009a).

Male FSL rats exhibit several neurochemical abnormalities, including serotonergic alterations, increased levels of catecholamines and their metabolites, and impaired communication between serotonergic and dopaminergic systems in limbic brain regions (Kokras et al. 2009a; Yadid et al. 2000). In male FSL rats at baseline, serotonergic activity, as indicated by serotonin turnover, is lower in limbic regions, compared to *Sprague-Dawley* controls, while it is modestly increased after chronic

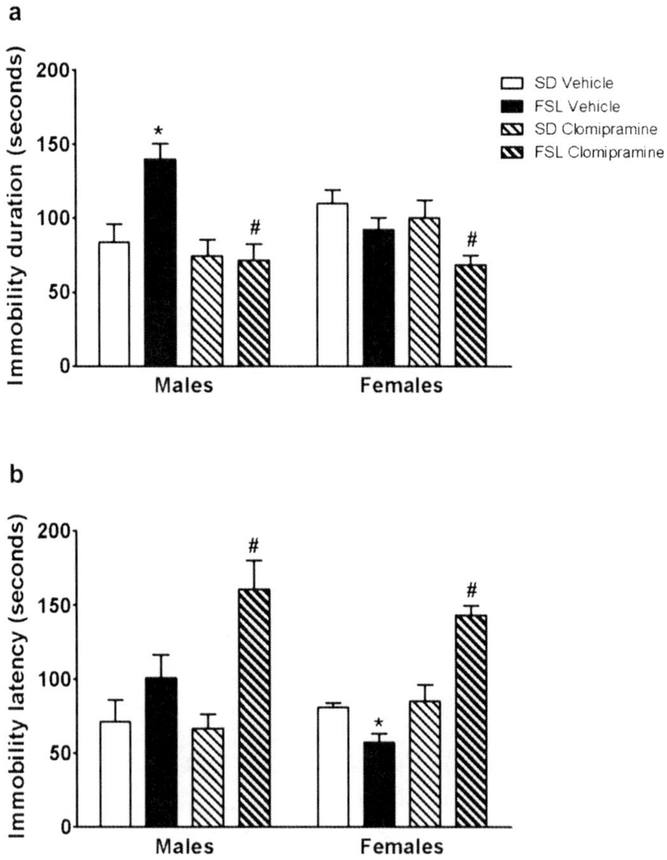

Fig. 5 *Behaviour of male and female Flinder rats during Forced swim test.* Male *Flinders Sensitive Line* rats (FSL) exhibited increased immobility duration during one session of FST, in comparison to *Sprague-Dawely* (SD) rats, which serve as controls. Female FSL rats exhibited reduced latency to become immobile during one session of FST, in comparison to SD rats, which serve as controls. Chronic clomipramine treatment reversed the immobility duration and latency in both sexes (*,# = $p < 0.05$) (Kokras et al. 2009a)

treatment with a tri-cyclic antidepressant (Kokras et al. 2009a). Interestingly, this is not the case for female FSL rats, which do not differ in serotonergic activity from their control *Sprague-Dawley* rats. However, clomipramine treatment results in a marked increase in serotonergic activity of female FSL rats in all brain regions tested (Kokras et al. 2009a). Also, clomipramine treatment increases cortical glutamate levels in both sexes and hippocampal glutamate only in female FSL rats (Kokras et al. 2009b). These findings indicate that antidepressant treatment may alleviate sex differences in the phenotype (i.e. converging effects on behaviour during FST) while maintaining or intensifying sex differences in the endophenotype (i.e. serotonergic activity).

6 General Conclusions

Overall, it seems that the type and the duration of stressor, as well as the behavioural parameters that we measure, influence the appearance and direction of sex differences in stress response. Depressive-like symptomatology is also expressed differently between the two sexes depending on the model. For example, although female rats appear to be more vulnerable in the FST and CMS models and following LPS administration, they do not exhibit learned helplessness behaviour in response to footshock.

In most cases, behavioural alterations can be linked with abnormalities in the endophenotype, such as in hormonal, neurochemical, immune and neuroplasticity indices. In particular, the HPA axis is activated in both sexes after stress exposure, but some components are differentiated between the two sexes pointing to its importance in the emergence of stress-related disorders, such as depression. Also, monoamines are generally activated after stress exposure, but in some cases, especially in females, we observe decreases in monoaminergic activity in response to stress, which may be linked with the higher incidence of depression in women (Deecher et al. 2008). Moreover, sex differences are mediated in a complex way by the effects of gonadal hormones during development of the brain (organizational effects), during puberty and in adulthood (activational effects).

Finally, it can be hypothesized that certain stress models can lead to decreases in serotonergic activity in the hippocampus of female rats, which in turn can be linked with decreased neuroplasticity associated with cognitive impairment and depressive-like symptomatology. However, more studies focusing on animal models validated for females are needed in order to evaluate this hypothesis.

Overall, it is clear that all the stress models have strengths and limitations that need to be recognized in order to use them effectively in the investigation of affective disorders. Limiting research on depression to male animals, may lead to inaccurate findings or hide important results that apply only to women. Also it is possible that sex hormone's actions on the brain from "womb to tomb" may affect the brain in a sexually dimorphic manner. This could lead to increased vulnerability to stressors, different coping strategies and differentiated responsivity to antidepressants in women.

Acknowledgements Dr. Christina Dalla was a Marie Curie International Fellow, funded from the European Commission within the 6th European Community Framework Programme. This work has been supported by the Greek General Secretariat of Research and Technology and by Research Grants from the Special Account for Research Grants of the University of Athens (Greece).

References

Amat J, Baratta MV, Paul E, Bland ST, Watkins LR, Maier SF (2005) Medial prefrontal cortex determines how stressor controllability affects behavior and dorsal raphe nucleus. Nat Neurosci 8:365–371

Anisman H, Matheson K (2005) Stress, depression, and anhedonia: caveats concerning animal models. Neurosci Biobehav Rev 29:525–546

Anisman H, Zacharko RM (1990) Multiple neurochemical and behavioral consequences of stressors: implications for depression. Pharmacol Ther 46:119–136

Avitsur R, Yirmiya R (1999) The immunobiology of sexual behavior: gender differences in the suppression of sexual activity during illness. Pharmacol Biochem Behav 64:787–796

Baker S, Bielajew C (2007) Influence of housing on the consequences of chronic mild stress in female rats. Stress 10:283–293

Baker SL, Kentner AC, Konkle AT, Santa-Maria Barbagallo L, Bielajew C (2006) Behavioral and physiological effects of chronic mild stress in female rats. Physiol Behav 87:314–322

Bangasser DA, Shors TJ (2007) The hippocampus is necessary for enhancements and impairments of learning following stress. Nat Neurosci 10:1401–1403

Bangasser DA, Shors TJ (2008) The bed nucleus of the stria terminalis modulates learning after stress in masculinized but not cycling females. J Neurosci 28:6383–6387

Bekris S, Antoniou K, Daskas S, Papadopoulou-Daifoti Z (2005) Behavioural and neurochemical effects induced by chronic mild stress applied to two different rat strains. Behav Brain Res 161:45–59

Beylin AV, Shors TJ (2003) Glucocorticoids are necessary for enhancing the acquisition of associative memories after acute stressful experience. Horm Behav 43:124–131

Bland ST, Schmid MJ, Greenwood BN, Watkins LR, Maier SF (2006) Behavioral control of the stressor modulates stress-induced changes in neurogenesis and fibroblast growth factor-2. Neuroreport 17:593–597

Bowman RE (2005) Stress-induced changes in spatial memory are sexually differentiated and vary across the lifespan. J Neuroendocrinol 17:526–535

Bowman RE, Beck KD, Luine VN (2003) Chronic stress effects on memory: sex differences in performance and monoaminergic activity. Horm Behav 43:48–59

Bowman RE, Micik R, Gautreaux C, Fernandez L, Luine VN (2009) Sex-dependent changes in anxiety, memory, and monoamines following one week of stress. Physiol Behav 97:21–29

Conrad CD, Jackson JL, Wieczorek L, Baran SE, Harman JS, Wright RL, Korol DL (2004) Acute stress impairs spatial memory in male but not female rats: influence of estrous cycle. Pharmacol Biochem Behav 78:569–579

Cryan JF, Slattery DA (2007) Animal models of mood disorders: recent developments. Curr Opin Psychiatry 20:1–7

Curtis AL, Bethea T, Valentino RJ (2006) Sexually dimorphic responses of the brain norepinephrine system to stress and corticotropin-releasing factor. Neuropsychopharmacology 31:544–554

Dalla C, Antoniou K, Drossopoulou G, Xagoraris M, Kokras N, Sfikakis A, Papadopoulou-Daifoti Z (2005) Chronic mild stress impact: are females more vulnerable? Neuroscience 135:703–714

Dalla C, Antoniou K, Kokras N, Drossopoulou G, Papathanasiou G, Bekris S, Daskas S, Papadopoulou-Daifoti Z (2008a) Sex differences in the effects of two stress paradigms on dopaminergic neurotransmission. Physiol Behav 93:595–605

Dalla C, Edgecomb C, Whetstone AS, Shors TJ (2008b) Females do not express learned helplessness like males do. Neuropsychopharmacology 33:1559–1569

Dalla C, Shors TJ (2009) Sex differences in learning processes of classical and operant conditioning. Physiol Behav 97:229–238

Dalla C, Pitychoutis PM, Kokras N, Papadopoulou-Daifoti Z (2010) Sex differences in the neurobiological substrate of depression and in antidepressant response. Basic and Clinical Pharmacology and Toxicology 106:226–233

Dalla C, Whetstone AS, Hodes GE, Shors TJ (2009) Stressful experience has opposite effects on dendritic spines in the hippocampus of cycling versus masculinized females. Neurosci Lett 449:52–56

Dantzer R, O'Connor JC, Freund GG, Johnson RW, Kelley KW (2008) From inflammation to sickness and depression: when the immune system subjugates the brain. Nat Rev Neurosci 9:46–56

Darmani NA (1996) Differential potentiation of L-tryptophan-induced head-twitch response in mice by cocaine and sertraline. Life Sci 59:1109–1119

DeBold CR, Sheldon WR, DeCherney GS, Jackson RV, Alexander AN, Vale W, Rivier J, Orth DN (1984) Arginine vasopressin potentiates adrenocorticotropin release induced by ovine corticotropin-releasing factor. J Clin Invest 73:533–538

Deecher D, Andree TH, Sloan D, Schechter LE (2008) From menarche to menopause: exploring the underlying biology of depression in women experiencing hormonal changes. Psychoneuroendocrinology 33:3–17

Drossopoulou G, Antoniou K, Kitraki E, Papathanasiou G, Papalexi E, Dalla C, Papadopoulou-Daifoti Z (2004) Sex differences in behavioral, neurochemical and neuroendocrine effects induced by the forced swim test in rats. Neuroscience 126:849–857

Duncko R, Kiss A, Skultetyova I, Rusnak M, Jezova D (2001) Corticotropin-releasing hormone mRNA levels in response to chronic mild stress rise in male but not in female rats while tyrosine hydroxylase mRNA levels decrease in both sexes. Psychoneuroendocrinology 26:77–89

Duncko R, Schwendt M, Jezova D (2003) Altered glutamate receptor and corticoliberin gene expression in brain regions related to hedonic behavior in rats. Pharmacol Biochem Behav 76:9–16

Engeland CG, Kavaliers M, Ossenkopp KP (2006) Influence of the estrous cycle on tolerance development to LPS-induced sickness behaviors in rats. Psychoneuroendocrinology 31:510–525

Franklin TB, Perrot-Sinal TS (2006) Sex and ovarian steroids modulate brain-derived neurotrophic factor (BDNF) protein levels in rat hippocampus under stressful and non-stressful conditions. Psychoneuroendocrinology 31:38–48

Frederic F, Oliver C, Wollman E, Delhaye-Bouchaud N, Mariani J (1993) IL-1 and LPS induce a sexually dimorphic response of the hypothalamo-pituitary-adrenal axis in several mouse strains. Eur Cytokine Netw 4:321–329

Galea LA, McEwen BS, Tanapat P, Deak T, Spencer RL, Dhabhar FS (1997) Sex differences in dendritic atrophy of CA3 pyramidal neurons in response to chronic restraint stress. Neuroscience 81:689–697

Garrett JE, Wellman CL (2009) Chronic stress effects on dendritic morphology in medial prefrontal cortex: sex differences and estrogen dependence. Neuroscience 162:195–207

Grippo AJ, Sullivan NR, Damjanoska KJ, Crane JW, Carrasco GA, Shi J, Chen Z, Garcia F, Muma NA, Van de Kar LD (2005) Chronic mild stress induces behavioral and physiological changes, and may alter serotonin 1A receptor function, in male and cycling female rats. Psychopharmacology (Berl) 179:769–780

Handa RJ, Nunley KM, Lorens SA, Louie JP, McGivern RF, Bollnow MR (1994) Androgen regulation of adrenocorticotropin and corticosterone secretion in the male rat following novelty and foot shock stressors. Physiol Behav 55:117–124

Heinrichs SC, Koob GF (2004) Corticotropin-releasing factor in brain: a role in activation, arousal, and affect regulation. J Pharmacol Exp Ther 311:427–440

Heinsbroek RP, van Haaren F, Feenstra MG, van Galen H, Boer G, van de Poll NE (1990) Sex differences in the effects of inescapable footshock on central catecholaminergic and serotonergic activity. Pharmacol Biochem Behav 37:539–550

Heinsbroek RP, van Haaren F, Feenstra MG, Boon P, van de Poll NE (1991) Controllable and uncontrollable footshock and monoaminergic activity in the frontal cortex of male and female rats. Brain Res 551:247–255

Holden C (2005) Sex and the suffering brain. Science 308:1574

Holsboer F (2000) The corticosteroid receptor hypothesis of depression. Neuropsychopharmacology 23:477–501

Holsboer F (2001) Stress, hypercortisolism and corticosteroid receptors in depression: implications for therapy. J Affect Disord 62:77–91

Kamper EF, Chatzigeorgiou A, Tsimpoukidi O, Kamper M, Dalla C, Pitychoutis PM, Papadopoulou-Daifoti Z (2009) Sex differences in oxidant/antioxidant balance under a chronic mild stress regime. Physiol Behav 98:215–222

Karandrea D, Kittas C, Kitraki E (2000) Contribution of sex and cellular context in the regulation of brain corticosteroid receptors following restraint stress. Neuroendocrinology 71:343–353

Karandrea D, Kittas C, Kitraki E (2002) Forced swimming differentially affects male and female brain corticosteroid receptors. Neuroendocrinology 75:217–226

Kendler KS, Thornton LM, Prescott CA (2001) Gender differences in the rates of exposure to stressful life events and sensitivity to their depressogenic effects. Am J Psychiatry 158:587–593

Khurana RC, Devaud LL (2007) Sex differences in neurotransmission parameters in response to repeated mild restraint stress exposures in intact male, female and ovariectomised female rats. J Neuroendocrinol 19:511–520

Kioukia N, Bekris S, Antoniou K, Papadopoulou-Daifoti Z, Christofidis I (2000) Effects of chronic mild stress (CMS) on thyroid hormone function in two rat strains. Psychoneuroendocrinology 25:247–257

Kioukia-Fougia N, Antoniou K, Bekris S, Liapi C, Christofidis I, Papadopoulou-Daifoti Z (2002) The effects of stress exposure on the hypothalamic-pituitary-adrenal axis, thymus, thyroid hormones and glucose levels. Prog Neuropsychopharmacol Biol Psychiatry 26:823–830

Kitay JI (1961) Sex differences in adrenal cortical secretion in the rat. Endocrinology 68:818–824

Kitraki E, Kremmyda O, Youlatos D, Alexis MN, Kittas C (2004) Gender-dependent alterations in corticosteroid receptor status and spatial performance following 21 days of restraint stress. Neuroscience 125:47–55

Klein LC, Corwin EJ (2002) Seeing the unexpected: how sex differences in stress responses may provide a new perspective on the manifestation of psychiatric disorders. Curr Psychiatry Rep 4:441–448

Kokras N, Antoniou K, Dalla C, Bekris S, Xagoraris M, Ovestreet D, Papadopoulou-Daifoti Z (2009a) Sex-related differential response to clomipramine treatment in a rat model of depression. J Psychopharmacol 23:945–956

Kokras N, Antoniou K, Polissidis A, Papadopoulou-Daifoti Z (2009b) Antidepressants induce regionally discrete, sex-dependent changes in brain's glutamate content. Neurosci Lett 464:98–102

Konkle AT, Baker SL, Kentner AC, Barbagallo LS, Merali Z, Bielajew C (2003) Evaluation of the effects of chronic mild stressors on hedonic and physiological responses: sex and strain compared. Brain Res 992:227–238

Leuner B, Mendolia-Loffredo S, Shors TJ (2004) Males and females respond differently to controllability and antidepressant treatment. Biol Psychiatry 56:964–970

Luine V (2002) Sex differences in chronic stress effects on memory in rats. Stress 5:205–216

Lupien SJ, McEwen BS, Gunnar MR, Heim C (2009) Effects of stress throughout the lifespan on the brain, behaviour and cognition. Nat Rev Neurosci 10:434–445

Maciejewski PK, Prigerson HG, Mazure CM (2001) Sex differences in event-related risk for major depression. Psychol Med 31:593–604

Maier SF (1984) Learned helplessness and animal models of depression. Prog Neuropsychopharmacol Biol Psychiatry 8:435–446

Maier SF, Watkins LR (2005) Stressor controllability and learned helplessness: the roles of the dorsal raphe nucleus, serotonin, and corticotropin-releasing factor. Neurosci Biobehav Rev 29:829–841

Malberg JE, Duman RS (2003) Cell proliferation in adult hippocampus is decreased by inescapable stress: reversal by fluoxetine treatment. Neuropsychopharmacology 28:1562–1571

Marcus SM, Young EA, Kerber KB, Kornstein S, Farabaugh AH, Mitchell J, Wisniewski SR, Balasubramani GK, Trivedi MH, Rush AJ (2005) Gender differences in depression: findings from the STAR*D study. J Affect Disord 87:141–150

Marriott I, Bost KL, Huet-Hudson YM (2006) Sexual dimorphism in expression of receptors for bacterial lipopolysaccharides in murine macrophages: a possible mechanism for gender-based differences in endotoxic shock susceptibility. J Reprod Immunol 71:12–27

Matuszewich L, Yamamoto BK (2003) Long-lasting effects of chronic stress on DOI-induced hyperthermia in male rats. Psychopharmacology (Berl) 169:169–175

McEwen BS (2002) Sex, stress and the hippocampus: allostasis, allostatic load and the aging process. Neurobiol Aging 23:921–939

Merali Z, Brennan K, Brau P, Anisman H (2003) Dissociating anorexia and anhedonia elicited by interleukin-1beta: antidepressant and gender effects on responding for "free chow" and "earned" sucrose intake. Psychopharmacology (Berl) 165:413–418

Miller WR, Seligman ME (1975) Depression and learned helplessness in man. J Abnorm Psychol 84:228–238

Mitsushima D, Yamada K, Takase K, Funabashi T, Kimura F (2006) Sex differences in the basolateral amygdala: the extracellular levels of serotonin and dopamine, and their responses to restraint stress in rats. Eur J Neurosci 24:3245–3254

Moore SA (2009) Cognitive abnormalities in posttraumatic stress disorder. Curr Opin Psychiatry 22:19–24

Nemeroff CB, Bremner JD, Foa EB, Mayberg HS, North CS, Stein MB (2006) Posttraumatic stress disorder: a state-of-the-science review. J Psychiatr Res 40:1–21

Nielsen CK, Arnt J, Sanchez C (2000) Intracranial self-stimulation and sucrose intake differ as hedonic measures following chronic mild stress: interstrain and interindividual differences. Behav Brain Res 107:21–33

Overstreet DH, Friedman E, Mathe AA, Yadid G (2005) The Flinders Sensitive Line rat: a selectively bred putative animal model of depression. Neurosci Biobehav Rev 29:739–759

Padilla E, Barrett D, Shumake J, Gonzalez-Lima F (2009) Strain, sex, and open-field behavior: factors underlying the genetic susceptibility to helplessness. Behav Brain Res 201:257–264

Palanza P (2001) Animal models of anxiety and depression: how are females different? Neurosci Biobehav Rev 25:219–233

Papp M, Klimek V, Willner P (1994) Effects of imipramine on serotonergic and beta-adrenergic receptor binding in a realistic animal model of depression. Psychopharmacology (Berl) 114:309–314

Pinos H, Collado P, Rodriguez-Zafra M, Rodriguez C, Segovia S, Guillamon A (2001) The development of sex differences in the locus coeruleus of the rat. Brain Res Bull 56:73–78

Pitychoutis PM, Griva E, Ioannou K, Tsitsilonis OE, Papadopoulou-Daifoti Z (2009a) Chronic antidepressant treatment exerts sexually dimorphic immunomodulatory effects in an experimental model of major depression: do females lack an advantage? Int J Neuropsychopharmacol 12:1–7

Pitychoutis PM, Nakamura K, Tsonis PA, Papadopoulou-Daifoti Z (2009b) Neurochemical and behavioral alterations in an inflammatory model of depression: sex differences exposed. Neuroscience 159:1216–1232

Pitychoutis PM, Papadopoulou-Daifoti Z (2010) Of Depression and Immunity: Does Sex matter? Int J Neuropsychopharmacol 13:675–689

Pitychoutis PM, Tsitsilonis OE, Papadopoulou-Daifoti Z (2010) Antidepressant Pharmacotherapy: Focus on Sex Differences in Neuroimmunopharmacological Crossroads. Future Neurology 5: 581–596

Seale JV, Wood SA, Atkinson HC, Bate E, Lightman SL, Ingram CD, Jessop DS, Harbuz MS (2004a) Gonadectomy reverses the sexually diergic patterns of circadian and stress-induced hypothalamic-pituitary-adrenal axis activity in male and female rats. J Neuroendocrinol 16: 516–524

Seale JV, Wood SA, Atkinson HC, Harbuz MS, Lightman SL (2004b) Gonadal steroid replacement reverses gonadectomy-induced changes in the corticosterone pulse profile and stress-induced hypothalamic-pituitary-adrenal axis activity of male and female rats. J Neuroendocrinol 16:989–998

Sherrill JT, Anderson B, Frank E, Reynolds CF III, Tu XM, Patterson D, Ritenour A, Kupfer DJ (1997) Is life stress more likely to provoke depressive episodes in women than in men? Depress Anxiety 6:95–105

Shors TJ (1998) Stress and sex effects on associative learning: for better or for worse. Neuroscientist 4:353–364

Shors TJ (2006) Stressful experience and learning across the lifespan. Annu Rev Psychol 57:55–85

Shors TJ, Seib TB, Levine S, Thompson RF (1989) Inescapable versus escapable shock modulates long-term potentiation in the rat hippocampus. Science 244:224–226

Shors TJ, Lewczyk C, Pacynski M, Mathew PR, Pickett J (1998) Stages of estrous mediate the stress-induced impairment of associative learning in the female rat. Neuroreport 9:419–423

Shors TJ, Chua C, Falduto J (2001) Sex differences and opposite effects of stress on dendritic spine density in the male versus female hippocampus. J Neurosci 21:6292–6297

Shors TJ, Falduto J, Leuner B (2004) The opposite effects of stress on dendritic spines in male vs. female rats are NMDA receptor-dependent. Eur J Neurosci 19:145–150

Shors TJ, Mathew J, Sisti HM, Edgecomb C, Beckoff S, Dalla C (2007) Neurogenesis and helplessness are mediated by controllability in males but not in females. Biol Psychiatry 62:487–495

Sotiropoulos I, Cerqueira JJ, Catania C, Takashima A, Sousa N, Almeida OF (2008) Stress and glucocorticoid footprints in the brain-the path from depression to Alzheimer's disease. Neurosci Biobehav Rev 32:1161–1173

Spinedi E, Salas M, Chisari A, Perone M, Carino M, Gaillard RC (1994) Sex differences in the hypothalamo-pituitary-adrenal axis response to inflammatory and neuroendocrine stressors. Evidence for a pituitary defect in the autoimmune disease-susceptible female Lewis rat. Neuroendocrinology 60:609–617

Sun MK, Alkon DL (2006) Differential gender-related vulnerability to depression induction and converging antidepressant responses in rats. J Pharmacol Exp Ther 316:926–932

Tanapat P, Hastings NB, Reeves AJ, Gould E (1999) Estrogen stimulates a transient increase in the number of new neurons in the dentate gyrus of the adult female rat. J Neurosci 19:5792–5801

Tonelli LH, Holmes A, Postolache TT (2008) Intranasal immune challenge induces sex-dependent depressive-like behavior and cytokine expression in the brain. Neuropsychopharmacology 33:1038–1048

Trentani A, Kuipers SD, te Meerman GJ, Beekman J, ter Horst GJ, den Boer JA (2003) Immunohistochemical changes induced by repeated footshock stress: revelations of gender-based differences. Neurobiol Dis 14:602–618

Willner P (1997) Validity, reliability and utility of the chronic mild stress model of depression: a 10-year review and evaluation. Psychopharmacology (Berl) 134:319–329

Willner P (2005) Chronic mild stress (CMS) revisited: consistency and behavioural-neurobiological concordance in the effects of CMS. Neuropsychobiology 52:90–110

Willner P, Towell A, Sampson D, Sophokleous S, Muscat R (1987) Reduction of sucrose preference by chronic unpredictable mild stress, and its restoration by a tricyclic antidepressant. Psychopharmacology (Berl) 93:358–364

Wood GE, Shors TJ (1998) Stress facilitates classical conditioning in males, but impairs classical conditioning in females through activational effects of ovarian hormones. Proc Natl Acad Sci U S A 95:4066–4071

Yadid G, Nakash R, Deri I, Tamar G, Kinor N, Gispan I, Zangen A (2000) Elucidation of the neurobiology of depression: insights from a novel genetic animal model. Prog Neurobiol 62:353–378

Zampeli E, Pitychoutis PM, Papadopoulou-Daifoti Z, Tiligada E (2009) Systemic challenge with lipopolysaccharide increases histamine levels in the conjunctiva and cartilage, but not hypothalamus of Sprague Dawley rats. Inflamm Res 58(Suppl 1):49–50

Importance of the COMT Gene for Sex Differences in Brain Function and Predisposition to Psychiatric Disorders

Elizabeth M. Tunbridge and Paul J. Harrison

Abstract As outlined elsewhere in this volume, sex differences can affect brain function and its dysfunction in psychiatric disorders. It is known that genetic factors contribute to these sex dimorphisms, but the individual genes have rarely been identified. The catechol-*O*-methyltransferase (COMT) gene, which encodes an enzyme that metabolises catechol compounds, including dopamine, is a leading candidate in this regard. COMT's enzyme activity, and the neurochemistry and behaviour of COMT knockout mice are both markedly sexually dimorphic. Furthermore, genetic associations between COMT and psychiatric phenotypes frequently show differences between men and women. Although many of these differences are unconfirmed or minor, some appear to be of reasonable robustness and magnitude and are reviewed in this chapter. Sexually dimorphic effects of COMT are usually attributed to transcriptional regulation by oestrogens; however, a careful examination of the literature suggests that additional mechanisms are likely to be at least as important. Here, we review the evidence for a sexually dimorphic influence of COMT upon psychiatric phenotypes and brain function, and discuss potential mechanisms by which this may occur. We conclude that despite the evidence being incomplete, there are accumulating and in places compelling data showing that COMT has markedly sexually dimorphic effects on brain function and its dysfunction in psychiatric disorders. Although oestrogenic regulation of COMT is probably partially responsible for these sex differences, other mechanisms are likely also involved. Since sex differences in the genetic architecture of brain function and psychiatric disorders are the rule not the exception, we anticipate that additional evidence will emerge for sexual dimorphisms, not only in COMT but also in many other autosomal genes.

E.M. Tunbridge (✉) and P.J. Harrison
Department of Psychiatry, University of Oxford, Neurosciences Building, Warneford Hospital, Oxford, OX3 7JX, UK
e-mail: elizabeth.tunbridge@psych.ox.ac.uk

J.C. Neill and J. Kulkarni (eds.), *Biological Basis of Sex Differences in Psychopharmacology*, 119
Current Topics in Behavioral Neurosciences 8, DOI 10.1007/7854_2010_97,

Keywords Dopamine · Gender · Genotype · Polymorphism · Sexual dimorphism

As the contributions to this volume make clear, sex is a key factor in neuroscience (Cahill 2006) and psychiatry (Cosgrove et al. 2007). Many aspects of brain function and structure vary between men and women (Gur et al. 1995; Murphy et al. 1996; De Courten-Myers 1999; De Bellis et al. 2001; Goldstein et al. 2001; Preece and Cairns 2003; De Vries 2004), and most psychiatric disorders show sex differences in one or more variables including incidence, age at onset, clinical features, and outcome (Lensi et al. 1996; Tamminga 1997; Piccinelli and Wilkinson 2000; Aleman et al. 2003; Baron-Cohen et al. 2005). These dimorphisms are usually assumed to result from the effects of sex hormones (Collaer and Hines 1995; Rubinow and Schmidt 1996; Seeman 1997; Kelly et al. 1999), as well as to the actions of genes located on the sex chromosomes (Vawter et al. 2004; Cutter et al. 2006; Davies and Wilkinson 2006). However, sex differences in epigenetic mechanisms, such as DNA methylation and chromatin modifications, relevant to brain function may also play a major role (Kaminsky et al. 2006). Furthermore, there is also evidence that autosomal genes contribute to sex differences in the genetic predisposition to psychiatric phenotypes. For example, sex differences have been found in the chromosomal loci implicated in the vulnerability to major depression (Holmans et al. 2004; Nash et al. 2004), neuroticism (Fullerton et al. 2003), obsessive–compulsive disorder ([OCD]; Nestadt et al. 2000), and autism (Stone et al. 2004). These differences occur in addition to sex differences in the heritability estimates for certain disorders; for example, major depression is estimated to be more heritable in women than men (Kendler et al. 2006). Taken together, these indications that there are differences in the genetic factors contributing to psychiatric phenotypes in men and women are consistent with the fact – not always appreciated – that the genetic basis of many, perhaps most, human traits is sexually dimorphic (Weiss et al. 2006).

Although, as described above, a diverse set of evidence demonstrates that there are sex differences in the genetic modulation of normal brain function and psychiatric disorders, few candidate genes for mediating these sex differences have been described. Here, we outline evidence that COMT is one such gene and consider the mechanisms by which its sexually dimorphic effects on brain function may be mediated. This chapter is an updated version of, and revised from, a recent review (Harrison and Tunbridge 2008).

1 Catechol-*O*-methyltransferase

The neurobiology and pharmacology of COMT have been reviewed in detail elsewhere (Weinberger et al. 2001; Tunbridge et al. 2006a; Mannisto and Kaakkola 1999); however, a brief introduction is given below. The COMT enzyme metabolises catechol-containing compounds, including dopamine and noradrenaline and, notably with respect to sex differences, catechol oestrogens. The COMT gene exists in at least two protein isoforms produced from alternate start codons: membrane-bound COMT (MB-COMT) and soluble COMT (S-COMT), which differ in their capacity and affinity for different catechols. In contrast to the dominance of S-COMT in most peripheral tissues, MB-COMT is the most abundant isoform in brain, consistent with its high affinity for the catecholamine neurotransmitters. The relevance for COMT in regulating brain function and risk for psychiatric disorders arises primarily from its key role in regulating cortical dopamine (Tunbridge et al. 2004a; Yavich et al. 2007). However, as outlined below, COMT's role in metabolising catechol oestrogens, a function which is generally overlooked from a neurobiological perspective, may be important for mediating some of its sexually dimorphic effects.

The COMT gene is located on chromosome 22q11. It contains several single nucleotide polymorphisms (SNPs) of known or suspected functional significance. The best studied of these is rs4680, a 472 G/A substitution at codon 158 in MB-COMT (codon 108 in S-COMT) that encodes either the ancestral valine (Val^{158}) or methionine (Met^{158}), which we refer to as $Val^{158}Met$. The $Val^{158}Met$ SNP directly affects the thermal stability of COMT; therefore, at 37°C the Val^{158} form is more stable, and thus more active, than the Met^{158} form. In human prefrontal cortex, this substitution results in an approximately 40% reduction in Met/Met homozygotes, compared with Val/Val homozygotes, with heterozygotes showing intermediate COMT enzyme activity (Chen et al. 2004). Earlier studies reported even greater differences in erythrocyte COMT activity between Val/Val and Met/Met carriers (Lachman et al. 1996), although this might, in part, result from methodological differences between the studies. Nevertheless, the $Val^{158}Met$-encoded enzyme activity differences, coupled with the importance of COMT for regulating cortical dopamine levels, suggest that cortical dopamine signalling is likely to be enhanced in Met^{158}- compared to Val^{158}-carrying individuals. Following the landmark study of Egan et al. (2001), a range of studies have shown that the $Val^{158}Met$ allele has a small but significant impact on working memory, attention and executive performance, and prefrontal efficiency, with Met^{158}-carrying individuals performing better and/or more efficiently than Val^{158}-carrying individuals (Barnett et al. 2007b; Mier et al. 2009). Conversely, the Val^{158} allele is associated with more "flexible" cognitive responses (Bilder et al. 2004), something which may be more adaptive during emotional processing (Smolka et al. 2005; Drabant et al. 2006). Apart from $Val^{158}Met$, most other common COMT SNPs are non-coding (synonymous, intronic, or in the 5′ or 3′-untranslated regions). As such, any functional impact they confer is presumably mediated by an effect on COMT expression. However, direct evidence for this hypothesis remains unclear (Bray et al. 2003; Chen et al. 2004; Tunbridge et al. 2004b; Dempster et al. 2006) and is likely complicated by non-linear interactions between SNPs within COMT (Meyer-Lindenberg et al. 2006; Nackley et al. 2006). An elegant demonstration of this was provided by Nackley et al. (2006), who demonstrated that different COMT haplotypes can result in different COMT mRNA secondary structure, thereby affecting its mRNA stability, protein expression, and enzyme activity in a manner that could not be predicted by examining any one of the haplotype-comprising SNPs individually. In any event, the majority of studies informing of sexually dimorphic effects of COMT have studied Val^{158}Met; therefore, this polymorphism is the main focus of this review.

2 Sexual Dimorphisms in COMT Function

COMT exhibits sexual dimorphisms in the normal brain that set the scene for, and may contribute to, differences in its involvement in psychiatric disorders between men and women. An important study by Chen et al. (2004), involving post-mortem brain tissue from 118 subjects, showed COMT activity in the prefrontal cortex to be 17% higher in men than women (independent of Val^{158}Met and other SNPs). These findings are in agreement with earlier studies that demonstrated approximately 30% higher enzyme activity in men in liver (Boudikova et al. 1990) and also in most (Fahndrich et al. 1980; Floderus and Wetterberg 1981; Philippu et al. 1981) but not all (Fitzgerald et al. 1980) studies of erythrocytes. It is not known at what stage in life the sex difference in brain COMT activity manifests, since the sole developmental study was conducted almost entirely in males (Tunbridge et al. 2007a). Interestingly, the higher brain COMT activity in men occurs despite levels of COMT protein and mRNA being similar in both sexes (Bray et al. 2003; Chen et al. 2004; Tunbridge et al. 2004b) or perhaps even higher in women (Dempster et al. 2006). This dissociation between expression and activity has implications for the likely mechanism underlying COMT's sexual dimorphisms, as we discuss later.

The other primary evidence for sexual dimorphism in normal COMT function comes from examination of neurochemical function and behaviour of the COMT knockout mouse (Gogos et al. 1998). Tissue dopamine levels in the frontal cortex are increased almost threefold in male COMT −/− mice (and twofold in +/− mice) compared to wild-type mice, confirming the importance of COMT in cortical dopamine metabolism; conversely, in female COMT +/− and −/− mice, dopamine levels were unchanged. The lack of effect of COMT deletion on frontal cortex dopamine in female mice presumably reflects the existence of sex-specific compensatory mechanisms (such as a higher activity of dopamine or noradrenaline transporters), although this has not been determined. Male, but not female, COMT knockout mice perform better on certain memory tests, showing higher levels of spontaneous alteration and better performance on the Barnes maze task than wild-type mice (Babovic et al. 2007). Similarly, male but not female knockout mice also have a significantly higher preference for ethanol, compared with their wild-type littermates (Tammimäki et al. 2008). In contrast, female but not male COMT null mice showed greater anxiety compared to wild-type mice (Gogos et al. 1998). These latter data suggest that COMT deletion is still of functional significance in female mice, despite the lack of an alteration in frontal dopamine. It is plausible that the anxiety changes in female COMT knockout mice results from the action of COMT in the hippocampus, a brain region which is critical for modulating anxiety phenotypes (Gray and McNaughton 2000) and in which COMT is highly expressed. However, the role of COMT in the hippocampus remains to be investigated. Further studies in the COMT knockout mice, using cocaine, GBR 12909 (a dopamine transporter inhibitor), levodopa, or amphetamine to pharmacologically modulate dopamine levels, showed that some but not all neurochemical and behavioural responses to these pharmacological challenges are sex-specific (Huotari et al. 2002a,b, 2004; see also O'Tuathaigh et al. 2007). For example, D-amphetamine administration affected dopamine metabolism equally in male and female knockout mice (Huotari et al. 2004), whereas hyperactivity in response to GBR 12909 was attenuated only in males (Huotari et al. 2002b).

A final noteworthy COMT sexual dimorphism concerns the potential for sex differences in allele frequencies. In 4,014 Ashkenazi Jews (control subjects participating in a genetic association study of schizophrenia discussed below), Shifman et al. (2002) found that the frequency of the A allele of the 3′-untranslated region SNP rs165599 was higher in women than men (65% vs. 61%, $p = 0.0009$); they state that a similar difference was seen in a second sample, giving a combined p value of 0.00009 (see also Shifman et al. 2004). The authors speculate that the finding may be due to a reduced viability of male foetuses carrying this allele. However, to our knowledge, this sex difference in rs165599 allele frequency has not been replicated (Sweet et al. 2005; see also Molero et al. 2007) – albeit no other study has reported on such a large sample – and so its significance remains unclear.

3 Sexually Dimorphic Effects of COMT Genotype on Brain Function and Psychiatric Disorders

Listed in Table 1 are many (but not all) of the studies that have reported a sex difference in the impact of COMT genotype on a cognitive or psychiatric phenotype. Patsopoulos et al. (2007) highlight that the majority of articles claiming sex differences fail to provide satisfactory evidence to support this claim. In part consistent with this assertion, the studies listed vary in the robustness and plausibility of the finding with regard to sex. Few have been statistically convincing, for example, by demonstrating a sex-by-genotype interaction in analysis of variance (ANOVA; e.g. Kates et al. 2006; Lang et al. 2007). Many studies simply found an association that reached significance in one sex but not in the other (e.g. Enoch et al. 2006). Findings of a significant difference in one sex and not in the other were often further compromised by the fact that the non-significant sex had a smaller sample size and therefore had less power to detect an effect, compared with the significant sex (e.g. Nolan et al. 2000; Ono et al. 2004; Stein et al. 2005). Other studies discuss sex-related effects despite reporting a non-significant sex-by-genotype ANOVA interaction (e.g. Olsson et al. 2005; Zinkstok et al. 2006), while a final category of studies, not included in the table, provide only trend-level findings of sex-related effects of COMT genotype (e.g. Eley et al. 2003; Sazci et al. 2004; Woo et al. 2004).

Table 1 Chronological list of studies reporting sexually dimorphic effects of COMT genotype on psychiatric disorders and allied phenotypes[a]

Citation	Phenotype/parameter	Male/females	Finding related to sex
Karayiorgou et al. (1997)[b]	OCD	117/104	Met^{158} allele associated in men ($p = 0.0002$) not women ($p = 0.066$)
Nolan et al. (2000)	Suicide attempts in schizophrenia	117/31	Met^{158} allele associated in men ($p = 0.028$) not women ($p > 0.5$)
Dauvilliers et al. (2001)	Sleep latency in narcolepsy	59/38	Longer latency associated with Met^{158} allele in men, Val^{158} allele in women
Shifman et al. (2002)[b]	Schizophrenia	3,980/1,643	G allele of rs165599 associated in women ($p < 0.00001$) not men ($p = 0.1$). Sex difference $p = 0.01$
Enoch et al. (2003)	Harm avoidance	160/241	Met^{158} homozygosity associated in women ($p \leq 0.03$) not men ($p \geq 0.79$)
Qian et al. (2003)	ADHD	170 case, 376 control/18 case, 17 control	Met^{158} associated with ADHD in males ($p = 0.05$; family-based study), while Val^{158} associated in females ($p = 0.044$; case control)
Domschke et al. (2004)	Panic disorder	82/148	Val^{158} allele associated in women ($p = 0.01$) not men ($p = 1.0$)
Ono et al. (2004)	Suicide	112/51	Val^{158} homozygosity protective in men ($p = 0.016$) not women ($p = 0.96$)
Olsson et al. (2005)	Persistent episodic anxiety	340/473	Met^{158} homozygosity associated in women ($p = 0.02$) not men ($p = 0.38$). No genotype × sex interaction
Poyurovsky et al. (2005)	OCD	109/141	Met^{158} allele associated in men ($p = 0.029$) not women ($p = 0.78$)
Stein et al. (2005)	Low extraversion	154/343[c]	Met^{158} homozygosity associated in women ($p = 0.001$) not men ($p = 0.6$)
Sweet et al. (2005)	Psychosis in Alzheimer's disease	130/243[c]	Val^{158} allele associated in women ($p = 0.005$) not men ($p = 0.383$)
Kates et al. (2006)[b]	Dorsal and orbital frontal volumes	25/26	Opposite effects in boys and girls with VCFS[d]. Sex x genotype interaction ($p < 0.001$)
Beuten et al. (2006)[b]	Nicotine dependence	668[e]/1369[c,e]	Met^{158} allele associated in women. Sex × genotype interactions ($0.006 < p < 0.02$)
Denys et al. (2006)	OCD	135/170	Met^{158} allele associated in men ($p = 0.036$) not women ($p = 0.23$)
Enoch et al. (2006)	Alcoholism and smoking	141/201	Val^{158} allele associated in women ($p = 0.011$) not men ($p = 0.186$)
Kim et al. (2006)	Harm avoidance	138/148	Val^{158} allele associated in women ($p = 0.003$) not men ($p = 0.36$)
O'Hara et al. (2006)	Delayed verbal recall tests	62[f]/101[f]	Val^{158} allele associated with 8 point WMS-R[g] advantage in men. Sex × genotype interaction on BNT[h]
Rothe et al. (2006)	Panic disorder	60/118[e]	Val^{158} allele associated in women ($p = 0.008$) not men ($p = 0.272$)
Rybakowski et al. (2006)	WCST[i] in schizophrenia	43/36	Val^{158} homozygosity associated with fewer errors by men ($p = 0.044$), more errors by women ($p = 0.042$)
Zinkstok et al. (2006)	Grey and white matter volumes	57/97	Ageing effects related to genotype in women only. No genotype × sex interactions
Barnett et al. (2007a)[b]	IQ and executive function	2,650[j,k]/2,650[j,k]	Effects in boys only. Genotype × sex interactions on attention and verbal IQ

Baud et al. (2007)	Trait anger in suicide attempters	211/536	Val^{158} homozygosity associated with higher trait anger in women ($p = 0.002$) but not men
Domschke et al. (2007)[b]	Panic disorder	209/319	Meta-analysis. Association in women only: Val allele in Caucasians, Met allele in Asians
Golimbet et al. (2007)	Novelty seeking	56/74	Met^{158} homozygosity associated with higher trait scores in women ($p = 0.018$) not men
Lang et al. (2007)[b]	Sensation seeking	214/218	Val^{158} homozygosity associated in women ($p \geq 0.005$) not men[l]. Sex × genotype interaction ($p = 0.005$)
Ma et al. (2007)	Schizotypy in healthy volunteers	231/234	Val^{158} homozygosity associated with lower schizotypy scores in men ($p = 0.009$) but not women
Pooley et al. (2007)[b]	OCD	580/718	Meta-analysis. Met^{158} allele associated in men ($p = .001$) not women ($p = 0.83$). Sex difference $p = 0.0001$
Weiss et al. (2007)	Emotional recognition	49/51	Met^{158} homozygosity associated with impaired recognition of sad faces in women ($p = 0.03$) not men
Zhang et al. (2007)	WISC[m]	142/163	Freedom from distractibility scores associated in girls (p-0.03; 0.06 after Bonferroni) but not boys
Barnett et al. (2008)	Meta-analysis of $Val^{158}Met$ and cognitive measures		N-back: negative association between study effect size and proportion of males ($p < 0.001$). Verbal recall: positive association between study effect size and number of male participants ($p = 0.007$)
Biederman et al. (2008)	ADHD[n]	308/166	Met^{158} allele associated in males ($p = 0.003$) but not females. Sex effect: $p = 0.071$; significance increases to $p = 0.007$ when combined with Qian et al. (2003)
Oosterhuis et al. (2008)	Opiate dependence	11 case, 50 control/18 case, 43 control	Met^{158} associated with dependence in female Hispanics ($p = 0.049$). Does not survive multiple testing correction and not found in other ethnicities
Pelayo-Terán et al. (2008)	Schizophrenia clinical characteristics	63/40	Val^{158} associated with longer duration of untreated psychosis in females ($p = 0.011$) but not males. Sex × genotype interaction: $p = 0.011$
Talkowski et al. (2008)[b]	Schizophrenia	Not stated (478 cases, 501 controls)	rs737865 associated in women ($p = 0.008$) but not men. Sex × genotype interaction: $p = 0.0007$
Barnett et al. (2009)[b]	IQ and cognition	4211/3962	COMT haplotype associated with verbal IQ in boys but not girls. Sex × genotype interaction ($p = 0.03$). Working memory span associated with rs165599 in boys but not girls. Sex × genotype interaction: $p = 0.02$
Domschke et al. (2009)	Therapeutic response to ECT[o]	33/71	Val^{158} allele associated with better response in females ($p = 0.016$) but not significant in males (although magnitude of change was similar in both groups)

(*continued*)

Table 1 (continued)

Citation	Phenotype/parameter	Male/females	Finding related to sex
Hoenicka et al. (2009)	Schizophrenia	226 cases, 117 controls/ 111 cases, 114 controls	Val^{158} homozygosity associated in men ($p = 0.022$) but not women
Katerberg et al. (2009)	OCD and patient phenotypic characteristics	151 cases, 235 controls/ 222 cases, 227 controls	Met^{158} associated with OCD in men ($p = 0.039$) but not women. Increased frequency of Met^{158} allele in women compared with men ($p = 0.012$). Sex × genotype interaction on OCD phenotype: heterozygous women showed a lower level of somatic and sensory symptoms compared with homozygotes ($p = 0.024$) but this relationship was absent in men
Kempton et al. (2009)	Brain activation during fearful affect processing	40/34	Val^{158}Met modulates brain activation in regions differing between men and women. Sex × genotype interaction significant in right temporal pole ($p = 0.028$): female Val^{158} activate > Met^{158}, while male Met^{158} deactivate > Val^{158}
Quednow et al. (2009)	PPI[p]	54/53	Met^{158} associated with higher PPI in men ($p < 0.05$) but not women
Tsai et al. (2009)	Fluoxetine response in major depression	138/196	Val^{158} associated with poorer response in men ($p = 0.035$) but not women

[a]Findings all relate to Val^{158}Met polymorphism except where stated
[b]Key studies, discussed in the text
[c]Haplotypes containing Val^{158}Met were also studied
[d]VCFS: velo-cardio-facial syndrome (22q11 deletion syndrome)
[e]Calculated from their Table 1
[f]Elderly adults
[g]WMS-R: Wechsler Memory Scale (Revised)
[h]BNT: Boston Naming Test
[i]WCST: Wisconsin Card Sorting Test
[j]Representative number; exact sample size differed between tests
[k]Children
[l]*p* value not stated
[m]Weschler Intelligence Scale for Children
[n]Attention-deficit hyperactivity disorder
[o]Electroconvulsive therapy
[p]Prepulse inhibition

It should be noted that some phenotypic associations with COMT are in the opposite direction in the two sexes, rather than just being limited to a significant finding in one or other sex (e.g. Dauvilliers et al. 2001; Rybakowski et al. 2006). Results of this nature may be correct; however, in the absence of a prior hypothesis, or replication, they may well be false positives.

These confounds and limitations mean that most of the studies in the table are merely suggestive and do not by themselves provide convincing evidence that COMT genetic variation has a sexually dimorphic influence on the phenotype in question. Nevertheless, the number of studies showing at least some evidence for sexual dimorphisms is intriguing, especially as among the studies are several more striking and robust findings of this kind. We discuss these more robust findings further here.

The first study to report a sex difference in the role of COMT in the genetic predisposition to a psychiatric disorder (Karayiorgou et al. 1997) showed that the low activity (Met^{158}) allele was associated with OCD in men, but not in women in a case–control study. The authors later replicated this finding in a family-based study (Karayiorgou et al. 1999), and it was further replicated in three of four subsequent case–control studies (Poyurovksy et al. 2005; Denys et al. 2006; Pooley et al. 2007). This sex-selective association between the COMT Met^{158} allele and OCD was confirmed in a meta-analysis (Pooley et al. 2007); the odds ratio associated with the Met^{158} allele in men was 1.88 ($p < 0.001$), with no effect in women (odds ratio 0.98, $p = 0.83$), and with a significant sex difference between the odds ratios ($p < 0.0001$). Although studies published since this meta-analysis have produced mixed results (Wray et al. 2008 found no association between COMT in men or women, while Katerberg et al. 2009 showed a trend-level association between Met^{158} and OCD in men but not in women, in line with the finding of Pooley et al.), the COMT OCD data are arguably the clearest evidence to date for a sexually dimorphic autosomal genetic association with a psychiatric disorder.

Barnett et al. (2007a, 2009) also demonstrated a sexually dimorphic association between COMT and IQ. They genotyped initially the $Val^{158}Met$ SNP (Barnett et al. 2007a), and more recently the functional haplotype described by Nackley et al. (2006) (Barnett et al. 2009), in over 5,000 participants in a longitudinal study of child development, and in whom a range of cognitive tests, including IQ, attention, and working memory, had been conducted between ages 8 and 10. In the initial study, the Met^{158} allele was associated with better function in several domains, with these effects greater in, or limited to, boys. More recently, they demonstrated that the strength of these associations was even greater when the functional haplotype described by Nackley et al. (2006) (which appears to have a larger effect on COMT enzyme activity than $Val^{158}Met$ SNP alone) was considered. For example, in boys, Met^{158} homozygotes had a verbal IQ 3 points higher than Val^{158} homozygotes, whereas the difference in girls was less than 1 point. Strikingly, in the later study, the effect of the COMT haplotypes was even greater: boys carrying the COMT diplotype that predicted the highest enzyme activity had a verbal IQ 6 points lower, on average, when compared with boys with the diplotype predicting the lowest activity; again, there was no significant association between the COMT haplotypes and verbal IQ in girls (Barnett et al. 2009). Intriguingly, in their initial study Barnett et al. (2007a) also showed that the COMT effect on verbal IQ was greater in pubertal than pre-pubertal boys (increasing to a 10 point difference between homozygote groups), suggesting that its influence increases with sexual maturation (they did not comment on the effect of age in their subsequent study). An increasing effect of COMT around the age of puberty is consistent with data linking COMT Val^{158}Met with verbal IQ in a longitudinal study of velocardiofacial syndrome (VCFS or 22q11 hemideletion syndrome, in which one copy of COMT is deleted; Gothelf et al. 2005). It is also consistent with significant maturational increases in COMT expression and activity in prefrontal cortex, although these changes were most marked post-adolescence and not peri-pubertally (Tunbridge et al. 2007a). Kates et al. (2006) also studied the role of COMT Val^{158}Met children with VCFS. Their finding, although in a small sample, is noteworthy because it concerns brain structure, and also because the dimorphism is statistically robust – a sex-by-genotype ANOVA interaction ($p < 0.001$), with an opposing effect of Val^{158}Met allele on frontal cortical volumes in boys and girls. Taken together, these data, together with those of Zinkstok et al. (2006), raise the possibility of a complex interaction between COMT genotype, sex, brain structure, and development.

Further evidence for sexually dimorphic associations between COMT and psychiatric phenotypes comes from hints that there may be sex differences in associations between COMT and schizophrenia. Shifman et al. (2002) reported that homozygosity for the G allele at rs165599 of COMT was strongly associated with schizophrenia in women ($p = 6.8 \times 10^{-6}$) but not in men ($p = 0.09$), with the sex difference in genotype effect being significant ($p < 0.01$). The authors concluded that there may be a sex-specific genetic component to schizophrenia, while acknowledging that twin studies had not predicted this. However, a sex difference in the genetic association with schizophrenia was not present for the other COMT SNPs they analysed, nor was it observed for haplotypes (that included rs165599). Furthermore, as noted by Craddock et al. (2006), their finding was driven by the sexually dimorphic allele frequency in the control group, described above. More recently, Talkowski et al. (2008) reported a female-specific association between COMT rs737865 and schizophrenia ($p = 0.008$) (but not in rs165599), in which the sex-by-genotype interaction was significant ($p = 0.0007$). They also replicated a female-specific association at this locus in a second cohort, although it was the other allele which was significantly associated in this second samples. Notably, this sex-specific association occurred in the absence of any sex differences in allele frequencies in the control group; thus, these findings are perhaps more convincing than those of Shifman and colleagues. However, despite the statistical significance of these two studies, the results remain difficult to interpret vis à vis schizophrenia, and in need of replication, especially given the different polymorphisms involved. The potential for sex differences in associations between COMT and schizophrenia is further complicated by the report of Sazci et al. (2004) of a female-predominant association (of Met^{158}) with schizophrenia, and by the fact that the meta-analyses of COMT Val^{158}Met with schizophrenia find no sex effect (Glatt et al. 2003; Fan et al. 2005). Thus, although these findings raise the intriguing possibility that different loci within COMT may be differentially associated with schizophrenia in men vs. women, they should be considered preliminary and are in need of further investigation.

A further sex-by-genotype interaction ($p < 0.0005$) was seen by Lang et al. (2007) who showed that the Val^{158} allele was associated with the personality trait of sensation seeking in women but not in men. This finding is the most statistically robust of a line of studies that show relationships between COMT genotype and personality traits in women but not in men. Of particular note, anxiety-related phenotypes (such as harm avoidance and neuroticism) have been repeatedly (though often weakly) associated in women with the Met^{158} allele (Eley et al. 2003; Enoch et al. 2003; Olsson et al. 2005; Stein et al. 2005), consistent with findings in anxiety disorder (Domschke et al. 2004; Woo et al. 2004; Rothe et al. 2006) and with the increased anxiety found in female, but not in male, COMT knockout mice, described above. A meta-analysis of the panic disorder studies (Domschke et al. 2007) confirmed the presence of a sex difference in the association with COMT. However, this study also revealed an additional complexity: the relationship interacted with ethnicity, such that panic disorder was associated with the Met^{158} allele in Caucasian women but with the Val^{158} allele in Asian women. The latter result is consistent with the findings of Kim et al. (2006), who showed a Val^{158} association with harm avoidance in Korean women. Two main explanations come to mind for the genotype-by-ethnicity interaction. First, given that the resulting subgroups are quite small and the p values modest ($p = 0.04$ in Caucasians, $p = 0.02$ in Asians; Domschke et al. 2007), the findings are quite likely to be false positives. Second, if true, the finding may relate to ethnic differences in the genetic background of the $Val^{158}Met$ polymorphism, such that the opposing allelic associations with panic disorder are both genuine (Lin et al. 2007). This possibility is quite plausible, given the complex manner in which SNPs in COMT have been shown to interact to regulate COMT activity (Nackley et al. 2006).

It is interesting to note that the data reviewed here suggest that the Met^{158} allele is associated with anxiety phenotypes in (Caucasian) women but with OCD – usually considered to be a type of anxiety disorder – in men. One interpretation is that Met^{158} is a risk factor for a shared predisposition to anxiety in both sexes, but that this manifests itself as different anxiety phenotypes in men and women because of other influences that are themselves sexually dimorphic, whether genetic, epigenetic, or environmental in origin.

4 Mechanisms Underlying Sex Differences in COMT's Effects on Brain Function and Its Associations with Psychiatric Disorders

Sexually dimorphic effects of COMT are normally explained by its regulation by oestrogens. Contemporary interest in oestrogenic COMT regulation can be traced to the work of Xie et al. (1999), although many years previously Axelrod had reported that 17-β-oestradiol (E2) administration decreased COMT activity in rat liver (Cohn and Axelrod 1971). Xie et al. (1999) identified two oestrogen response elements in the COMT promoter and showed that E2 at physiological concentrations inhibits COMT mRNA expression in cells expressing oestrogen receptors, but not in those which do not. The same group later showed that the oestrogen-mediated decrease in COMT mRNA was accompanied by a proportional decrease in COMT immunoreactivity and activity (Jiang et al. 2003). This inhibitory regulation by oestrogens is consistent not only with the normal sex differences in COMT activity noted above, but also with the evidence that women with high oestrogen states (e.g. on the combined oral contraceptive, or in the third trimester of pregnancy) have lower COMT activity than other women (Briggs and Briggs 1973). In post-menopausal women, oestrogen levels fall dramatically, to substantially below that of men of the same age (Bjornerem et al. 2004); it would be of interest to know whether sex differences in COMT activity (and its genetic associations) present in younger adults are lost or reversed in the elderly.

In addition to itself being regulated by oestrogens, COMT in turn plays an important role in metabolising catechol oestrogens and thereby lowering levels of these potential carcinogens (Creveling 2003). There is both in vivo (Worda et al. 2003) and in vitro (Dawling et al. 2001; but see Goodman et al. 2002) evidence that the Val^{158}Met SNP influences this pathway with greater E2 metabolism in those with the high activity Val^{158} allele, a finding that may also explain the associations reported in some studies between the COMT Met^{158} allele and oestrogenic cancers (Goodman et al. 2002). These complex, reciprocal, and partly genotype-influenced interactions between COMT and oestrogens may be relevant to the question of sexual dimorphism. COMT genotype may modulate the role that oestrogens play in brain function and dysfunction (Seeman 1997), while oestrogens affect COMT activity and therefore its associations with behaviour and psychiatric disorders by virtue of their influence on COMT gene expression.

However, despite the focus on oestrogenic regulation of COMT gene expression, it is probably only a partial explanation for sex differences in COMT's function. Although COMT mRNA expression is lower in peripheral tissue in women (Wang et al. 2009), Cohn and Axelrod (1971) found that E2 did not affect rat brain COMT activity despite its robust down-regulation in the liver. Consistent with these findings, Jiang et al. (2003) demonstrated oestrogenic regulation of COMT in a breast cancer cell line but not in a glioblastoma cell line, again suggesting that this relationship may not pertain in the brain. The suggestion that oestrogenic regulation of COMT mRNA abundance may be less important for mediating COMT's sex differences in the brain, compared with the periphery, is supported by several recent human post-mortem brain studies showing that COMT activity is higher in men (Chen et al. 2004) even though COMT mRNA abundance is the same (Tunbridge et al. 2004b; Chen et al. 2004) or even higher (Dempster et al. 2006) in women. Hence, for a given level of COMT mRNA abundance, COMT activity in the male brain is higher than it is in females, and therefore the sexual dimorphism cannot be readily explained by transcriptional regulation. This fact also argues against potential explanations for sex differences in COMT based on epigenetic regulation of COMT mRNA (Kaminsky et al. 2006); in any event, there is no evidence for a sex difference in COMT promoter methylation status (Abdolmaleky et al. 2006). Finally, the finding that COMT protein and activity levels rise considerably in men between the third and fifth decade of life (Tunbridge et al. 2007a), despite steady oestradiol levels across this period (Bjornerem et al. 2004) emphasises that oestrogens are not the only factor responsible for regulating COMT activity in the brain.

The mechanism mediating a sex difference in COMT activity in the absence of a difference in COMT mRNA and protein abundance is unknown. It is possible that the discrepancy is in fact spurious, since localised or phasic sex differences in COMT mRNA or protein abundance in brain – e.g. in specific cell types, or in relation to menstrual cycle – are likely, based on evidence from peripheral tissue (Salih et al. 2008). However, the fact that a robust sex difference in COMT enzyme activity was seen with no hint of a sex difference in COMT protein measured in the same samples (Chen et al. 2004) suggests that it is genuine. As such, one possibility is that it reflects a sex difference in the relative abundance of novel mRNA (Tunbridge et al. 2007b) or protein (Tunbridge et al. 2006b) isoforms of COMT, which could differ in their enzyme activity and might not affect the amount of total COMT mRNA or protein. Or, there could be a sex difference in the abundance of co-factors, endogenous inhibitors (e.g. *S*-adenosylmethionine [SAM]; Zhu 2002), or interacting proteins that modulate COMT activity. In support of the hypothesis that SAM levels may be rate-limiting in the female brain, there are hints from a small pilot study that SAM augmentation may be therapeutically beneficial for depressive symptoms for women but not for men (Strous et al. 2009)

More broadly, sexual dimorphisms in COMT occur against a background of sex differences in its biochemical pathways. In addition to differences between females and males in terms of catechol oestrogen function, the dopamine system is markedly sexually dimorphic (Di Paolo 1994; Becker 1999; Andersen and Teicher 2000; Carroll et al. 2004). For example, women have higher striatal [^{18}F]fluorodopa uptake, suggesting greater presynaptic dopamine synthesis (Laakso et al. 2002), and a lower D2 receptor affinity (Pohjalainen et al. 1998) than men, which the authors hypothesise reflects higher dopamine levels in the female brain at baseline. Conversely, women have lower amphetamine-stimulated dopamine release (Munro et al. 2006), and a greater dopamine transporter uptake, suggestive of more rapid clearance of dopamine from the synaptic cleft (Mozley et al. 2001). Thus, compared with men, women appear to have elevated basal, but decreased stimulated, striatal dopamine levels. Taken together, these findings may indicate a difference in the balance between tonic and phasic dopamine between women and men, a parameter which is also hypothesised to be regulated by COMT activity (Bilder et al. 2004). However, oestrogenic state (e.g. phase of menstrual cycle) has not been fully taken into account in these studies and may be a significant confounder: rodent studies have shown marked fluctuations in multiple dopaminergic parameters in the striatum across the oestrus cycle (Jori and Cecchetti 1973, Favis et al. 1977, Crowley et al. 1978; Fernandez-Ruiz et al. 1991; Morissette and Di Paolo 1993; Xiao and Becker 1994). Sex differences in dopamine function are less well studied outside the striatum; therefore little information is available to inform on relative dopaminergic function in the sex differences in cortical regions where COMT is likely to have its primary impact (Karoum et al. 1994; Gogos et al. 1998; Tunbridge et al. 2004a; Yavich et al. 2007). Cortical tissue dopamine concentrations are reportedly similar in men and women (Robinson et al. 1977), although women may have a higher extra-striatal D2 receptor-binding potential (Kaasinen et al. 2001). In summary, there are diverse data suggesting substantial sex differences in central dopamine parameters (Cosgrove et al. 2007). However, it is not clear what their net effect is upon dopaminergic function in men compared to women, nor how the dopaminergic sexual dimorphisms impinge upon the inverted U-shaped relationship between dopamine activity and prefrontal function (Goldman-Rakic et al. 2000) known to be regulated by COMT activity (Mattay et al. 2003; Tunbridge et al. 2006a; Williams-Gray et al. 2007) and likely to be sensitive to interactive effects of COMT with other genes (Talkowski et al. 2008; Tan et al. 2007) and environmental factors (Caspi et al. 2005).

5 Conclusions

Genetic epidemiological and other studies show that sex differences in the genetic architecture of many human phenotypes, including psychiatric disorders, are common. There is also a substantial literature attesting to male–female differences across many domains of brain function, structure, and development. The data reviewed here suggest that COMT is one of the genes that contribute to these sexual dimorphisms. In addition to a difference in COMT enzyme activity between men and women, there is evidence that the involvement of COMT in predisposition to OCD and anxiety phenotypes is sex-specific, and weaker evidence for sex differences in its roles in several other phenotypes.

Not all COMT genotypic associations are demonstrably sexually dimorphic; for example, its influence on Wisconsin Card Sort Test performance (Barnett et al. 2007b) and on homocysteine metabolism (Tunbridge et al. 2008). Neither is COMT the only autosomal gene for which sexually dimorphic genetic associations with psychiatric phenotypes have been reported; using schizophrenia as an example, reelin is reportedly associated only in women (Shifman et al. 2008), while GNB1L (Williams et al. 2008) and MTHFR have been associated only in men (Sazci et al. 2005; Kempisty et al. 2006). However, the data do appear more extensive, and in places statistically more convincing, for COMT than for any other autosomal gene we could determine. Furthermore, as we have discussed, the findings parallel sex differences found in the COMT knockout mouse, and are complemented by the sexual dimorphism in COMT activity and by a plausible, if incomplete, mechanistic explanation in terms of oestrogenic regulation. As such, COMT may well contribute to the genetic basis for sexual dimorphisms in human brain, behaviour, and -related disorders, although it is clearly but one of many genes acting in this way, and in isolation explains only a tiny proportion of the variance. Future studies are needed not only to establish the range and magnitude of the COMT-related sexual dimorphisms, but also to identify other genes, their epistatic and gene–environment interactions, and the underlying biological mechanisms.

Acknowledgements We thank Daniel Weinberger for insightful discussions and fruitful collaborations. Valerie West kindly provided secretarial assistance. E.M.T. is supported by a Royal Society Research Fellowship. We apologise to authors of relevant studies that we have inadvertently omitted.

References

Abdolmaleky HM, Cheng KH, Faraone SV et al (2006) Hypomethylation of MB-COMT promoter is a major risk factor for schizophrenia and bipolar disorder. Hum Mol Genet 15:3132–3145

Aleman A, Kahn RS, Selten JP (2003) Sex differences in the risk of schizophrenia – evidence from meta-analysis. Arch Gen Psychiatry 60:565–571

Andersen SL, Teicher MH (2000) Sex differences in dopamine receptors and their relevance to ADHD. Neurosci Biobehav Rev 24:137–141

Babovic D, O'Tuathaigh CM, O'Sullivan GJ et al (2007) Exploratory and habituation phenotype of heterozygous and homozygous COMT knockout mice. Behav Brain Res 183:236–239

Barnett JH, Heron J, Ring SM et al (2007a) Gender-specific effects of the catechol-O-methyltransferase Val(108)/(158)Met polymorphism on cognitive function in children. Am J Psychiatry 164:142–149

Barnett JH, Jones PB, Robbins TW et al (2007b) Effects of the catechol-O-ethyltransferase Val158Met polymorphism on executive function: a meta-analysis of the Wisconsin Card Sort Test in schizophrenia and healthy controls. Mol Psychiatry 12:502–509

Barnett JH, Scoriels L, Munafò MR (2008) Meta-analysis of the cognitive effects of the catechol-O methyltransferase gene Val158/108Met polymorphism. Biol Psychiatry 64:137–144

Barnett JH, Heron J, Goldman D et al (2009) Effects of catechol-O-methyltransferase on normal variation in the cognitive function of children. Am J Psychiatry 166:909–916

Baron-Cohen S, Knickmeyer RC, Belmonte MK (2005) Sex differences in the brain: implications for explaining autism. Science 310:819–823

Baud P, Courtet P, Perroud N et al (2007) Catechol-O-methyltransferase polymorphism (COMT) in suicide attempters: a possible gender effect on anger traits. Am J Med Genet B Neuropsychiatr Genet 144B:1042–1047

Becker JB (1999) Gender differences in dopaminergic function in striatum and nucleus accumbens. Pharmacol Biochem Behav 64:803–812

Beuten J, Payne T, Ma J et al (2006) Significant association of catechol-O-methyltransferase (COMT) haplotypes with nicotine dependence in male and female smokers of two ethnic populations. Neuropsychopharmacology 31:675–684

Biederman J, Kim JW, Doyle AE et al (2008) Sexually dimorphic effects of four genes (COMT, SLC6A2, MAOA, SLC6A4) in genetic associations of ADHD: a preliminary study. Am J Med Genet B Neuropsychiatr Genet 147B:1511–1518

Bilder RM, Volavka J, Lachman HM et al (2004) The catechol-O-methyltransferase polymorphism: relations to the tonic-phasic dopamine hypothesis and neuropsychiatric phenotypes. Neuropsychopharmacology 29:1943–1961

Bjornerem A, Straume B, Midtby M et al (2004) Endogenous sex hormones in relation to age, sex, lifestyle factors, and chronic diseases in a general population: the Tromso Study. J Clin Endocrinol Metab 89:6039–6047

Boudikova B, Szumlanski C, Maidak B et al (1990) Human liver catechol-O-methyltransferase pharmacogenetics. Clin Pharmacol Therap 48:381–389

Bray N, Buckland P, Williams N et al (2003) A haplotype implicated in schizophrenia susceptibility is associated with reduced COMT expression in human brain. Am J Hum Genet 73:152–161

Briggs MH, Briggs M (1973) Hormonal influences on erythrocyte catechol-O-methyl transferase activity in humans. Experientia 29:278–280

Cahill L (2006) Why sex matters for neuroscience. Nat Rev Neurosci 7:477–484

Carroll ME, Lynch WJ, Roth ME et al (2004) Sex and estrogen influence drug abuse. Trends Pharmacol Sci 25:273–279

Caspi A, Moffitt TE, Cannon M et al (2005) Moderation of the effect of adolescent-onset cannabis use on adult psychosis by a functional polymorphism in the catechol-O-methyltransferase gene: longitudinal evidence of a gene X environment interaction. Biol Psychiatry 57:1117–1127

Chen JS, Lipska BK, Halim N et al (2004) Functional analysis of genetic variation in catechol-o-methyltransferase (COMT): effects on mRNA, protein, and enzyme activity in postmortem human brain. Am J Hum Genet 75:807–821

Cohn CK, Axelrod J (1971) The effect of estradiol on catechol-O-methyltransferase activity in rat liver. Life Sci 10:1351–1354

Collaer ML, Hines M (1995) Human behavioral sex differences: a role for gonadal hormones during early development? Psychol Bull 118:55–107

Cosgrove KP, Mazure CM, Staley JK (2007) Evolving knowledge of sex differences in brain structure, function, and chemistry. Biol Psychiatry 62:847–855. doi:10.1016/j.biopsych.2007.03.001

Craddock N, Owen MJ, O'Donovan MC (2006) The catechol-O-methyl transferase (COMT) gene as a candidate for psychiatric phenotypes: evidence and lessons. Mol Psychiatry 11:446–58

Creveling CR (2003) The role of catechol-O-methyltransferase in the inactivation of catecholestrogen. Cell Mol Neurobiol 23:289–291

Crowley WR, O'Donohue TL, Jacobowitz DM (1978) Changes in catecholamine content in discrete brain nuclei during the estrous cycle of the rat. Brain Res 147:315–326

Cutter WJ, Daly EM, Robertson DMW et al (2006) Influence of X chromosome and hormones on human brain development: a magnetic resonance imaging and proton magnetic resonance spectroscopy study of Turner syndrome. Biol Psychiatry 59:273–283

Dauvilliers Y, Neidhart E, Lecendreux M et al (2001) MAO-A and COMT polymorphisms and gene effects in narcolepsy. Mol Psychiatry 6:367–372

Davies W, Wilkinson LS (2006) It is not all hormones: alternative explanations for sexual differentiation of the brain. Brain Res 1126:36–45

Dawling S, Roodi N, Mernaugh RL et al (2001) Catechol-O-methyltransferase (COMT)-mediated metabolism of catechol estrogens: comparison of wild-type and variant COMT isoforms. Cancer Res 61:6716–6722

De Bellis J, Keshavan MS, Beers SR et al (2001) Sex differences in brain maturation during childhood and adolescence. Cereb Cortex 11:552–557

De Courten-Myers GM (1999) The human cerebral cortex: gender differences in structure and function. J Neuropathol Exp Neurol 58:217–226

De Vries GJ (2004) Sex differences in adult and developing brain: compensation, compensation, compensation. Endocrinol 145:1063–1068

Dempster EL, Mill J, Craig IW et al (2006) The quantification of COMT mRNA in post mortem cerebellum tissue: diagnosis, genotype, methylation and expression. BMC Med Genet 7:10

Denys D, Van Nieuwerburgh F, Deforce D, Westenberg H (2006) Association between the dopamine D2 receptor TaqI A2 allele and low activity COMT allele with obsessive-compulsive disorder in males. Eur Neuropsychopharmacol 16:446–50

Di Paolo T (1994) Modulation of brain dopamine transmission by sex steroids. Rev Neurosci 5:27–41

Domschke K, Freitag CM, Kuhlenbaumer G et al (2004) Association of the functional V158M catechol-O-methyltransferase polymorphism with panic disorder in women. Int J Neuropsychopharmacol 7:183–188

Domschke K, Deckert J, O'Donovan MC et al (2007) Meta-analysis of COMT val158met in panic disorder: ethnic heterogeneity and gender specificity. Am J Med Genet B Neuropsychiatr Genet 144B:667–673. doi:10.1002/ajmg.b.30494

Domschke K, Zavorotnyy M, Diemer J et al (2009) COMT val158met influence on electroconvulsive therapy response in major depression. Am J Med Genet B Neuropsychiatr Genet. doi:10.1002/ajmg.b.30949

Drabant EM, Hariri AR, Meyer-Lindenberg A et al (2006) Catechol O-methyltransferase val(158) met genotype and neural mechanisms related to affective arousal and regulation. Arch Gen Psychiatry 63:1396–1406

Egan MF, Goldberg TE, Kolachana BS et al (2001) Effect of COMT Val$^{108/158}$ Met genotype on frontal lobe function and risk for schizophrenia. Proc Natl Acad Sci USA 98:6917–6922

Eley TC, Tahir E, Angleitner A et al (2003) Association analysis of MAOA and COMT with neuroticism assessed by peers. Am J Med Genet 120B:90–96

Enoch MA, Xu K, Ferro E et al (2003) Genetic origins of anxiety in women: a role for a functional catechol-O-methyltransferase polymorphism. Psychiatr Genet 13:33–41

Enoch MA, Waheed JF, Harris CR et al (2006) Sex differences in the influence of COMT Val158Met on alcoholism and smoking in plains American Indians. Alcohol Clin Exp Res 30:399–406

Fahndrich E, Coper H, Christ W et al (1980) Erythrocyte COMT-activity in patients with affective disorders. Acta Psychiatrica Scand 61:427–437

Fan JB, Zhang CS, Gu NF et al (2005) Catechol-O-methyltransferase gene Val/Met functional polymorphism and risk of schizophrenia: a large scale association study plus meta-analysis. Biol Psychiatry 57:139–144

Favis CF, Davis BF, Halaris AE (1977) Variations in the uptake of [3H] dopamine during the estrous cycle. Life Sci 20:1319–1332

Fernandez-Ruiz JJ, Hernandez ML, De Miguel R et al (1991) Nigrostriatal and mesolimbic dopaminergic activities were modified throughout the ovarian cycle of female rats. J Neural Transm Gen Sect 85:223–229

Fitzgerald GA, Hamilton CA, Jones DH et al (1980) Erythrocytes catechol-O-methyltransferase activity and indices of sympathetic activity in man. Clin Sci 58:423–425

Floderus Y, Wetterberg L (1981) The inheritance of human erythrocyte catechol-O-methyltransferase activity. Clin Genet 19:392–393

Fullerton J, Cubin M, Tiwari H et al (2003) Linkage analysis of extremely discordant and concordant sibling pairs identifies quantitative trait loci that influence variation in the human personality trait neuroticism. Am J Hum Genet 72:879–890

Glatt SJ, Faraone SV, Tsuang MT (2003) Association between a functional catechol O-methyltransferase gene polymorphism and schizophrenia: meta-analysis of case-control and family-based studies. Am J Psychiatry 160:469–476

Gogos JA, Morgan M, Luine V et al (1998) Catechol-O-methyltransferase-deficient mice exhibit sexually dimorphic changes in catecholamine levels and behavior. Proc Natl Acad Sci USA 95:9991–9996

Goldman-Rakic PS, Muly EC III, Williams GV (2000) D1 receptors in prefrontal cells and circuits. Brain Res Rev 31:295–301

Goldstein JM, Seidman LJ, Horton NJ et al (2001) Normal sexual dimorphism of the adult human brain assessed by in vivo magnetic resonance imaging. Cereb Cortex 11:490–497

Golimbet VE, Alfimova MV, Gritsenko IK et al (2007) Relationship between dopamine system genes and extraversion and novelty seeking. Neurosci Behav Physiol 37:601–606

Goodman JE, Jensen LT, He P et al (2002) Characterization of human soluble high and low activity catechol-O-methyltransferase catalyzed catechol estrogen methylation. Pharmacogenetics 12:517–528

Gothelf D, Eliez S, Thompson T et al (2005) COMT genotype predicts longitudinal cognitive decline and psychosis in 22q11.2 deletion syndrome. Nat Neurosci 8:1500–1502

Gray JA, McNaughton N (2000) The neuropsychology of anxiety: an enquiry into the functions of the septo-hippocampal system, 2nd edn. OUP, Oxford

Gur RC, Mozley LH, Mozley PD et al (1995) Sex differences in regional cerebral glucose metabolism during a resting state. Science 267:528–531

Harrison PJ, Tunbridge EM (2008) Catechol-O-methyltransferase: a gene contributing to sex differences in brain function, and the sexual dimorphisms in the predisposition to psychiatric disorders. Neuropsychopharmacology 33:3037–3045

Hoenicka J, Garrido E, Martínez I et al (2009) Gender-specific COMT Val158Met polymorphism association in Spanish schizophrenic patients. Am J Med Genet B Neuropsychiatr Genet. doi:10.1002/ajmg.b.30957

Holmans P, Zubenko G, Crowe R et al (2004) Genomewide significant linkage to recurrent, early-onset major depressive disorder on chromosome 15q. Am J Hum Genet 74:1154–1167

Huotari M, Gogos JA, Karayiorgou M et al (2002a) Brain catecholamine metabolism in catechol-O-methyltransferase (COMT)-deficient mice. Eur J Neurosci 15:246–256

Huotari M, Santha M, Lucas LR et al (2002b) Effect of dopamine uptake inhibition on brain catecholamine levels and locomotion in catechol-O-methyltransferase-disrupted mice. J Pharmacol Exp Ther 303:1309–1316

Huotari M, Garcia-Horsman JA, Karayiorgou M et al (2004) D-amphetamine responses in catechol-O-methyltransferase (COMT) disrupted mice. Psychopharmacology 172:1–10

Jiang H, Xie T, Ramsden DB et al (2003) Human catechol-O-methyltransferase down-regulation by estradiol. Neuropharmacology 45:1011–1018

Jori A, Cecchetti G (1973) Homovanillic acid levels in rat striatum during the oestrus cycle. J Endocrinol 58:341–342

Kaasinen V, Nogren K, Hietala J et al (2001) Sex differences in extrastriatal dopamine D_2-like receptors in the human brain. Am J Psychiatry 158:308–311

Kaminsky Z, Wang SC, Petronis A (2006) Complex disease, gender and epigenetics. Ann Med 38:530–544

Karayiorgou M, Altemus M, Galke BL et al (1997) Genotype determining low catechol-O-methyltransferase activity as a risk factor for obsessive-compulsive disorder. Proc Natl Acad Sci USA 94:4572–4575

Karayiorgou M, Sobin C, Blundell ML et al (1999) Family-based association studies support a sexually dimorphic effect of COMT and MAOA on genetic susceptibility to obsessive-compulsive disorder. Biol Psychiatry 45:1178–1189

Karoum F, Chrapusta SJ, Egan MF (1994) 3-Methoxytyramine is the major metabolite of released dopamine in the rat frontal cortex: reassessment of the effects of antipsychotics on the dynamics of dopamine release and metabolism in the frontal cortex, nucleus accumbens, and striatum by a simple two pool model. J Neurochem 63:972–979

Katerberg H, Cath DC, Denys DA et al (2009) The role of the COMT Val(158)Met polymorphism in the phenotypic expression of obsessive-compulsive disorder. Am J Med Genet B Neuropsychiatr Genet. doi:10.1002/ajmg.b.30971

Kates WR, Antshel KM, AbdulSabur N et al (2006) A gender-moderated effect of a functional COMT polymorphism on prefrontal brain morphology and function in velo-cardio-facial syndrome (22q11.2 deletion syndrome). Am J Med Genet 141B:274–280

Kelly SJ, Ostrowski NL, Wilson MA (1999) Gender differences in brain and behavior: hormonal and neural bases. Pharmacol Biochem Behav 64:655–664

Kempisty A, Mostowska A, G¢rska I et al (2006) Association of 677C>T polymorphism of methylenetetrahydrofolate reductase (MTHFR) gene with bipolar disorder and schizophrenia. Neurosci Lett 400:267–271

Kempton MJ, Haldane M, Jogia J et al (2009) The effects of gender and COMT Val158Met polymorphism on fearful facial affect recognition: a fMRI study. Int J Neuropsychopharmacol 12:371–381

Kendler KS, Gatz M, Gardner CO et al (2006) A Swedish national twin study of lifetime major depression. Am J Psychiatry 163:109–114
Kim SJ, Kim YS, Kim SY et al (2006) An association study of catechol-O-methyltransferase and monoamine oxidase A polymorphisms and personality traits in Koreans. Neurosci Lett 401:154–158
Laakso A, Vilkman H, Bergman J et al (2002) Sex differences in striatal presynaptic dopamine synthesis capacity in healthy subjects. Biol Psychiatry 52:759–763
Lachman HM, Papolos DF, Saito T et al (1996) Human catechol-o-methyltransferase polymorphism and its potential application to neuropsychiatric disorders. Pharmacogenetics 6:243–250
Lang UE, Bajbouj M, Sander T et al (2007) Gender-dependent association of the functional catechol-O-methyltransferase Val158Met genotype with sensation seeking personality trait. Neuropsychopharmacology 32:1950–1955
Lensi P, Cassano GB, Correddu G et al (1996) Obsessive-compulsive disorder. Familial-developmental history, symptomatology, comorbidity and course with special reference to gender-related differences. Br J Psychiatry 169:101–107
Lin PI, Vance JM, Pericak-Vance MA et al (2007) No gene is an island: the flip-flop phenomenon. Am J Hum Genet 80:531–538
Ma X, Sun J, Yao J et al (2007) A quantitative association study between schizotypal traits and COMT, PRODH and BDNF genes in a healthy Chinese population. Psychiatry Res 153:7–15
Mannisto PT, Kaakkola S (1999) Catechol-O-methyltransferase (COMT): biochemistry, molecular biology, pharmacology, and clinical efficacy of the new selective COMT inhibitors. Pharmacol Rev 51:593–628
Mattay VS, Goldberg TE, Fera F et al (2003) Catechol O-methyltransferase val158-met genotype and individual variation in the brain response to amphetamine. Proc Natl Acad Sci USA 100:6186–6191
Meyer-Lindenberg A, Nichols T, Callicott JH et al (2006) Impact of complex genetic variation in COMT on human brain function. Mol Psychiatry 11:877–882
Mier D, Kirsch P, Meyer-Lindenberg A (2009) Neural substrates of pleiotropic action of genetic variation in COMT: a meta-analysis. Mol Psychiatry. doi:10.1038/mp. 2009.36
Molero P, Ortuno F, Zalacain M et al (2007) Clinical involvement of catechol-O-methyltransferase polymorphisms in schizophrenia spectrum disorders: influence on the severity of psychotic symptoms and on the response to neuroleptic treatment. Pharmacogenomics J 7:418–426
Morissette M, Di Paolo T (1993) Sex and estrous cycle variations of rat striatal dopamine uptake sites. Neuroendocrinology 58:16–22
Mozley LH, Gur RC, Mozley PD et al (2001) Striatal dopamine transporters and cognitive functioning in healthy men and women. Am J Psychiatry 158:1492–1499
Munro CA, McCaul ME, Wong DF et al (2006) Sex differences in striatal dopamine release in healthy adults. Biol Psychiatry 59:966–974
Murphy DGM, DeCarli C, McIntosh AR et al (1996) Sex differences in human brain morphometry and metabolism: an in vivo quantitative magnetic resonance imaging and positron emission tomography study on the effect of aging. Arch Gen Psychiatry 53:585–594
Nackley A, Shabalina S, Tchivileva I et al (2006) Human catechol-O-methyl-transferase haplotypes modulate protein expression by altering mRNA secondary structure. Science 314:1930–1933
Nash MW, Huezo-Diaz P, Williamson RJ et al (2004) Genome-wide linkage analysis of a composite index of neuroticism and mood-related scales in extreme selected sibships. Hum Mol Genet 13:2173–2182
Nestadt G, Lan T, Samuels J et al (2000) Complex segregation analysis provides compelling evidence for a major gene underlying obsessive-compulsive disorder and for heterogeneity by sex. Am J Hum Genet 67:1611–1616
Nolan KA, Volavka J, Czobor P et al (2000) Suicidal behavior in patients with schizophrenia is related to COMT polymorphism. Psychiatr Genet 10:117–124

O'Hara R, Miller E, Liao CP et al (2006) COMT genotype, gender and cognition in community-dwelling, older adults. Neurosci Lett 409:205–209
Olsson CA, Anney RJ, Lotfi-Miri M et al (2005) Association between the COMT Val158Met polymorphism and propensity to anxiety in an Australian population-based longitudinal study of adolescent health. Psychiatr Genet 15:109–115
Ono H, Shirakawa O, Nushida H et al (2004) Association between catechol-O-methyltransferase functional polymorphism and male suicide completers. Neuropsychopharmacology 29: 1374–1377
Oosterhuis BE, LaForge KS, Proudnikov D et al (2008) Catechol-O-methyltransferase (COMT) gene variants: possible association of the Val158Met variant with opiate addiction in Hispanic women. Am J Med Genet B Neuropsychiatr Genet 147B:793–798
O'Tuathaigh CMP, Babovic D, O'Meara G et al (2007) Susceptibility genes for schizophrenia: characterisation of mutant mouse models at the level of phenotypic behaviour. Neurosci Biobehav Rev 31:60–78
Patsopoulos NA, Tatsioni A, Ioannidis JP (2007) Claims of sex differences: an empirical assessment in genetic associations. JAMA 298:880–893
Pelayo-Terán JM, Crespo-Facorro B, Carrasco-Marín E et al (2008) Catechol-O-methyltransferase Val158Met polymorphism and clinical characteristics in first episode non-affective psychosis. Am J Med Genet B Neuropsychiatr Genet 147B:550–556
Philippu G, Hoo JJ, Milech U et al (1981) Catechol-O-methyltransferase of erythrocytes in patients with endogenous psychoses. Psychiatry Res 4:139–146
Piccinelli M, Wilkinson G (2000) Gender differences in depression – critical review. Br J Psychiatry 177:486–492
Pohjalainen T, Rinne JO, Nogren K et al (1998) Sex differences in the striatal dopamine D_2 receptor binding characteristics in vivo. Am J Psychiatry 155:768–773
Pooley EC, Fineberg N, Harrison PJ (2007) The met^{158} allele of catechol-O-methyltransferase (COMT) is associated with obsessive-compulsive disorder in men: case-control study and meta-analysis. Mol Psychiatry 12:556–561
Poyurovsky M, Michaelovsky E, Frisch A, Knoll G, Amir I, Finkel B, Buniak F, Hermesh H, Weizman R (2005) COMT Val158Met polymorphism in schizophrenia with obsessive-compulsive disorder: a case-control study. Neurosci Lett. 389:21–4
Preece P, Cairns NJ (2003) Quantifying mRNA in postmortem human brain: influence of gender, age at death, postmortem interval, brain pH, agonal state and inter-lobe mRNA variance. Mol Brain Res 118:60–71
Qian Q, Wang Y, Zhou R, Li J, Wang B, Glatt S, Faraone SV (2003) Family-based and case-control association studies of catechol-O-methyltransferase in attention deficit hyperactivity disorder suggest genetic sexual dimorphism. Am J Med Genet B Neuropsychiatr Genet. 118B:103–9
Quednow BB, Schmechtig A, Ettinger U et al (2009) Sensorimotor gating depends on polymorphisms of the serotonin-2A receptor and catechol-O-methyltransferase, but not on neuregulin-1 Arg38Gln genotype: a replication study. Biol Psychiatry 66:614–620
Robinson DS, Sourkes TL, Nies A et al (1977) Monoamine metabolism in human brain. Arch Gen Psychiatry 34:89–92
Rothe C, Koszycki D, Bradwejn J et al (2006) Association of the Val158Met caetchol-O-methyltransferase genetic polymorphism with panic disorder. Neuropsychopharmacol 31:2237–2242
Rubinow DR, Schmidt PJ (1996) Androgens, brain, and behavior. Am J Psychiatry 153:974–984
Rybakowski JK, Borkowska A, Czerski PM et al (2006) Performance on the Wisconsin Card Sorting Test in schizophrenia and genes of dopaminergic inactivation (COMT, DAT, NET). Psychiatry Res 143:13–19
Salih SM, Salama SA, Jamaluddin M et al (2008) Progesterone-mediated regulation of catechol-O-methyl transferase expression in endometrial cancer cells. Reprod Sci 15(2):210–220

Sazci A, Ergul E, Kucukali I et al (2004) Catechol-O-methyltransferase gene Val108/158Met polymorphism, and susceptibility to schizophrenia: association is more significant in women. Mol Brain Res 132:51–56
Sazci A, Ergul E, Kucukali I et al (2005) Association of the C677T and A1298C polymorphisms of methylenetetrahydrofolate reductase gene with schizophrenia: association is significant in men but not in women. Progr Neuropsychopharmacol Biol Psychiatry 29:1113–1123
Seeman MV (1997) Psychopathology in women and men: focus on female hormones. Am J Psychiatry 154:1641–1647
Shifman S, Bronstein M, Sternfeld M et al (2002) A highly significant association between a COMT haplotype and schizophrenia. Am J Hum Genet 71:1296–1302
Shifman S, Bronstein M, Sternfeld M et al (2004) COMT: a common susceptibility gene in bipolar disorder and schizophrenia. Am J Med Genet Neuropsychiatr Genet 128B:61–64
Shifman S, Johannesson M, Bronstein M et al (2008) Genome-wide association identifies a common variant in the reelin gene that increases the risk of schizophrenia only in women. PLoS Genet 4:e28
Smolka MN, Schumann G, Wrase J et al (2005) Catechol-O-methyltransferase val^{158}met genotype affects processing of emotional stimuli in the amygdala and prefrontal cortex. J Neurosci 25:836–842
Stein MB, Fallin MD, Schork NJ et al (2005) COMT polymorphisms and anxiety related personality traits. Neuropsychopharmacology 30:2092–2102
Stone JL, Merriman B, Cantor RM et al (2004) Evidence for sex-specific risk alleles in autism spectrum disorder. Am J Hum Genet 75:1117–1123
Strous RD, Ritsner MS, Adler S et al (2009) Improvement of aggressive behavior and quality of life impairment following S-adenosyl-methionine (SAM-e) augmentation in schizophrenia. Eur Neuropsychopharmacol 19:14–22
Sweet RA, Devlin B, Pollock BG et al (2005) Caetchol-O-methyltransferase haplotypes are associated with psychosis in Alzheimer disease. Mol Psychiatry 10:1026–1036
Talkowski ME, Kirov G, Bamne M et al (2008) A network of dopaminergic gene variations implicated as risk factors for schizophrenia. Hum Mol Genet 17:747–758
Tammimäki A, Forsberg MM, Karayiorgou M et al (2008) Increase in free choice oral ethanol self-administration in catechol-o-methyltransferase gene-disrupted male mice. Basic Clin Pharmacol Toxicol 103:297–304
Tamminga CA (1997) Gender and schizophrenia. J Clin Psychiatry 58(suppl 15):33–37
Tan HY, Chen Q, Sust S et al (2007) Epistasis between catechol-O-methyltransferase and type II metabotropic glutamate receptor 3 genes on working memory brain function. Proc Natl Acad Sci USA 104:12536–12541
Tsai SJ, Gau YT, Hong CJ et al (2009) Sexually dimorphic effect of catechol-O-methyltransferase val158met polymorphism on clinical response to fluoxetine in major depressive patients. Affect Disord 113:183–187
Tunbridge EM, Bannerman DM, Sharp T et al (2004a) Catechol-O-methyltransferase inhibition improves set-shifting performance and elevates stimulated dopamine release in the rat prefrontal cortex. J Neurosci 24:5331–5335
Tunbridge E, Burnet PWJ, Sodhi MS et al (2004b) Catechol-o-methyltransferase (COMT) and proline dehydrogenase (PRODH) mRNAs in the dorsolateral prefrontal cortex in schizophrenia, bipolar disorder, and major depression. Synapse 51:112–118
Tunbridge EM, Harrison PJ, Weinberger DR (2006a) Catechol-o-methyltransferase, cognition, and psychosis: Val^{158}Met and beyond. Biol Psychiatry 60:141–151
Tunbridge EM, Weinberger DR, Harrison PJ (2006b) A novel protein isoform of catechol O-methyltransferase (COMT): brain expression analysis in schizophrenia and bipolar disorder and effect of Val^{158}Met genotype. Mol Psychiatry 11:116–117
Tunbridge EM, Weickert CS, Kleinman JE et al (2007a) Catechol-o-methyltransferase enzyme activity and protein expression in human prefrontal cortex across the postnatal lifespan. Cereb Cortex 17:1206–1212

Tunbridge EM, Lane TA, Harrison PJ (2007b) Expression of multiple catechol-o-methyltransferase (COMT) mRNA variants in human brain. Am J Med Genet Neuropsychiatr Genet 144B: 834–839

Tunbridge EM, Harrison PJ, Warden D et al (2008) Human plasma homocysteine levels are associated with the catechol-O-methyltransferase Val^{158}Met polymorphism. Am J Med Genet B Neuropsychiatr Genet 147B:996–999

Vawter MP, Evans S, Choudary P et al (2004) Gender-specific gene expression in post-mortem human brain: localization to sex chromosomes. Neuropsychopharmacology 29:373–384

Wang Z, Xiao Z, Inslicht SS et al (2009) Low expression of catecholamine-O-methyl-transferase gene in obsessive-compulsive disorder. J Anxiety Disord 23:660–664

Weinberger DR, Egan MF, Bertolino A et al (2001) Prefrontal neurons and the genetics of schizophrenia. Biol Psychiatry 50:825–844

Weiss LA, Pan L, Abney M et al (2006) The sex-specific genetic architecture of quantitative traits in humans. Nat Genet 38:218–222

Weiss EM, Stadelmann E, Kohler CG et al (2007) Differential effect of catechol-O-methyltransferase Val158Met genotype on emotional recognition abilities in healthy men and women. J Int Neuropsychol Soc 13:881–887

Williams NM, Glaser B, Norton N et al (2008) Strong evidence that GNB1L is associated with schizophrenia. Hum Mol Genet 17:555–566

Williams-Gray CH, Hampshire A, Robbins TW et al (2007) Catechol O-methyltransferase val$^{158\text{-}}$met genotype influences frontoparietal activity during planning in patients with Parkinson's disease. J Neurosci 27:4832–4838

Woo J-M, Yoon K-S, Choi Y-H et al (2004) The association between panic disorder and the L/L genotype of catechol-O-methyltransferase. J Psychiatr Res 38:365–370

Worda C, Sator MO, Schneeberger C et al (2003) Influence of the catechol-O-methyltransferase (COMT) codon 158 polymorphism on estrogen levels in women. Hum Reprod 18:262–266

Wray NR, James MR, Dumenil T et al (2008) Association study of candidate variants of COMT with neuroticism, anxiety and depression. Am J Med Genet B Neuropsychiatr Genet 147B:1314–1318

Xiao L, Becker JB (1994) Quantitative microdialysis determination of extracellular striatal dopamine concentration in male and female rats: effects of estrous cycle and gonadectomy. Neurosci Lett 180:155–158

Xie T, Ho SL, Ramsden D (1999) Characterization and implications of estrogenic down-regulation of human catechol-O-methyltransferase gene transcription. Mol Pharmacol 56:31–38

Yavich L, Forsberg MM, Karayiorgou M et al (2007) Site-specific role of catechol-O-methyltransferase in dopamine overflow within prefrontal cortex and dorsal striatum. J Neurosci 27:10196–10209

Zhang K, Zheng Z, Gao X et al (2007) Possible relationship between the COMT gene ValMet polymorphism and psychometric IQ in girls of the Qinba region in China. Neuropsychobiology 56:98–103

Zhu BT (2002) Catechol-O-methyltranferase (COMT)-mediated methylation metabolism of endogenous bioactive catechols and modulation by endobiotics and xenobiotics: importance in pathophysiology and pathogenesis. Curr Drug Metab 3:321–329

Zinkstok J, Schmitz N, van Amelsvoort T et al (2006) The COMT val158met polymorphism and brain morphometry in healthy young adults. Neurosci Lett 405:34–39

Sex Differences and Hormonal Influences in Human Sensorimotor Gating: Implications for Schizophrenia

Veena Kumari

Contents

Abstract Prepulse inhibition (PPI) of the startle response serves to prevent the interruption of ongoing perceptual and early sensory analysis and provides a simple operational measure of sensorimotor gating. In line with postulated deficits in early stages of information processing, PPI is disrupted in schizophrenia. PPI is considered a valid candidate for an endophenotypic marker in genetic studies of schizophrenia and has also been extensively used in translational research. Importantly, there are well-replicated sex differences and menstrual phase effects in prepulse-elicited startle modulation of nonclinical young populations. Lack of knowledge about the precise roles of sex differences and hormonal effects in prepulse-elicited startle modulation and in the schizophrenia disease process presents a stumbling block to continuous progress in this field. This chapter reviews a wealth of data demonstrating sex and hormonal influences in prepulse-elicited startle modulation and considers their implications for our understanding of the pathophysiology, genetics, and potential treatments of schizophrenia.

V. Kumari
Department of Psychology, Institute of Psychiatry, King's College London, De Crespigny Park, London SE5 8AF, UK
e-mail: veena.kumari@kcl.ac.uk

J.C. Neill and J. Kulkarni (eds.), *Biological Basis of Sex Differences in Psychopharmacology*,
Current Topics in Behavioral Neurosciences 8, DOI 10.1007/7854_2010_117,

Keywords Estrogen · Hormones · Prepulse facilitation · Prepulse inhibition · Progesterone · Schizophrenia · Sex · Startle response

1 Introduction

Schizophrenia is a severe neuropsychiatric disorder of unknown etiology. The condition is clinically heterogeneous and often results in disabling cognitive, perceptual, and emotional symptoms. The symptoms are generally classified as positive (e.g., hallucinations and delusions), negative (e.g., anhedonia, thought paucity), and cognitive (e.g., thought disorder, bizarre thinking). There are well-established sex differences in onset, prognosis, and course of this disorder (Hafner 2003). Current pharmacological treatments, largely involving blockade of dopamine D2 receptors (Kapur and Remington 2001; Guillin et al. 2007), are effective in reducing the acute symptoms but provide no cure. Accordingly, the need for a better understanding of the etiology and pathophysiology of schizophrenia and the development of novel treatments persists. The future of drug development in schizophrenia depends on many experimental strategies as well as on serendipity (Carpenter and Koenig 2008; Javitt et al. 2008). Over the last 30–40 years, translational models, such as prepulse inhibition (PPI) of the startle response, have played an important role in the strategic effort to develop and characterize new treatments. The aim of this chapter is to review sex differences and menstrual phase effects in PPI of nonclinical young populations and discuss their implications with a view to advance our understanding of the pathophysiology, genetics, and treatment of schizophrenia.

2 Prepulse-Elicited Startle Modulation (PESM) as a Measure of Sensorimotor Gating

The startle reflex consists of a set of reflexive, involuntary responses to a sudden, intense stimulus. It is known to exhibit several forms of plasticity which are remarkably similar in animals and humans, examples being habituation (Hoffman and Searle 1968) and fear potentiation (Brown et al. 1951). A further example of startle plasticity is that the startle response can be modified reliably by presenting a more innocuous stimulus (prepulse) before the strong startle-eliciting stimulus (pulse) (Graham 1975). When the time from prepulse onset to pulse onset, or stimulus onset asynchrony (SOA), is between 30 and 500 ms, such modification will appear as inhibition (PPI), evident as the attenuation of the startle response (Fig. 1). However, with a longer interval between the prepulse and the pulse, such modification will appear as facilitation of the startle response (PPF) (Fig. 1). PESM is increasingly being used as a measure of early and late information processing in experimental animals as well as in clinical and nonclinical human populations.

PPI is considered to provide an operational index of sensorimotor gating: while information processing resources are targeted at the prepulse, any incoming

Fig. 1 PPI and PPF effects

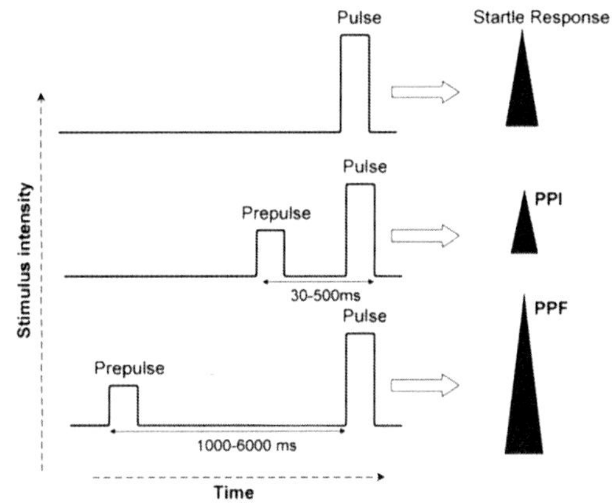

information (i.e., the pulse) is attended to a reduced level, thereby protecting the processing of the initial stimulus (i.e., the prepulse). A reduced ability to gate (or filter out) such stimulus interference (i.e., reduced PPI) has been associated with sensory overstimulation and confusion (Geyer et al. 1990). Animal studies show that PPI is mediated by brain stem circuits involving the inferior colliculus, pedunculopontine tegmental nucleus, laterodorsal tegmental nucleus, substantia nigra pars reticulata, and caudal pontine reticular nucleus (Fendt et al. 2001) and modulated by forebrain circuits involving the prefrontal cortex, thalamus, hippocampus, amygdala, nucleus accumbens, striatum, ventral pallidum, globus pallidus, and subpallidal efferents to the pedunculopontine nucleus (Swerdlow et al. 2001, 2008). Although the neural substrates of PPI may vary somewhat between animals and humans (Swerdlow et al. 2008), imaging studies (Hazlett and Buchsbaum 2001; Kumari et al. 2003b, 2005a, 2007a; Postma et al. 2006; Hazlett et al. 2008) support the involvement of similar brain regions in modulation of human PPI. PPF is a relatively less well-studied phenomenon. It may reflect sustained attention (Dawson et al. 1997), sensory enhancement linked with modality (Anthony 1985), or a different aspect of the same mechanism underlying PPI (Kumari et al. 2003a). The neural substrates of PPF are not well studied even in animals. In the only study (Neuner et al. 2010) to have investigated the neural substrates of PPF (in healthy men only), many regions, consistent with previous PPI studies, were found to be activated for both PPI and PPF.

3 PESM Deficits in Schizophrenia

Graham (1975) suggested that "startle modulation might provide a powerful technique for probing what underlies the normal processing of information and especially for probing processing characteristics of relatively inaccessible subjects" (p. 238). Applying this to schizophrenia patients, Braff et al. (1978) demonstrated impaired PPI confirming theorized deficits in early stages of information processing in this population. Many subsequent studies have replicated and refined this finding (Braff et al. 2001; Kumari et al. 2004, 2007b, 2008c; Swerdlow et al. 2006a; Takahashi et al. 2008). It has been hypothesized that a breakdown in the gating system causes sensory overload (Gottschalk et al. 1972) with secondary cognitive fragmentation, thought disorder, and possibly other psychotic symptoms in this clinical population (Braff and Geyer 1990). PPF is not as widely studied as PPI but is also impaired in schizophrenia, especially in female patients compared to female controls (Kumari et al. 2004).

PPI, given its high temporal stability (Abel et al. 1998; Cadenhead et al. 1999) and amenability to cross-species comparisons (Swerdlow et al. 1994), is one of the leading animal paradigms dominating research on antipsychotic drug activity and in other comparative psychopharmacological studies (Geyer et al. 2001; Swerdlow et al. 2006b; Talledo et al. 2009). Attesting the usefulness of PPI as a translational model, an emerging body of evidence suggests that second generation antipsychotics may attenuate the PPI deficit in schizophrenia (Kumari et al. 2000, 2007a; Swerdlow et al. 2006a).

PPI has also been recommended as a valid candidate for an endophenotypic marker in genetic studies of schizophrenia for several reasons. First, PPI is heritable with genetic factors contributing to 30–50% of the variance (Greenwood et al. 2007; Hasenkamp et al. 2010). Second, PPI is reduced not only in patients but also in schizophrenia-spectrum populations, for example, in unaffected biological relatives of schizophrenia patients (Cadenhead et al. 2000; Kumari et al. 2005b) and people with schizotypal personality disorder (Cadenhead et al. 2000) or high scores on psychometric measures of psychosis-proneness (Kumari et al. 1997, 2008b; Evans et al. 2005). Third, PPI is influenced by schizophrenia-relevant single nucleotide polymorphisms (SNPs) within the dopamine, serotonin, and acetylcholine systems (Quednow et al. 2008, 2009; Roussos et al. 2008a, b; Petrovsky et al. 2010). However, some of the genetically mediated effects in PPI found reliably in healthy men may not be present in healthy women who are tested without regard to their menstrual cycle phase. For example, catechol *O*-methyltransferase Val158Met SNP has been reported to significantly influence PPI in independent samples of healthy men (Roussos et al. 2008b; Quednow et al. 2009) but not of healthy women (Montag et al. 2008; Quednow et al. 2009). The most likely reason for this is menstrual cycle-related variability in PPI of healthy young women (next section). It is becoming increasing obvious that sex differences and hormonal influences need to be considered in future applications of PESM models in the context of schizophrenia and other disorders that are characterized by disrupted sensorimotor gating.

4 Sex Differences and Menstrual Phase Effects in PESM

PPI is sexually dimorphic (Fig. 2) with several studies reporting less PPI in healthy young women, when tested regardless of where they are in their menstrual cycle, than in healthy young men (Swerdlow et al. 1993, 1999; Abel et al. 1998;

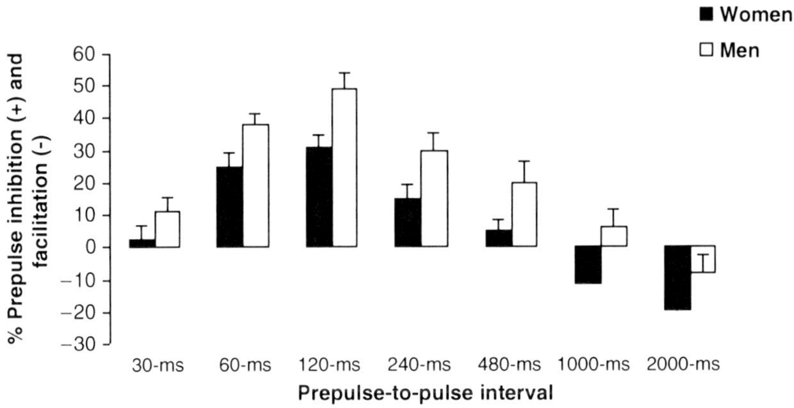

Fig. 2 Sex differences in PPI and PPF (data from Aasen et al. 2005). PPI and PPF = ([$a - b$])/$a \times 100$, where "a" = amplitude over pulse-alone trials, and "b" = amplitude over prepulse trials (PPF is expressed as a negative value). PPI is seen to increase from 30-ms through 60-ms to 120-ms prepulse-to-pulse intervals, then to decrease and turn into PPF at 1,000-ms prepulse-to-pulse interval, especially in women. More PPI is present in men and more PPF in women

Kumari et al. 2003a, 2004, 2008a; Aasen et al. 2005). The sex difference in PPI of nonclinical young people remains true after possible confounds, such as cigarette smoking and personality, are controlled for (Swerdlow et al. 1999; Aasen et al. 2005). A sex effect in PPI (females < males) has also been reported in rats (Koch 1998; Faraday et al. 1999) and mice (Ison and Allen 2007).

PPI is sensitive to menstrual cycle status in healthy women, with more PPI observed during the early follicular phase relative to the luteal phase in both cross-sectional (Swerdlow et al. 1997) and within-subjects studies (Jovanovic et al. 2004; Kumari et al. 2010). Healthy pregnant women in their third trimester, when the levels of both estrogen and progesterone are roughly 50 times and 10 times (respectively) the levels in normally cycling women, have lower PPI in comparison with healthy postpartum women (Kask et al. 2008). Luteal phase young women show lower PPI in comparison with postmenopausal women (Bannbers et al. 2010). The sex effect in PPI is not present in children under 8 years of age (Ornitz et al. 1991), in postmenopausal women compared to age-matched men (Kumari et al. 2008a) or aged mice (Ison and Allen 2007).

Sex differences in PPF are less widely studied. Previous studies from our laboratory suggest that healthy young menstruating women display higher PPF than men (Kumari et al. 2003a, 2004, 2008a; Aasen et al. 2005) (Fig. 2) and greater PPF during the luteal relative to the early follicular phase (Kumari et al. 2010) (Fig. 3).

5 Hormonal Influences in PESM

Lower PPI during the luteal phase, compared to the early follicular phase, in healthy young women has been proposed to be caused by high levels of the ovarian hormone, estrogen (Swerdlow et al. 1997; Jovanovic et al. 2004). PPI is also

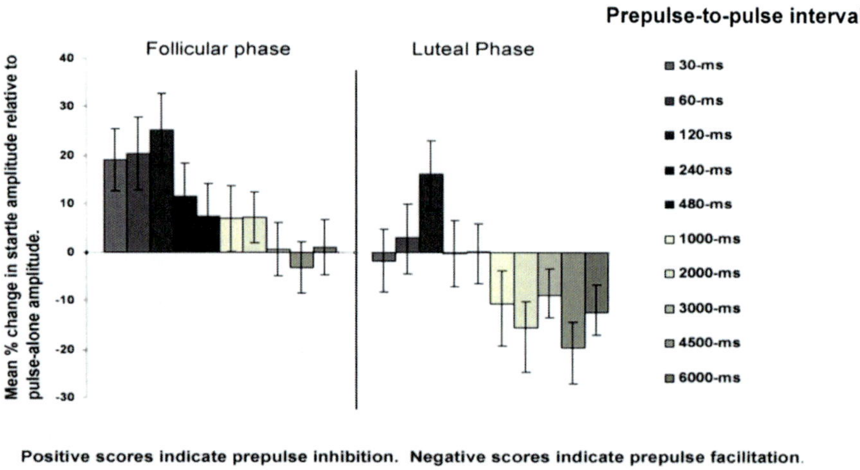

Fig. 3 Menstrual phase effects in PPI and PPF (data from Kumari et al. 2010). PPI increases from 30-ms through 60-ms to 120-ms prepulse intervals during both the follicular and the luteal phases, then decreases and turns into PPF at 1,000-ms especially during the luteal phase. Significantly more PPI is seen during the follicular phase and significantly more PPF during the luteal phase

reduced during periods of high estrogen in experimental animals (Vaillancourt et al. 2002). Estrogen influences various neurochemical activities including dopaminergic activity in the nucleus accumbens, an area critical for PPI both in experimental animals (Swerdlow et al. 2001) and in humans (Kumari et al. 2003b, 2005a). Administration of dopamine agonists, such as amphetamine or apomorphine, disrupts PPI in both rodents and human participants (Geyer et al. 2001; Swerdlow et al. 2003). Estradiol administration to ovariectomized rats has been reported to induce a decrease in the number of inhibitory synaptic inputs, an increase in the number of excitatory synapses, and an enhancement of the frequency of neuronal firing (Parducz et al. 2002). Since these findings fit the observed pattern of sex differences in PPI and PPF, we earlier suggested (Aasen et al. 2005) that women might show reduced PPI during the high estrogen phase of the menstrual cycle because estrogen elevates excitatory neuronal firing at the gate, inhibiting the gate to close as efficiently as men at short prepulse-to-pulse intervals and express increased PPF since the gate opens more efficiently at the long prepulse-to-pulse intervals compared to men. However, recent studies from our (Kumari et al. 2008a, 2010) and other laboratories (Talledo et al. 2009) have failed to detect a direct relationship between PPI and varying estrogen levels in healthy women, while in an earlier study estrogen administration (2 mg) prevented the disruption of PPI by buspirone, the serotonin-1A (5-HT1A) receptor partial agonist, but had no influence on PPI when given on its own to early follicular healthy women (Gogos et al. 2006).

Perhaps progesterone, another ovarian hormone, plays a role in menstrual cycle-related variability in PPI as suggested by findings of our most recent study (Kumari et al. 2010). This study showed a smaller decrease in PPI from the follicular phase to the luteal phase in women who had a larger increase in progesterone.

Progesterone shows marked fluctuations over the menstrual cycle (Marshall 2001) and possesses psychotropic properties in addition to its role in reproductive endocrinology (Rupprecht 2003). It is known to modulate the release of dopamine (Dluzen and Ramirez 1990; Ramirez and Zheng 1996) with biphasic effects, initially increasing but ultimately decreasing basal and amphetamine-stimulated dopamine release (Dluzen and Ramirez 1984). In rodents, progesterone reduces amphetamine-induced stereotypy (Michanek and Meyerson 1982). It is also implicated in the modulation of PPI (Rupprecht et al. 1999; Gogos and Van den Buuse 2004) and reverses apomorphine-induced disruption of PPI in rodents (Rupprecht et al. 1999). In healthy women, estrogen enhances the response to stimulant drugs, but this effect is masked in the presence of progesterone in healthy women (Justice and de Wit 1999). PPI, however, is sensitive not only to dopaminergic but also to serotonergic, glutamatergic, and cholinergic systems (Geyer et al. 2001; Swerdlow et al. 2008). Progesterone too, in addition to dopamine, affects other neurotransmitter systems. It is known to act as a functional antagonist at 5-HT3 receptors (Wetzel et al. 1998) and to have a role in the control of nicotinic cholinergic receptors (Valera et al. 1992). At present, it is unclear which of these systems might be most pertinent to the effect of progesterone in PPI of healthy young females.

6 Consideration of Sex and Hormonal Influences in PSEM in the Context of Schizophrenia

The first issue deserving consideration is the roles of ovarian hormones in the schizophrenia disease process. A later age of illness onset, less severe forms of the illness, a superior response to antipsychotics, as well as better functional and social outcomes are reported for women than men with schizophrenia (Castle and Murray 1991; Faraone et al. 1994; Castle et al. 1995), supposedly due to a neuroprotective role of estrogen in women (Hafner et al. 1998; Kulkarni 2009). Female schizophrenia patients display greater symptom severity during the periods of low estrogen (e.g., postpartum) and lower symptom severity during the periods of high estrogen (e.g., pregnancy) (Riecher-Rossler et al. 1994). These observations showing "reduced symptoms" with "high estrogen" appear inconsistent with the earlier proposal (Swerdlow et al. 1997) that high estrogen during the luteal phase causes lower PPI in healthy women. It is, of course, possible that estrogen–PPI relationship follows an inverted U pattern with both very low and very high levels producing lower PPI or, as mentioned earlier, is modulated by another ovarian hormone, progesterone. Our recent finding did suggest a role for progesterone, more specifically an antipsychotic-like PPI-restorative action of progesterone, during the luteal phase in PPI of healthy young women (Kumari et al. 2010). Considering this finding in the context of schizophrenia, women are more susceptible to the onset of schizophrenia after menopause and during the postpartum period (Hafner et al. 1993). This effect can be attributed to a drop in progesterone levels

(Shulman and Tibbo 2005). Studies have also reported high progesterone levels in unmedicated chronic patients in response to metabolic stress (Breier and Buchanan 1992) but normal progesterone levels in medicated patients with early psychosis (Oades and Schepker 1994) as well medicated chronic schizophrenia patients (Taherianfard and Shariaty 2004). Progesterone has been proposed to act as an endogenous antipsychotic and serve to restore normal function during times of stress (Shulman and Tibbo 2005). Further support for this notion comes from the reports that the progesterone metabolite 3α, 5α-THP produces a behavioral profile similar to that of dopamine receptor antagonists by increasing GABAergic tone in rodents (Khisti et al. 1998, 2002). Other data in rodents show increases in cortical progesterone and/or 3α, 5α-THP concentrations with olanzapine (Marx et al. 2000, 2003) and clozapine (Barbaccia et al. 2001; Marx et al. 2003), but not with haloperidol (Barbaccia et al. 2001). Atypical antipsychotic-induced increases in progesterone have been suggested to contribute to clinical benefits of these drugs (Barbaccia 2004; Marx et al. 2006). Studies so far have been incapable of disentangling the effect of estrogen and progesterone in human PESM.

The second issue is whether the "normal" reduction in PPI of healthy young women during the luteal phase means the same as "deficient" PPI in schizophrenia and related populations. As suggested earlier (Kumari et al. 2003a), reduced PPI in healthy women during the luteal phase "may not be a simple reduction but rather a shift of the inhibition/facilitation curve in the direction of facilitation in women, relative to men." At present, very few studies have examined both PPI and PPF in schizophrenia patients and not much is known about the pharmacology and neuroanatomy of PPF even in animals. It is plausible that schizophrenia is associated not only with impaired PPI but also with a defect in mechanisms that underlie a smooth transition of PPI into PPF with increasing prepulse-to-pulse intervals in women during the luteal phase and appear to be modulated by ovarian hormones.

7 Conclusions and Directions for Future Research

Given the importance of PPI as an animal model of schizophrenia and known sex differences in prognosis and course of this order, it is vital to uncover the biological basis of sex differences and menstrual cycle-related variations in PESM. While rodent studies in this area provide valuable information for hypothesis development, extrapolating their findings to humans is challenging because neuroanatomically "the further forward one moves in the brain, the greater the anatomical and functional differences between rodents and humans" (Swerdlow et al. 2008). Due to poor understanding of sex and menstrual phase effects in human PESM, the majority of imaging and pharmacological studies so far have used mostly, if not exclusively, men. Their findings are unlikely to be applicable to healthy or ill women. Future studies investigating pharmacological and treatment effects using a prepulse modification paradigm in normal and clinical populations of both sexes would benefit from consideration of sex differences and menstrual phase effects and assessments of both PPI and PPF. Future research should also aim to clarify the roles of varying ovarian hormones, especially estrogen and progesterone, in menstrual phase-related variability in PESM and establish whether menstrual phase-related variability in PPI and PPF is mediated by estrogen alone, progesterone alone, or via their interaction. The knowledge gained from such research will enable further studies examining the therapeutic potential of hormones for treatment as well as prevention (e.g., postpartum psychosis) for disorders characterized by disrupted sensorimotor gating and, wherever relevant, combining them with genetic profiles.

Acknowledgment The author acknowledges funding support by The British Academy (SG-39919) for her studies of sex differences and menstrual phase effects in PESM.

References

Aasen I, Kolli L, Kumari V (2005) Sex effects in prepulse inhibition and facilitation of the acoustic startle response: implications for pharmacological and treatment studies. J Psychopharmacol 19:39–45

Abel K, Waikar M, Pedro B, Hemsley D, Geyer M (1998) Repeated testing of prepulse inhibition and habituation of the startle reflex: a study in healthy human controls. J Psychopharmacol 12:330–337

Anthony BJ (1985) In the blink of an eye: implications of reflex modification for information processing. In: Ackles P et al (eds) Advances in psychophysiology, vol 1. JAI Press, Greenwich, CT, pp 167–218

Bannbers E, Kask K, Wikstrom J, Sundstrom Poromaa I (2010) Lower levels of prepulse inhibition in luteal phase cycling women in comparison with postmenopausal women. Psychoneuroendocrinology 35:422–429

Barbaccia ML (2004) Neurosteroidogenesis: relevance to neurosteroid actions in brain and modulation by psychotropic drugs. Crit Rev Neurobiol 16:67–74

Barbaccia ML, Affricano D, Purdy RH, Maciocco E, Spiga F, Biggio G (2001) Clozapine, but not haloperidol, increases brain concentrations of neuroactive steroids in the rat. Neuropsychopharmacology 25:489–497

Braff D, Stone C, Callaway E, Geyer M, Glick I, Bali L (1978) Prestimulus effects on human startle reflex in normals and schizophrenics. Psychophysiology 15(4):339–343

Braff DL, Geyer MA (1990) Sensorimotor gating and schizophrenia. Human and animal model studies. Arch Gen Psychiatry 47:181–188

Braff DL, Geyer MA, Swerdlow NR (2001) Human studies of prepulse inhibition of startle: normal subjects, patient groups, and pharmacological studies. Psychopharmacology (Berl) 156:234–258

Breier A, Buchanan RW (1992) The effects of metabolic stress on plasma progesterone in healthy volunteers and schizophrenic patients. Life Sci 51:1527–1534

Brown JS, Kalish HI, Farber IE (1951) Conditioned fear as revealed by magnitude of startle response to an auditory stimulus. J Exp Psychol 41:317–328

Cadenhead KS, Carasso BS, Swerdlow NR, Geyer MA, Braff DL (1999) Prepulse inhibition and habituation of the startle response are stable neurobiological measures in a normal male population. Biol Psychiatry 45:360–364

Cadenhead KS, Swerdlow NR, Shafer KM, Diaz M, Braff DL (2000) Modulation of the startle response and startle laterality in relatives of schizophrenic patients and in subjects with schizotypal personality disorder: evidence of inhibitory deficits. Am J Psychiatry 157:1660–1668

Carpenter WT, Koenig JI (2008) The evolution of drug development in schizophrenia: past issues and future opportunities. Neuropsychopharmacology 33:2061–2079
Castle DJ, Murray RM (1991) The neurodevelopmental basis of sex differences in schizophrenia. Psychol Med 21:565–575
Castle DJ, Abel K, Takei N, Murray RM (1995) Gender differences in schizophrenia: hormonal effect or subtypes? Schizophr Bull 21:1–12
Dawson ME, Schell AM, Swerdlow NR, Filion DL (1997) Cognitive, clinical, and neuropsychological implications of startle modulation. In: Lang P et al (eds) Attention and orienting: sensory and motivational processes. Lawrence Erlbaum Associates, Hillsdale, NJ, pp 257–279
Dluzen DE, Ramirez VD (1984) Bimodal effect of progesterone on in vitro dopamine function of the rat corpus striatum. Neuroendocrinology 39:149–155
Dluzen DE, Ramirez VD (1990) In vitro progesterone modulates amphetamine-stimulated dopamine release from the corpus striatum of castrated male rats treated with estrogen. Neuroendocrinology 52:517–520
Evans LH, Gray NS, Snowden RJ (2005) Prepulse inhibition of startle and its moderation by schizotypy and smoking. Psychophysiology 42:223–231
Faraday MM, O'Donoghue VA, Grunberg NE (1999) Effects of nicotine and stress on startle amplitude and sensory gating depend on rat strain and sex. Pharmacol Biochem Behav 62:273–284
Faraone SV, Chen WJ, Goldstein JM, Tsuang MT (1994) Gender differences in age at onset of schizophrenia. Br J Psychiatry 164:625–629
Fendt M, Li L, Yeomans JS (2001) Brain stem circuits mediating prepulse inhibition of the startle reflex. Psychopharmacology (Berl) 156:216–224
Geyer MA, Swerdlow NR, Mansbach RS, Braff DL (1990) Startle response models of sensorimotor gating and habituation deficits in schizophrenia. Brain Res Bull 25:485–498
Geyer MA, Krebs-Thomson K, Braff DL, Swerdlow NR (2001) Pharmacological studies of prepulse inhibition models of sensorimotor gating deficits in schizophrenia: a decade in review. Psychopharmacology (Berl) 156:117–154
Gogos A, Van den Buuse M (2004) Estrogen and progesterone prevent disruption of prepulse inhibition by the serotonin-1A receptor agonist 8-hydroxy-2-dipropylaminotetralin. J Pharmacol Exp Ther 309:267–274
Gogos A, Nathan PJ, Guille V, Croft RJ, van den Buuse M (2006) Estrogen prevents 5-HT1A receptor-induced disruptions of prepulse inhibition in healthy women. Neuropsychopharmacology 31:885–889
Gottschalk LA, Haer JL, Bates DE (1972) Effect of sensory overload on psychological state. Changes in social alienation-personal disorganization and cognitive-intellectual impairment. Arch Gen Psychiatry 27:451–457
Graham FK (1975) Presidential Address, 1974. The more or less startling effects of weak prestimulation. Psychophysiology 12:238–248
Greenwood TA, Braff DL, Light GA, Cadenhead KS, Calkins ME, Dobie DJ, Freedman R, Green MF, Gur RE, Gur RC, Mintz J, Nuechterlein KH, Olincy A, Radant AD, Seidman LJ, Siever LJ, Silverman JM, Stone WS, Swerdlow NR, Tsuang DW, Tsuang MT, Turetsky BI, Schork NJ (2007) Initial heritability analyses of endophenotypic measures for schizophrenia: the consortium on the genetics of schizophrenia. Arch Gen Psychiatry 64:1242–1250
Guillin O, Abi-Dargham A, Laruelle M (2007) Neurobiology of dopamine in schizophrenia. Int Rev Neurobiol 78:1–39
Hafner H (2003) Gender differences in schizophrenia. Psychoneuroendocrinology 28(Suppl 2):17–54
Hafner H, Maurer K, Loffler W, Riecher-Rossler A (1993) The influence of age and sex on the onset and early course of schizophrenia. Br J Psychiatry 162:80–86
Hafner H, an der Heiden W, Behrens S, Gattaz WF, Hambrecht M, Loffler W, Maurer K, Munk-Jorgensen P, Nowotny B, Riecher-Rossler A, Stein A (1998) Causes and consequences of the gender difference in age at onset of schizophrenia. Schizophr Bull 24:99–113

Hasenkamp W, Epstein MP, Green A, Wilcox L, Boshoven W, Lewison B, Duncan E (2010) Heritability of acoustic startle magnitude, prepulse inhibition, and startle latency in schizophrenia and control families. Psychiatry Res 178:236–243

Hazlett EA, Buchsbaum MS (2001) Sensorimotor gating deficits and hypofrontality in schizophrenia. Front Biosci 6:D1069–D1072

Hazlett EA, Buchsbaum MS, Zhang J, Newmark RE, Glanton CF, Zelmanova Y, Haznedar MM, Chu KW, Nenadic I, Kemether EM, Tang CY, New AS, Siever LJ (2008) Frontal-striatal-thalamic mediodorsal nucleus dysfunction in schizophrenia-spectrum patients during sensorimotor gating. Neuroimage 42:1164–1177

Hoffman HS, Searle JL (1968) Acoustic and temporal factors in the evocation of startle. J Acoust Soc Am 43:269–282

Ison JR, Allen PD (2007) Pre- but not post-menopausal female CBA/CaJ mice show less prepulse inhibition than male mice of the same age. Behav Brain Res 185:76–81

Javitt DC, Spencer KM, Thaker GK, Winterer G, Hajos M (2008) Neurophysiological biomarkers for drug development in schizophrenia. Nat Rev Drug Discov 7:68–83

Jovanovic T, Szilagyi S, Chakravorty S, Fiallos AM, Lewison BJ, Parwani A, Schwartz MP, Gonzenbach S, Rotrosen JP, Duncan EJ (2004) Menstrual cycle phase effects on prepulse inhibition of acoustic startle. Psychophysiology 41:401–406

Justice AJ, de Wit H (1999) Acute effects of d-amphetamine during the follicular and luteal phases of the menstrual cycle in women. Psychopharmacology (Berl) 145:67–75

Kapur S, Remington G (2001) Dopamine D(2) receptors and their role in atypical antipsychotic action: still necessary and may even be sufficient. Biol Psychiatry 50:873–883

Kask K, Backstrom T, Gulinello M, Sundstrom-Poromaa I (2008) Lower levels of prepulse inhibition of startle response in pregnant women compared to postpartum women. Psychoneuroendocrinology 33:100–107

Khisti RT, Mandhane SN, Chopde CT (1998) The neurosteroid 3 alpha-hydroxy-5 alpha-pregnan-20-one induces catalepsy in mice. Neurosci Lett 251:85–88

Khisti RT, Deshpande LS, Chopde CT (2002) The neurosteroid 3 alpha-hydroxy-5 alpha-pregnan-20-one affects dopamine-mediated behavior in rodents. Psychopharmacology (Berl) 161:120–128

Koch M (1998) Sensorimotor gating changes across the estrous cycle in female rats. Physiol Behav 64:625–628

Kulkarni J (2009) Oestrogen – a new treatment approach for schizophrenia? Med J Aust 190:S37–S38

Kumari V, Antonova E, Geyer MA, Ffytche D, Williams SC, Sharma T (2007a) A fMRI investigation of startle gating deficits in schizophrenia patients treated with typical or atypical antipsychotics. Int J Neuropsychopharmacol 10:463–477

Kumari V, Antonova E, Geyer MA (2008b) Prepulse inhibition and "psychosis-proneness" in healthy individuals: an fMRI study. Eur Psychiatry 23:274–280

Kumari V, Aasen I, Papadopoulos A, Bojang F, Poon L, Halari R, Cleare AJ (2008a) A comparison of prepulse inhibition in pre- and postmenopausal women and age-matched men. Neuropsychopharmacology 33:2610–2618

Kumari V, Aasen I, Sharma T (2004) Sex differences in prepulse inhibition deficits in chronic schizophrenia. Schizophr Res 69:219–235

Kumari V, Antonova E, Zachariah E, Galea A, Aasen I, Ettinger U, Mitterschiffthaler MT, Sharma T (2005a) Structural brain correlates of prepulse inhibition of the acoustic startle response in healthy humans. Neuroimage 26:1052–1058

Kumari V, Das M, Zachariah E, Ettinger U, Sharma T (2005b) Reduced prepulse inhibition in unaffected siblings of schizophrenia patients. Psychophysiology 42(5):588–594

Kumari V, Fannon D, Sumich AL, Sharma T (2007b) Startle gating in antipsychotic-naive first episode schizophrenia patients: one ear is better than two. Psychiatry Res 151: 21–28

Kumari V, Gray J, Gupta P, Luscher S, Sharma T (2003a) Sex differences in prepulse inhibition of the acoustic startle response. Person Indiv Diff 34:733–742

Kumari V, Gray JA, Geyer MA, ffytche D, Soni W, Mitterschiffthaler MT, Vythelingum GN, Simmons A, Williams SC, Sharma T (2003b) Neural correlates of tactile prepulse inhibition: a functional MRI study in normal and schizophrenic subjects. Psychiatry Res 122:99–113

Kumari V, Konstantinou J, Papadopoulos A, Aasen I, Poon L, Halari R, Cleare AJ (2010) Evidence for a role of progesterone in menstrual cycle-related variability in prepulse inhibition in healthy young women. Neuropsychopharmacology 35:929–937

Kumari V, Peters ER, Fannon D, Premkumar P, Aasen I, Cooke MA, Anilkumar AP, Kuipers E (2008c) Uncontrollable voices and their relationship to gating deficits in schizophrenia. Schizophr Res 101:185–194

Kumari V, Soni W, Mathew VM, Sharma T (2000) Prepulse inhibition of the startle response in men with schizophrenia: effects of age of onset of illness, symptoms, and medication. Arch Gen Psychiatry 57:609–614

Kumari V, Toone B, Gray JA (1997) Habituation and prepulse inhibition of the acoustic startle reflex: Effects of smoking status and psychosis-proneness. Pers Individ Dif 23:183–191

Marshall JC (2001) Hormonal regulation of the menstrual cycle and mechanisms of ovulation. In: Leslie J et al (eds) Endocrinology. W.B. Saunders Company, Phliladelphia, pp 2073–2085

Marx CE, Duncan GE, Gilmore JH, Lieberman JA, Morrow AL (2000) Olanzapine increases allopregnanolone in the rat cerebral cortex. Biol Psychiatry 47:1000–1004

Marx CE, VanDoren MJ, Duncan GE, Lieberman JA, Morrow AL (2003) Olanzapine and clozapine increase the GABAergic neuroactive steroid allopregnanolone in rodents. Neuropsychopharmacology 28:1–13

Marx CE, Shampine LJ, Duncan GE, VanDoren MJ, Grobin AC, Massing MW, Madison RD, Bradford DW, Butterfield MI, Lieberman JA, Morrow AL (2006) Clozapine markedly elevates pregnenolone in rat hippocampus, cerebral cortex, and serum: candidate mechanism for superior efficacy? Pharmacol Biochem Behav 84:598–608

Michanek A, Meyerson BJ (1982) Influence of estrogen and progesterone on behavioral effects of apomorphine and amphetamine. Pharmacol Biochem Behav 16:875–879

Montag C, Hartmann P, Merz M, Burk C, Reuter M (2008) D2 receptor density and prepulse inhibition in humans: negative findings from a molecular genetic approach. Behav Brain Res 187:428–432

Neuner I, Stocker T, Kellermann T, Ermer V, Wegener HP, Eickhoff SB, Schneider F, Shah NJ (2010) Electrophysiology meets fMRI: neural correlates of the startle reflex assessed by simultaneous EMG-fMRI data acquisition. Hum Brain Mapp 31:1675–1685

Oades RD, Schepker R (1994) Serum gonadal steroid hormones in young schizophrenic patients. Psychoneuroendocrinology 19:373–385

Ornitz EM, Guthrie D, Sadeghpour M, Sugiyama T (1991) Maturation of prestimulation-induced startle modulation in girls. Psychophysiology 28:11–20

Parducz A, Hoyk Z, Kis Z, Garcia-Segura LM (2002) Hormonal enhancement of neuronal firing is linked to structural remodelling of excitatory and inhibitory synapses. Eur J Neurosci 16:665–670

Petrovsky N, Quednow BB, Ettinger U, Schmechtig A, Mössner R, Collier DA, Kühn KU, Maier W, Wagner M, Kumari V (2010) Sensorimotor gating is associated with CHRNA3 polymorphisms in schizophrenia and healthy volunteers. Neuropsychopharmacology 35:1429–1439

Postma P, Gray JA, Sharma T, Geyer M, Mehrotra R, Das M, Zachariah E, Hines M, Williams SC, Kumari V (2006) A behavioural and functional neuroimaging investigation into the effects of nicotine on sensorimotor gating in healthy subjects and persons with schizophrenia. Psychopharmacology (Berl) 184:589–599

Quednow BB, Kuhn KU, Mossner R, Schwab SG, Schuhmacher A, Maier W, Wagner M (2008) Sensorimotor gating of schizophrenia patients is influenced by 5-HT2A receptor polymorphisms. Biol Psychiatry 64:434–437

Quednow BB, Schmechtig A, Ettinger U, Petrovsky N, Collier DA, Vollenweider FX, Wagner M, Kumari V (2009) Sensorimotor gating depends on polymorphisms of the serotonin-2A receptor and catechol-O-methyltransferase, but not on neuregulin-1 Arg38Gln genotype: a replication study. Biol Psychiatry 66:614–620

Ramirez VD, Zheng J (1996) Membrane sex-steroid receptors in the brain. Front Neuroendocrinol 17:402–439

Riecher-Rossler A, Hafner H, Stumbaum M, Maurer K, Schmidt R (1994) Can estradiol modulate schizophrenic symptomatology? Schizophr Bull 20:203–214

Roussos P, Giakoumaki SG, Bitsios P (2008a) The dopamine D(3) receptor Ser9Gly polymorphism modulates prepulse inhibition of the acoustic startle reflex. Biol Psychiatry 64:235–240

Roussos P, Giakoumaki SG, Rogdaki M, Pavlakis S, Frangou S, Bitsios P (2008b) Prepulse inhibition of the startle reflex depends on the catechol O-methyltransferase Val158Met gene polymorphism. Psychol Med 38:1651–1658

Rupprecht R (2003) Neuroactive steroids: mechanisms of action and neuropsychopharmacological properties. Psychoneuroendocrinology 28:139–168

Rupprecht R, Koch M, Montkowski A, Lancel M, Faulhaber J, Harting J, Spanagel R (1999) Assessment of neuroleptic-like properties of progesterone. Psychopharmacology (Berl) 143:29–38

Shulman Y, Tibbo PG (2005) Neuroactive steroids in schizophrenia. Can J Psychiatry 50:695–702

Swerdlow NR, Auerbach P, Monroe SM, Hartston H, Geyer MA, Braff DL (1993) Men are more inhibited than women by weak prepulses. Biol Psychiatry 34:253–260

Swerdlow NR, Braff DL, Taaid N, Geyer MA (1994) Assessing the validity of an animal model of deficient sensorimotor gating in schizophrenic patients. Arch Gen Psychiatry 51:139–154

Swerdlow NR, Hartman PL, Auerbach PP (1997) Changes in sensorimotor inhibition across the menstrual cycle: implications for neuropsychiatric disorders. Biol Psychiatry 41:452–460

Swerdlow NR, Geyer MA, Hartman PL, Sprock J, Auerbach PP, Cadenhead K, Perry W, Braff DL (1999) Sex differences in sensorimotor gating of the human startle reflex: all smoke? Psychopharmacology (Berl) 146:228–232

Swerdlow NR, Geyer MA, Braff DL (2001) Neural circuit regulation of prepulse inhibition of startle in the rat: current knowledge and future challenges. Psychopharmacology (Berl) 156:194–215

Swerdlow NR, Stephany N, Wasserman LC, Talledo J, Shoemaker J, Auerbach PP (2003) Amphetamine effects on prepulse inhibition across-species: replication and parametric extension. Neuropsychopharmacology 28:640–650

Swerdlow NR, Light GA, Cadenhead KS, Sprock J, Hsieh MH, Braff DL (2006a) Startle gating deficits in a large cohort of patients with schizophrenia: relationship to medications, symptoms, neurocognition, and level of function. Arch Gen Psychiatry 63:1325–1335

Swerdlow NR, Talledo J, Sutherland AN, Nagy D, Shoemaker JM (2006b) Antipsychotic effects on prepulse inhibition in normal 'low gating' humans and rats. Neuropsychopharmacology 31:2011–2021

Swerdlow NR, Weber M, Qu Y, Light GA, Braff DL (2008) Realistic expectations of prepulse inhibition in translational models for schizophrenia research. Psychopharmacology (Berl) 199:331–388

Taherianfard M, Shariaty M (2004) Evaluation of serum steroid hormones in schizophrenic patients. Indian J Med Sci 58:3–9

Takahashi H, Iwase M, Ishii R, Ohi K, Fukumoto M, Azechi M, Ikezawa K, Kurimoto R, Canuet L, Nakahachi T, Iike N, Tagami S, Morihara T, Okochi M, Tanaka T, Kazui H, Yoshida T, Tanimukai H, Yasuda Y, Kudo T, Hashimoto R, Takeda M (2008) Impaired prepulse inhibition and habituation of acoustic startle response in Japanese patients with schizophrenia. Neurosci Res 62:187–194

Talledo JA, SutherlandOwens AN, Schortinghuis T, Swerdlow NR (2009) Amphetamine effects on startle gating in normal women and female rats. Psychopharmacology (Berl) 204:165–175

Vaillancourt C, Cyr M, Rochford J, Boksa P, Di Paolo T (2002) Effects of ovariectomy and estradiol on acoustic startle responses in rats. Pharmacol Biochem Behav 74:103–109

Valera S, Ballivet M, Bertrand D (1992) Progesterone modulates a neuronal nicotinic acetylcholine receptor. Proc Natl Acad Sci U S A 89:9949–9953

Wetzel CH, Hermann B, Behl C, Pestel E, Rammes G, Zieglgansberger W, Holsboer F, Rupprecht R (1998) Functional antagonism of gonadal steroids at the 5-hydroxytryptamine type 3 receptor. Mol Endocrinol 12:1441–1451

Estrogens and Gonadal Function in Schizophrenia and Related Psychoses

Anita Riecher-Rössler and Jayashri Kulkarni

Contents

Abstract Recent research has increasingly pointed to the importance of estrogens and the hypothalamic–pituitary–gonadal axis in schizophrenia. Specifically, there is mounting evidence from clinical, epidemiological, and basic research that estradiol, the main component of estrogens, exerts protective effects in schizophrenia and related psychoses. Possible modes of action of this hormone in the brain have been suggested, and clinical intervention studies have reported the first positive results.

A. Riecher-Rössler (✉)
Psychiatric University Clinic Basel, University Psychiatric Outpatient Department, c/o University Hospital Basel, Petersgraben 4, Basel 4031, Switzerland
e-mail: anita.riecher@upkbs.ch

J. Kulkarni
Monash Alfred Psychiatry Research Centre, The Alfred Hospital and School of Psychology and Psychiatry, Monash University, Commercial Road, Melbourne 3004, Australia
e-mail: jayashri.kulkarni@med.monash.edu.au

J.C. Neill and J. Kulkarni (eds.), *Biological Basis of Sex Differences in Psychopharmacology*, 155
Current Topics in Behavioral Neurosciences 8, DOI 10.1007/7854_2010_100,

Furthermore, there are an increasing number of reports on gonadal dysfunction and states of estrogen deficiency in women with schizophrenia. These findings could have important implications for clinicians and researchers alike.

Keywords Estradiol · Estrogens · Gonadal function · Psychoses · Schizophrenia

Abbreviations

ESR1	Endogen receptor-α
HERS	Heart and Estrogen/Progestin Replacement Study
HRT	Hormone replacement therapy
PANSS	Positive and negative syndrome scale
RNA	Ribonucleic acid
WHI	Women's Health Initiative Study
WHI-M	Women's Health Initiative Memory Study

1 Introduction

Recent research increasingly points to the importance of estrogens and the hypothalamic–pituitary–gonadal axis in schizophrenia and related psychoses.

On one hand, there are reports of gonadal dysfunction and states of estrogen deficiency in women with schizophrenia (the hypothesis of hypoestrogenism). On the other hand, there is mounting evidence from clinical as well as from epidemiological and basic research that estradiol, the main component of estrogens, exerts protective effects in schizophrenia and related psychoses (the estrogen protection hypothesis) (Riecher-Rössler and Häfner 1993).

2 Estrogens: A Protective Factor in Schizophrenia and Related Psychoses? (The "Estrogen Protection Hypothesis")

2.1 *Historical Findings*

As long ago as at the beginning of the last century, psychiatrists recognized the possible association between schizophrenia and estrogens (for review, see Riecher-Rössler and Häfner 1993). There are longstanding observations indicating an association between lowered estrogen blood levels and *acute* psychotic symptomatology. Early clinicians such as Kraepelin and Kretschmer described signs of *chronic* "hypoestrogenism" in women with schizophrenia.

Kraft-Ebing was among the first to describe women becoming psychotic before or during menstruation, i.e., when blood levels of estrogen are relatively low. Kraepelin even created a separate diagnostic category, labeled "menstrual psychosis". Kretschmer reported cases where the outbreak of schizophrenia and related psychoses had a temporal relationship with "surgery of ovaries, pregnancy, delivery, and puerperium". Finally, Manfred Bleuler noted that late-onset schizophrenia with onset after age 40 years was much more frequent in women than in men, a finding he attributed to the "loss of ovarian function" starting at around that age (for review, see Riecher-Rössler and Häfner 1993).

2.2 *Basic Research Findings*

Important findings from basic research were the identification of estrogen receptors in the limbic system of the brain, and the observation that the effects of estrogens in rodents are, in some respects, similar to those of antipsychotic medications. Furthermore, it was shown that estrogens can modulate the sensitivity and number of dopamine receptors. It was therefore hypothesized that estrogens exert their antipsychotic effects in a manner similar to that of traditional antipsychotic medications at least partly by blockade of dopaminergic transmission (Riecher-Rössler and Häfner 1993).

We now know that estrogens, especially 17-β-estradiol (the natural estrogen that is most active in the brain), have many other neuroprotective and psychoprotective effects. For example, they appear to improve cerebral blood flow and glucose metabolism, promote neuronal sprouting and myelination, enhance synaptic density and plasticity, facilitate neuronal connectivity, act as antioxidants, and inhibit neuronal cell death.

Estrogens have also been shown to exert profound effects on brain differentiation during development, particularly during late gestation and the early postnatal period, and are important in normal maintenance of brain function during aging (Cyr et al. 2002; Goldstein et al. 2002; Oesterlund 2002; Vedder and Behl 2005). In a well-controlled magnetic resonance imaging study, Goldstein et al. (2002) showed that normal patterns of sexual brain dimorphism (brain regions found to be structurally different in normal men and women) are disrupted in schizophrenia and related psychoses, especially in the cortex. Apart from later "activational" effects of circulating hormones (e.g., during puberty), those investigators suggested that these early "organizational" effects of gonadal hormones that occur during the developmental period (which is probably critical for at least some forms of schizophrenia and related psychoses) could be partly responsible for that finding.

The mechanisms of action of estrogens are now known not only to depend on the classical genomic pathway but also to involve nongenomic, rapid interactions, which explain the differing latency of effects. They clearly modulate the dopaminergic and other neurotransmitter systems that are believed to be relevant to schizophrenia and related psychoses, such as the serotonergic and glutamatergic system, but also the noradrenergic and cholinergic system (for reviews, see Cyr et al. 2002; Oesterlund 2002; Garcia-Segura et al. 2001; McEwen 2002; Stahl 2001a, b). Recently it has even been suggested that 17-β-estradiol in the brain might rather be regarded as a neurotransmitter itself than as a hormone (Balthazart and Ball 2006).

There are at least two subtypes of estrogen receptors, namely estrogen receptor-α and estrogen receptor-β, which are transcribed from two distinct genes (Oesterlund 2002). Autopsy studies showed that estrogen receptor-α messenger RNA is expressed in discrete areas of the human brain such as the amygdala, hypothalamus, cerebral cortex, and hippocampus; these areas are associated with neuroendocrine function, as well as emotion, memory, and cognition (Oesterlund et al. 2000).

Recently Weickert et al. (2008) reported a variation in the endogen receptor-α (ESR1) gene to be associated with schizophrenia and speculated that the mechanism of this association may involve alternative gene regulation and transcript processing.

Regarding the therapeutic effect of estrogens, it must also be noted that both the numerous direct effects on the brain and indirect effects may play a role. For example, estrogens may also increase blood levels of antipsychotic drugs via their actions on liver metabolism (Yonkers et al. 1992).

2.3 Epidemiological Findings

Epidemiological studies into sex differences in schizophrenic disorders suggest that the physiologically high estradiol production in young fertile women contributes to the later age of onset of schizophrenia in women as compared with men, to the second peak of onsets in women around the menopause, and to the better course of the disease in young women (Häfner et al. 1993; Riecher-Rössler et al. 1997). Thus, in an *epidemiological study* on a representative sample of 392 first admitted patients with schizophrenia, the ABC Study, we found that schizophrenic women have a later peak of illness onset in comparison with schizophrenic men (Häfner et al. 1991a, b; Riecher et al. 1991). They also exhibit an additional, smaller peak after age 45. We postulated that estrogens raise the vulnerability threshold for the outbreak of the disease. According to this hypothesis, women would be protected against schizophrenia between puberty and menopause to some extent by their relatively high gonadal estrogen production during this time. Then, around age 45, several years before menopause sets in at a mean age of 51.4 years, estrogen production begins to fall (Labhart 1978). Thus, women would lose the protection estrogens give, which could account for their second peak of illness onset after age 45.

A number of risk factors appear to counteract the protective effect of estrogens. Thus, the sex difference in the age of onset diminishes in the subgroup of cases with a genetic risk and in patients with perinatal complications (Häfner 2005; Könnecke et al. 2000).

Recent results regarding the age of menarche further support the hypothesis that physiological estrogens play a protective role against the development of the disease. We demonstrated a significantly later age of menarche in a representative group of first admitted women with schizophrenia and related psychoses as compared with a healthy control group (Riecher-Rössler 2002). Seeman and co-workers (Cohen et al. 1999; Hayeems and Seeman 2005) found that later menarche was associated with an earlier onset of the illness, an association that was independent of factors such as family history and obstetric complications.

2.4 Clinical Findings

Clinically, psychotic symptomatology has often been found to correlate with the estrogenic state of women (for review, see Riecher-Rössler and Häfner 1993; Seeman 1996). For example, during high estrogen phases such as pregnancy, chronic psychoses appear to improve, whereas there is an excess of psychoses after delivery.

Psychosis associated with estrogen withdrawal due to conditions other than the puerperium was recently reviewed by Mahe and Dumaine (2001). Those investigators reported cases of premenstrual psychosis, post abortion psychosis, and psychoses associated with removal of hydatiform mole, cessation of oral contraceptives, clomiphene and tamoxifen administration (both estrogen receptor antagonists), and gonadorelin agonist administration (which blocks pituitary stimulation of endogenous estrogen secretion). Psychotic episodes were acute, short, and with a wide range of psychotic, but also affective, symptomatology. Recurrences were often reported when estrogen levels were normalized, and puerperal psychosis was frequent in the history of patients who were affected.

Psychotic symptoms in schizophrenic patients have also often been shown to deteriorate premenstrually or perimenstrually (i.e., in the low estrogen phase of the cycle; for review, see Riecher-Rössler and Häfner 1993; Riecher-Rössler 2002; Seeman 1996). Thus, Riecher-Rössler et al. (1994a, b) could show an inverse correlation of estradiol blood levels with psychopathology. They examined 32 acutely admitted women with schizophrenia, who gave a history of regular menstrual cycles, and found a significant excess of admissions during the perimenstrual low estrogen phase of the cycle ($p < 0.005$). During the hospital stay of the 32 women, there emerged a significant association between estradiol levels on one hand, and psychopathology scores on the other hand. Psychopathology seemed to improve when estradiol blood levels rose and vice versa. This was true not only for the total score of the Brief Psychiatric Rating Scale (BPRS; Overall and Gorham 1962) and for almost all the subscores of this scale such as anergia, thought disturbance, activation, and hostile suspiciousness, but also for the general behavior on ward as rated by the nurses (NOSIE; Honigfeld et al. 1976) and for general well-being and paranoid feelings as rated by the patients themselves (BfS and paranoid subscore of PDS, both by von Zerssen and Koeller 1976).

Also, Bergemann et al. (2007a) found a significant effect of the menstrual cycle phase and 17-β-estradiol levels on positive and negative symptoms of 125 women with schizophrenia and related psychoses. The same authors (Bergemann et al. 2008) could furthermore show a significant effect of estrogen on the comprehension of metaphoric speech and/or concretism, a main feature of schizophrenic thought and language disturbance. Ko et al. (2006) similarly found that 35 women with chronic schizophrenia and related psychoses had low levels of estrogen which was associated with severe negative symptomatology and reduced cognitive functioning, especially in the domains of verbal performance and executive tasks.

Most studies, however, did not examine the correlation between psychosis symptomatology and estradiol serum levels directly. For example, a more recent study (Choi et al. 2001) reported behavioral, affective, and somatic symptoms of schizophrenia (not psychotic ones) to be associated with the menstrual cycle phase. Estradiol was not measured.

Rather than examining cyclic fluctuations, Hoff et al. (1996) assessed the relationship between average estrogen levels from four consecutive weeks sampling with psychopathology and cognitive functioning in 22 female inpatients (aged between 22 and 63 years) with chronic schizophrenia. There was no significant association between average estrogen levels and psychopathology, but higher average estrogen levels were strongly associated with better cognitive abilities. However, this finding may be due in part to the effects of aging.

It has to be noted that an elevated number of admissions during the perimenstrual period has also been identified in other disorders (Althaus et al. 2000), and exacerbation of many psychiatric symptoms (not only psychotic ones) during the perimenstrual period was observed in patients with schizophrenia and related psychoses (Riecher-Rössler et al. 1994b; Harris 1997). In theory, this lack of specificity is to be expected because of the multiple effects of estrogen on mental functioning.

Studies on gender differences conducted in the area of late-onset schizophrenia emphasize the significance of hormone changes related to the menopause. Riecher-Rössler and colleagues (Riecher-Rössler et al. 1997; Riecher-Rössler 2002) showed that there are twice as many women as men with the onset of schizophrenia and related psychoses beyond 40 years of age, and that female patients with late-onset illness suffered from unexpectedly severe disease in terms of symptomatology and course. One explanation for this could be the fluctuation and sharp decline in estrogen levels just before and during the menopause. In support of this proposed explanation are results from long-term studies in women with schizophrenia showing that the course of illness in women tends to deteriorate rapidly during menopause and thereafter (for review, see Riecher-Rössler and Häfner 1993; Riecher-Rössler et al. 1998).

2.5 *Intervention Studies*

Intervention studies have also been conducted over long periods, mainly with positive results (Korhonen et al. 1995; Lindamer et al. 1997) (for review, see Riecher-Rössler and Häfner 1993). Thus, as early as the 1940s, Manfred Bleuler (Bleuler 1943) reported the first unsystematic trials using a combination of ovarian and pituitary hormones. Mall (1960), a German psychiatrist in charge of a large hospital, examined 167 women suffering from schizophrenia with respect to estrogen excretion in 24-h urine samples, basal temperature, and vaginal cytology. Based on his findings, he divided the psychoses into two groups: hypofollicular and hyperfollicular. In the former group, he replaced estrogens and found that "hypofollicular psychosis can be healed relatively easily by this substitution therapy." Unfortunately, Mall does not give many details about these interesting studies.

In the first systematic trial conducted in 1996, Kulkarni et al. (1996) found that women with schizophrenia receiving estradiol as an adjunct to antipsychotic medication treatment exhibited rapid and greater improvement in psychotic symptoms than women receiving antipsychotics only. In 2001, the same group performed a double-blind, 28-day, placebo-controlled study (Kulkarni et al. 2001) in which 12 women were administered transdermal 17-β-estradiol (patches) 50 μg/24 h, another 12 women received 100 μg/24 h patches, and the third group received placebo patches. The 100 μg group experienced greater improvement than either the 50 μg or placebo groups, with striking improvements observed in the key psychotic symptoms.

Akhondzadeh et al. (2003) published a randomized study of 32 women of childbearing age with chronic schizophrenia. They administered ethinyl estradiol as an adjunct to haloperidol over 8 weeks and the control group received haloperidol only. The combination with estradiol showed a significant superiority over haloperidol alone regarding positive and negative symptom response. Furthermore, the estrogen group needed significantly less adjunctive anticholinergic medication to treat extrapyramidal side effects caused by haloperidol treatment. This finding is in line with other studies suggesting that estrogen treatment can also reduce the severity of antipsychotic medication-induced extrapyramidal side effects (Thompson et al. 2000).

Louza et al. (2004) did not find a positive response to estrogen treatment in their study, which they correctly discussed as possibly being due to having used conjugated estrogens rather than 17-β-estradiol. In a Cochrane review in 2005, Chua et al. (2005) surveyed data from only five randomized double-blind intervention studies with appropriate methodology. They concluded that the effects of estrogen as sole treatment or adjunctive therapy for those with schizophrenia and related psychoses were still unclear, but that further, larger clinical trials were needed. Studies in this review with negative results used conjugated estrogens and not 17-β-estradiol, although the latter has been shown to be the estrogen type with the most potent activity in the brain. Furthermore, to prevent endometrial hyperplasia, the estrogens were usually combined with progestogens which can counteract the positive effects of estradiol in the brain.

Kulkarni (2009) recently conducted a proof-of-concept study of 102 women with DSM-IV schizophrenia. In this double-blind randomized controlled 28-day study, women received either an active 100 μg estradiol skin patch treatment ($n = 56$) or an identical placebo patch ($n = 46$). All patients received antipsychotic drug treatment according to a standardized protocol. Progesterone was not given during the study; hence, the trial measured the impact of unopposed estradiol.

Psychopathology was assessed using the PANSS rating scale. Serum levels of estrogen, progesterone, prolactin, luteinizing hormone, and follicle-stimulating hormone were measured. Several cognitive tests were also administered. They found that patients who received the 100 μg estradiol adjunct made a significantly better recovery in their total positive, negative, and general symptoms of schizophrenia than the patients who received standard antipsychotic medication only ($p < 0.01$). Women who received the estradiol patch also showed significant improvement in cognition ($p < 0.01$). By measuring luteinizing hormone, they could also demonstrate that there is a direct effect on the pituitary gland, which suggests that this dose and type of unconjugated estrogen directly affects the hypothalamic–pituitary–gonadal axis.

Overall, these studies provide strong evidence for the estrogen protection hypothesis. The addition of transdermally delivered estradiol seems to be associated with significant abatement of psychotic symptoms in women with schizophrenia compared with standardized antipsychotic drug treatment alone.

Most of the estrogen treatment studies conducted so far have been in young, reproductive age women and not in peri- or postmenopausal women with estrogen deficiency. Theoretically, the greatest effect of estradiol would be expected when it is replaced in a woman in a hypoestrogenic state. Good et al. (1999) conducted a study in *postmenopausal* patients. He administered estradiol and progesterone to 14 women with schizophrenia, schizophreniform disorder, or schizoaffective disorder and found a significant improvement of negative symptoms over 6 months.

There are also some case reports regarding positive results of hormone replacement therapy (HRT) in postmenopausal women with schizophrenia and related psychoses. Bergemann et al. (2007b) reported a case study of a woman with first onset of schizophrenia in the perimenopause period. The patient experienced severe acute psychosis symptoms over several months, but refused antipsychotic treatment. As she was diagnosed to be in a periclimacteric state based on clinical symptoms and hormone analysis, she was started on transdermal estradiol in combination with norethisterone acetate and had an impressive remission of the psychotic symptoms. Lindamer et al. (1997) reported details about a postmenopausal woman, whose psychotic symptoms improved with estradiol treatment as an adjunct to her antipsychotic.

Lindamer et al. (2001) studied a community sample of postmenopausal women with schizophrenia and related psychoses. Twenty-four women received standard HRT, and 28 women had never received hormone treatment. Interestingly, the users of HRT needed a relatively lower average dose of antipsychotic medication and suffered fewer severe negative symptoms.

Ahokas et al. (2000) described positive effects of estrogen treatment in women with postpartum psychosis. In those women who exhibited sustained estrogen deficiency states, the addition of 17-β-estradiol, without any further medication, yielded a dramatic antipsychotic effect within 1 week. However, the proportion of schizophrenia-like psychoses in the sample was not given.

Finally, Kulkarni (2005) also tested the use of adjunctive estradiol in a small sample of 11 men with schizophrenia. They gave 2 mg oral estradiol valerate as an adjunct to six men who were taking antipsychotic drugs. Five men received oral placebo plus their standard antipsychotic medication for 7 days. This small study was only conducted for 7 days to avoid feminization and other side effects in the men. The groups were matched for age, illness severity, and duration. Oral, rather than transdermal, estradiol was used to ensure treatment adherence in men with acute psychosis. Psychopathology was assessed using the standardized rating scales, PANSS and the BPRS, Brief Psychiatric Rating Scale. By day 5, the estradiol group showed significant abatement of psychotic symptoms compared with the placebo group, and by day 7, the estradiol group made further improvements. This study, although small in sample size and short in duration, raises the possibility that nonfeminizing estrogen may provide useful treatment possibilities for men with schizophrenia.

3 Hypoestrogenism in Women with Schizophrenia (The Hypothesis of Hypoestrogenism)

Several studies have recently confirmed earlier findings of disturbed gonadal function and hypoestrogenism in women with schizophrenia (Riecher-Rössler and Häfner 1993; Riecher-Rössler et al. 1994b, 1998; Choi et al. 2001; Kulkarni et al. 1996; Bergemann et al. 2002; Canuso et al. 2002; Hoff et al. 2001; Huber et al. 2001; Smith et al. 2002; Zhang-Wong and Seeman 2002). They described menstrual irregularities and reduced blood levels of estradiol, progesterone, and gonadotropins (follicle-stimulating hormone, luteinizing hormone) throughout the menstrual cycle, plus anovulation in the majority of women with schizophrenia and related psychoses. Reduced fertility was also reported.

There appear to be multiple reasons for these disturbances including the consequences of emotional stress and/or antipsychotic medication-induced hyperprolactinaemia, which is known to suppress gonadal function (Maguire 2002). However, these are probably not the only causes, because women experiencing other psychiatric disorders with similar emotional stress do not have the same hypothalamic–pituitary–gonadal axis hormone changes, at least not to the same degree (Riecher-Rössler et al. 1998; Huber et al. 2001). Furthermore, hypoestrogenism was observed long before the introduction of antipsychotics.

Smith et al. (2002) found the dose of typical antipsychotics to correlate with prolactin levels especially in women and prolactin to correlate inversely with estradiol serum levels. In contrast to those findings, Huber et al. (2001) were unable to identify a significant association of prolactin and estradiol in 43 women with acute psychosis, 14 women with other diagnoses, and 9 healthy control women. Nevertheless, the women with schizophrenia and related psychoses had significantly lower estradiol serum levels than the control women. Women with other psychiatric diagnoses fell in between the psychotic and the healthy group with regard to estradiol and prolactin levels. Also, Canuso et al. (2002) found a high rate of ovarian dysfunction and estradiol levels below normal, irrespective of medication type or prolactin status in 16 premenopausal women with schizophrenia and schizoaffective disorders. Interestingly, Warner et al. (2001) found prolactin levels in unmedicated schizophrenic patients to be even lower than in control individuals. Those investigators suggested that this was due to a disordered dopaminergic system because dopamine tonically inhibits prolactin.

Taken together, these results imply that the hypothalamic–pituitary–gonadal axis is disturbed in many women with schizophrenia and related psychoses, and that the reasons for this are far from clear yet. An interesting research question in this context is whether gonadal dysfunction with estrogen deficiency could even be part of the underlying pathogenetic process, at least in a subgroup of women (Riecher-Rössler 2002).

4 Implications for Clinicians and Researchers

Further research into the impact of gonadal function and estrogen on schizophrenia and related psychoses is warranted because new diagnostic and therapeutic strategies could emerge that would benefit the many women worldwide who suffer from this disorder.

4.1 Assessment and Therapy of Gonadal Dysfunction

As there is growing evidence that many, even younger women, with schizophrenia and related psychoses are in a state of estrogen deficiency, in future estrogens and the gonadal axis should be considered more seriously in the treatment of women with schizophrenia and related psychoses. Psychiatric history taking should always include questions regarding menstrual irregularities, amenorrhoea, and galactorrhoea. Also, prolactin and estrogen serum levels should be tested, if necessary. Gonadal dysfunction and hypoestrogenic states can often be found even in menstruating women (Riecher-Rössler et al. 1994b, 1998; Smith et al. 2002). In addition, hyperprolactinaemia is clearly underdiagnosed (Maguire 2002). Some authors have therefore suggested routine laboratory tests (Smith et al. 2002).

Most antipsychotics can cause hyperprolactinaemia and – especially if they are taken over a number of years – theoretically induce "iatrogenic early menopause" via suppression of physiological estradiol production. The concomitant risks include both short-term effects, such as hot flushes and sexual dysfunction, and long-term consequences, including osteoporosis and potentially cardiovascular disease or cognitive deterioration (Oesterlund 2002; Maguire 2002). In schizophrenia patients, these risks are further increased by additional risk factors such as smoking, poor diet, and reduced exercise (Smith et al. 2002). Furthermore, menopausal complaints may lead to compliance problems.

In the case of hyperprolactinaemia with secondary estrogen deficiency, prolactin-sparing antipsychotics (e.g., clozapine, quetiapine, aripiprazole, or maybe olanzapine;

Maguire 2002) should therefore be preferred. If a switch to these antipsychotics is not possible for clinical reasons or if hypoestrogenism persists despite switching, then estrogen can be added to the treatment. Issues regarding contraception must be taken into account in such cases because, when switching to prolactin-sparing antipsychotics, the menstrual cycle often normalizes and fertility is regained, with high risk for unplanned pregnancy (Neumann and Frasch 2001).

4.2 *Estradiol as a Therapeutic Agent?*

First trials of estrogens in schizophrenia and related psychoses indicate that estradiol could be used as an adjunct to antipsychotic medication. However, further replications of these findings in larger control studies by different groups are needed before recommendations for broad clinical application can be made.

In women who suffer from frequent perimenstrual psychotic relapses, "cycle modulated" antipsychotic medication therapy or, if contraception is needed at the same time, continuous use of oral contraceptives without hormone-free intervals may be strategies worthy of research (Riecher-Rössler 2002; Braendle et al. 2001).

Even more promising could be hormonal replacement with estrogens in women with schizophrenia in peri- and postmenopause, because estrogens in other disorders such as depression have proven to be especially helpful when they are used to restore hormonal balance.

Hormonal replacement with estrogens for women with schizophrenia *during and after the perimenopause* could be recommended as an augmentation strategy respectively an adjunct to antipsychotic medication. Possibly, the dose of antipsychotics could then be reduced and corresponding side effects minimized. The replacement of estrogens in these women could also attenuate perimenopausal complaints such as hot flushes, night sweats with sleep disturbances, and general irritability (see Table 1), which can contribute to a general deterioration of the mental state and, in vulnerable women, potentially provoke a psychotic episode. In contrast to this recommendation, it has been reported that women with schizophrenia are less likely to ever use HRT as compared to women without psychiatric diagnoses (Lindamer et al. 2003).

Estrogen replacement therapy for women of this age group has been recommended anyway for many reasons, for example prophylaxis of osteoporosis, and also delay of age-dependent cognitive deterioration or Alzheimer's dementia (Sherwin 2005) (for review, see Riecher-Rössler and de Geyter 2007; Table 1). Further research into estrogen treatment as an *additional* indication in peri- and postmeno-pausal women schizophrenia and related psychoses is urgently needed.

The use of estrogen has been questioned in the context of perimenopausal estrogen replacement by studies such as the WHI, Women's Health Initiative Study (Rossouw et al. 2002), the WHI-M, Women's Health Initiative Memory Study (Craig et al. 2005), and the HERS, Heart and Estrogen/Progestin Replacement Study (Hlatky et al. 2002). These studies have highlighted the side effects of

Table 1 Some important effects of estrogen replacement

Positive	Negative
Perimenopausal complaints ↓ Physical: hot flushes, genital discomfort, aging of collagen (skin, joints, intervertebral discs) ↓ Mental: depression, irritability, emotional lability ↓	Endometrial carcinoma ↑ if unopposed estrogens are administered (→ in women without hysterectomy always combine with progestogens!)
Risk of osteoporosis ↓	Risk of breast cancer ↑? (→ do not use in patients with a familiar or own risk of breast cancer and usually not longer than 7 years!)
Delay of cognitive decline/Morbus Alzheimer?	Risk of thrombosis and cerebral insult ↑? (→ no prescription for patients at risk!)
Cardiovascular protection? (if started right after menopause)	Other cardiovascular risks (coronary heart disease, arteriosclerosis) ↑? (→ start only within the first 10 years after menopause and not in patients with cardiovascular disease!)

Sources: Riecher-Rössler and de Geyter (2007), Birkhäuser et al. (2008), Rossouw et al. (2007)

HRT, which has provoked an ongoing controversy about the advantages and disadvantages of this regime.

However, the WHI study has been criticized by many experts and by the International Menopause Society (Birkhäuser et al. 2008) because of the advanced age of the study population (mean age at inclusion was 63 years) who had a high prevalence of cardiovascular risk factors. Many of the complications noted for participants in the WHI study such as stroke, pulmonary embolism, and myocardial infarction, which were attributed to the vascular effects of estrogens, could well have been due to pre-existing arteriosclerosis.

The WHI conclusions have now been partially counteracted by a reanalysis (Rossouw et al. 2007), which showed that the cardiovascular complications can be reduced using replacement therapy early in the perimenopause. This has opened a window of opportunity in which possibly even a cardiovascular benefit can be obtained in healthy menopausal women when replacement therapy is started early after the menopause (for review, see Riecher-Rössler and de Geyter 2007).

Overall, the use of estrogens for *therapeutic* reasons must be distinguished from their *preventative* use. The WHI investigated the latter use and as described above, concerns about the conclusions have been made. The potential *therapeutic* use in women with psychosis or other mental disorders has been discussed here, and while care must be given to monitoring potential side effects, the benefits for women with poor quality of life due to intractable schizophrenia often outweigh the risk of side effects.

Further research needs to be done into the best mode of HRT for psychiatric patients. To date, the natural 17-β-estradiol has been shown to have the best neuropsychoprotective effects compared with other estrogens that are often prescribed. Transdermal application in the form of patches or gel is preferred because of smoother metabolism and potentially fewer side effects. Progestogens are usually added to estrogens to prevent endometrial cancer, but they can antagonize the positive effects of estrogens with respect to mental state (Cyr et al. 2002; Braendle et al. 2001). These systemic effects of progestogens should be minimized by careful selection of the progestogen, and also, potentially by other forms of application.

Furthermore, alternatives to conventional HRT, i.e., compounds with more specific and potent estrogenic activity in the brain as opposed to other tissues, need to be investigated (Riecher-Rössler 2002; Halbreich 2002). Such compounds would both minimize the side effects of hormonal therapy and permit new therapeutic strategies in men. Possible candidates are selective estrogen receptor modulators, which have agonistic or antagonistic properties that depend on the target tissue. However, the effects of the available selective estrogen receptor modulators on the brain remain to be clarified. Raloxifene, for example, appears to exert its main effects on the bone, although recent data suggest that it also acts on different brain receptors (Craig et al. 2005). A recent study by Kulkarni et al. has shown promising antipsychotic effects for adjunctive raloxifene in postmenopausal women with schizophrenia (Kulkarni et al. 2010). Also, the synthetic steroid tibolone appears to cause less endometrial proliferation, but its effects on the central nervous system are still not clear, apart from the fact that it appears to have an androgenic effect and increases β-endorphin levels, with improvement in mood and libido (Davis 2002). Further studies on the brain-specific effects of selective estrogen receptor modulators and other estrogenic compounds (e.g., phyto-estrogens, xeno-estrogens, and dihydroepiandrosterone) are urgently needed.

5 Summary and Conclusions

In summary, there is emerging evidence that estrogens are very useful neuroprotective and psychoprotective adjunctive therapies, which could complement and enhance the traditional drug therapies for people with schizophrenia and related psychoses. However, it must be emphasized that the role for estrogen treatment strategies still requires further research. In particular, results from larger, well-controlled studies are needed before estrogens may be recommended as adjunct therapy in standard clinical practice for the treatment of women with schizophrenia, without proven estrogen deficiency.

In contrast, other strategies should already be part of standard clinical care (Grigoriadis and Seeman 2002). These include examination of the gonadal axis, with therapeutic actions, if indicated. In peri- and postmenopause HRT can be used, but the decision to use estrogen replacement therapy must always be made on the basis of an individual risk–benefit assessment (Brikhäuser et al. 2008) and in close collaboration with a gynecologist.

For future research, many questions remain unresolved, regarding not only new therapeutic strategies and compounds but also the poorly understood disturbances of estrogens and the hypothalamic–pituitary–gonadal axis in women with schizophrenia and related psychoses. Further research in this area may substantially contribute to our understanding of the pathogenesis of this disease, at least in a subgroup of women.

References

Ahokas A, Aito M, Rimon R (2000) Positive treatment effect of estradiol in postpartum psychosis: a pilot study. J Clin Psychiatry 61:166–9

Akhondzadeh S, Nejatisafa AA, Amini H, Mohammadi MR, Larijani B, Kashani L et al (2003) Adjunctive estrogen treatment in women with chronic schizophrenia: a double-blind, randomized, and placebo-controlled trial. Prog Neuropsychopharmacol Biol Psychiatry 27:1007–12

Althaus G, Pfuhlmann B, Franzek E (2000) The effect of the menstruation cycle on manifestations of pychiatric diseases. Fortschr Neurol Psychiatr 68:357–62

Balthazart J, Ball GF (2006) Is brain estradiol a hormone or a neurotransmitter? Trends Neurosci 29:241–9

Bergemann N, Parzer P, Nagl I, Salbach B, Runnebaum B, Mundt C et al (2002) Acute psychiatric admission and menstrual cycle phase in women with schizophrenia. Arch Womens Ment Health 5:119–26

Bergemann N, Parzer P, Runnebaum B, Resch F, Mundt C (2007a) Estrogen, menstrual cycle phases, and psychopathology in women suffering from schizophrenia. Psychol Med 37: 1427–36

Bergemann N, Abu-Tair F, Strowitzki T (2007b) Estrogen in the treatment of late-onset schizophrenia. J Clin Psychopharmacol 27:718–20

Bergemann N, Parzer P, Jaggy S, Auler B, Mundt C, Maier-Braunleder S (2008) Estrogen and comprehension of metaphoric speech in women suffering from schizophrenia: results of a double-blind, placebo-controlled trial. Schizophr Bull 34:1172–81

Birkhäuser MH, Panay N, Archer DF, Barlow D, Burger H, Gambacciani M et al (2008) Updated practical recommendations for hormone replacement therapy in the peri- and postmenopause. Climacteric 11:108–23

Bleuler M (1943) Die spätschizophrenen Krankheitsbilder. Neurology 15:259–90

Braendle W, Breckwoldt M, Kuhl H (2001) Sexualhormone und Psyche – Ergebniosse des 2. Interdisziplinären Frankfurter Gesprächs zur Kontrazeption. Frauenarzt 42:154–60

Canuso CM, Goldstein JM, Wojcik J, Dawson R, Brandman D, Klibanski A et al (2002) Antipsychotic medication, prolactin elevation, and ovarian function in women with schizophrenia and schizoaffective disorder. Psychiatry Res 111:11–20

Choi SH, Kang SB, Joe SH (2001) Changes in premenstrual symptoms in women with schizophrenia: a prospective study. Psychosom Med 63:822–9

Chua W, De Izquierdo S, Kulkarni J, Mortimer AM (2005) Estrogen for schizophrenia. Cochrane Database Syst Rev. doi:10.1002/14651858.CD0044719.pub2

Cohen RZ, Seeman MV, Gotowiec A, Kopala L (1999) Earlier puberty as a predictor of later onset of schizophrenia in women. Am J Psychiatry 156:1059–64

Craig MC, Maki PM, Murphy DG (2005) The Women's Health Initiative Memory Study: findings and implications for treatment. Lancet Neurol 4:190–4

Cyr M, Calon F, Morissette M, Di Paolo T (2002) Estrogenic modulation of brain activity: implications for schizophrenia and Parkinson's disease. J Psychiatry Neurosci 27:12–27

Davis SR (2002) The effects of tibolone on mood and libido. Menopause 9:162–70

Garcia-Segura L, Azcoitia I, Doncarlos L (2001) Neuroprotection by estradiol. Prog Neurobiol 63:29–60

Goldstein JM, Seidman LJ, O'Brien LM, Horton NJ, Kennedy DN, Makris N et al (2002) Impact of normal sexual dimorphisms on sex differences in structural brain abnormalities in schizophrenia assessed by magnetic resonance imaging. Arch Gen Psychiatry 59:154–64

Good KP, Kopala LC, Martzke JS, Fluker M, Seeman MV, Parish B et al (1999) Hormone replacement therapy in postmenopausal women with schizophrenia: preliminary findings. Schizophr Res 12:131

Grigoriadis S, Seeman MV (2002) The role of estrogen in schizophrenia: implications for schizophrenia practice guidelines for women. Can J Psychiatry 47:437–42

Häfner H (2005) Gender differences in schizophrenia. In: Bergemann N, Riecher-Rössler A (eds) Estrogen effects in psychiatric disorders. Springer, Wien, pp 53–94

Häfner H, Behrens S, De Vry J, Gattaz WF, Löffler W, Maurer K et al (1991a) Warum erkranken Frauen später an Schizophrenie? Erhöhung der Vulnerabilitätsschwelle durch Östrogen. Nervenheilkunde 10:154–63

Häfner H, Riecher-Rössler A, Hambrecht M, Maurer K, Meissner S, Schmidtke A et al (1991b) Geschlechtsunterschiede bei schizophrenen Erkrankungen. Fortschr Neurol Psychiatr 59: 343–60

Häfner H, Riecher-Rössler A, An Der Heiden W, Maurer K, Fatkenheuer B, Löffler W (1993) Generating and testing a causal explanation of the gender difference in age at first onset of schizophrenia. Psychol Med 23:925–40

Halbreich U (2002) The spectrum of estrogens, estrogen agonists and serms. Int J Neuropsychopharmacol 5(Suppl 1):12

Harris AH (1997) Menstrually related symptom changes in women with schizophrenia. Schizophr Res 27:93–9

Hayeems R, Seeman MV (2005) Puberty and schizophrenia onset. In: Bergemann N, Riecher-Rössler A (eds) Estrogen effects in psychiatric disorders. Springer, Wien, p 95

Hlatky MA, Boothroyd D, Vittinghoff E, Sharp P, Whooley MA (2002) Quality-of-life and depressive symptoms in postmenopausal women after receiving hormone therapy: results from the Heart and Estrogen/Progestin Replacement Study (HERS) trial. JAMA 287:591–7

Hoff AL, Harris D, Faustman WO, Beal M, DeVilliers D, Mone RD et al (1996) A neuropsychological study of early onset schizophrenia. Schizophr Res 20:21–8

Hoff AL, Kremen WS, Wieneke MH, Lauriello J, Blankfeld HM, Faustman WO et al (2001) Association of estrogen levels with neuropsychological performance in women with schizophrenia. Am J Psychiatry 158:1134–9

Honigfeld G, Gillis R, Klett C (eds) (1976) NOSIE: nurses' observation scale for inpatient evaluation. NIMH, Rockville

Huber TJ, Rollnik J, Wilhelms J, von zur Muhlen A, Emrich HM, Schneider U (2001) Estradiol levels in psychotic disorders. Psychoneuroendocrinology 26:27–35

Ko YH, Joe SH, Cho W, Park JH, Lee JJ, Jung IK et al (2006) Estrogen, cognitive function and negative symptoms in female schizophrenia. Neuropsychobiology 53:169–75

Könnecke R, Häfner H, Maurer K, Löffler W, An der Heiden W (2000) Main risk factors for schizophrenia: increased familial loading and pre- and peri-natal complications antagonize the protective effect of oestrogen in women. Schizophr Res 44:81–93

Korhonen S, Saarijarvi S, Aito M (1995) Successful estradiol treatment of psychotic symptoms in the premenstrual phase: a case report. Acta Psychiatr Scand 92:237–8

Kulkarni J (2005) Clinical estrogen trials in patients with schizophrenia. In: Bergemann N, Riecher-Rössler A (eds) Estrogen effects in psychiatric disorders. Springer, Vienna, pp 107–23

Kulkarni J (2009) Oestrogen – a new treatment approach for schizophrenia? Med J Aust 190: S37–8

Kulkarni J, de Castella A, Smith D, Taffe J, Keks N, Copolov D (1996) A clinical trial of the effects of estrogen in acutely psychotic women. Schizophr Res 20:247–52

Kulkarni J, Riedel A, de Castella AR, Fitzgerald PB, Rolfe TJ, Taffe J et al (2001) Estrogen – a potential treatment for schizophrenia. Schizophr Res 48:137–44

Kulkarni J, Gurvich C, Lee SJ, Gilbert H, Gavrilidis E, de Castella A et al (2010) Piloting the effective therapeutic dose of adjunctive selective estrogen receptor modulator treatment in postmenopausal women with schizophrenia. Psychoneuroendocrinology 35(8):1142–1147

Labhart A (1978) Klinik der inneren Sekretion. Springer, Berlin

Lindamer LA, Lohr JB, Harris MJ, Jeste DV (1997) Gender, estrogen, and schizophrenia. Psychopharmacol Bull 33:221–8

Lindamer LA, Buse DC, Lohr JB, Jeste DV (2001) Hormone replacement therapy in postmenopausal women with schizophrenia: positive effect on negative symptoms? Biol Psychiatry 49:47–51

Lindamer LA, Buse DC, Auslander L, Unutzer J, Bartels SJ, Jeste DV (2003) A comparison of gynecological variables and service use among older women with and without schizophrenia. Psychiatr Serv 54:902–4

Louza MR, Marques AP, Elkis H, Bassitt D, Diegoli M, Gattaz WF (2004) Conjugated estrogens as adjuvant therapy in the treatment of acute schizophrenia: a double-blind study. Schizophr Res 66:97–100

Maguire GA (2002) Prolactin elevation with antipsychotic medications: mechanisms of action and clinical consequences. J Clin Psychiatry 63(Suppl 4):56–62

Mahe V, Dumaine A (2001) Oestrogen withdrawal associated psychoses. Acta Psychiatr Scand 104:323–31

Mall G (1960) Diagnostik und Therapie ovarieller Psychosen. Zentralbl Gesamte Neurol Psychiatr 155:250

McEwen B (2002) Interplay between membrane and genomic actions of estrogens. Int J Neuropsychopharmacol 5(Suppl 1):12

Neumann NU, Frasch K (2001) Olanzapine and pregnancy. 2 case reports. Nervenarzt 72:876–8

Oesterlund M (2002) The role of estrogens in neuropsychiatric disorders. Curr Opin Psychiatry 15:307–12

Oesterlund MK, Keller E, Hurd YL (2000) The human forebrain has discrete estrogen receptor alpha messenger RNA expression: high levels in the amygdaloid complex. Neuroscience 95:333–42

Overall J, Gorham D (1962) The brief psychiatric rating scale. Psychol Rep 10:799–812

Riecher A, Maurer K, Löffler W, Fätkenheuer B, An Der Heiden W, Munk-Jorgensen P (1991) Gender differences in age at onset and course of schizophrenic disorders. In: Häfner H, Gattaz W (eds) Search for the causes of schizophrenia. Springer, Berlin, pp 14–33

Riecher-Rössler A (2002) Oestrogen effects in schizophrenia and their potential therapeutic implications – review. Arch Womens Ment Health 5:111–8

Riecher-Rössler A, de Geyter C (2007) The forthcoming role of treatment with oestrogens in mental health. Swiss Med Wkly 137:565–72

Riecher-Rössler A, Häfner H (1993) Schizophrenia and oestrogens – is there an association? Eur Arch Psychiatry Clin Neurosci 242:323–8

Riecher-Rössler A, Häfner H, Dutsch-Strobel A, Oster M, Stumbaum M, van Gulick-Bailer M et al (1994a) Further evidence for a specific role of estradiol in schizophrenia? Biol Psychiatry 36:492–4

Riecher-Rössler A, Häfner H, Stumbaum M, Maurer K, Schmidt R (1994b) Can estradiol modulate schizophrenic symptomatology? Schizophr Bull 20:203–14

Riecher-Rössler A, Löffler W, Munk-Jorgensen P (1997) What do we really know about late-onset schizophrenia? Eur Arch Psychiatry Clin Neurosci 247:195–208

Riecher-Rössler A, Häfner H, Dütsch-Strobel A, Stumbaum M (1998) Gonadal function and its influence on psychopathology. A comparison of schizophrenic and non-schizophrenic female inpatients. Arch Womens Ment Health 1:15–26

Rossouw JE, Anderson GL, Prentice RL, LaCroix AZ, Kooperberg C, Stefanick ML et al (2002) Risks and benefits of estrogen plus progestin in healthy postmenopausal women: principal results From the Women's Health Initiative randomized controlled trial. JAMA 288:321–33

Rossouw JE, Prentice RL, Manson JE, Wu L, Barad D, Barnabei VM et al (2007) Postmenopausal hormone therapy and risk of cardiovascular disease by age and years since menopause. JAMA 297:1465–77

Seeman MV (1996) Schizophrenia, gender, and affect. Can J Psychiatry 41:263–4

Sherwin BB (2005) Estrogen and memory in women: how can we reconcile the findings? Horm Behav 47:371–5

Smith S, Wheeler MJ, Murray R, O'Keane V (2002) The effects of antipsychotic-induced hyperprolactinaemia on the hypothalamic-pituitary-gonadal axis. J Clin Psychopharmacol 22:109–14

Stahl SM (2001a) Effects of estrogen on the central nervous system. J Clin Psychiatry 62:317–8

Stahl SM (2001b) Why drugs and hormones may interact in psychiatric disorders. J Clin Psychiatry 62:225–6
Thompson KN, Kulkarni J, Sergejew AA (2000) Extrapyramidal symptoms and oestrogen. Acta Psychiatr Scand 101:130–4
Vedder H, Behl C (2005) Estrogens in neuropsychiatric disorders: from physiology to pathophysiology. In: Bergemann N, Riecher-Rössler A (eds) Estrogen effects in psychiatric disorders. Springer, Wien, pp 1–30
von Zerssen D, Koeller D (1976) Klinische Selbstbeurteilungs-Skalen (KSb-Si) aus dem Münchener Psychiatrischen Informations-System (PSYCHIS München). Manual. Beltz, Weinheim
Warner MD, Walker AM, D'Souza DC, Lee D, Nasseri D, Peabody CA (2001) Lower prolactin bioactivity in unmedicated schizophrenic patients. Psychiatry Res 102:249–54
Weickert CS, Miranda-Angulo AL, Wong J, Perlman WR, Ward SE, Radhakrishna V et al (2008) Variants in the estrogen receptor alpha gene and its mRNA contribute to risk for schizophrenia. Hum Mol Genet 17:2293–309
Yonkers KA, Kando JC, Cole JO, Blumenthal S (1992) Gender differences in pharmacokinetics and pharmacodynamics of psychotropic medication. Am J Psychiatry 149:587–95
Zhang-Wong JH, Seeman MV (2002) Antipsychotic drugs, menstrual regularity and osteoporosis risk. Arch Womens Ment Health 5:93–8

Oestradiol and Psychosis: Clinical Findings and Biological Mechanisms

Angelika Wieck

Contents

Abstract Female sex steroids easily access the central nervous system and modulate a number of intracerebral processes via their specific receptors. Oestradiol is the biologically dominant female sex steroid and has been implicated in the aetiology and course of psychotic illnesses. There is evidence for interaction between oestradiol and several neurobiological systems that have been implicated in the pathogenesis of psychotic illnesses. Clinical studies have indicated that psychosis, and in particular schizophrenia, is associated with reduced ovarian function and that this may be inherent to the illness itself. In schizophrenia several studies have suggested a therapeutic effect of oestradiol and selective oestrogen modulators although research is still at an early stage. In bipolar disorder, the relationship between childbirth and first onsets or recurrences is one of the most reproducible findings in psychiatric research. Whether or not the rapid fall of oestrogens is the mediating mechanism is not yet clear but preliminary oestrogen treatment studies commenced immediately after childbirth are promising. Outside the perinatal context, tamoxifen, a selective oestrogen receptor modulator, has

A. Wieck
Laureate House, Wythenshawe Hospital, Manchester Mental Health and Social Care Trust, University of Manchester, Southmoor Road, Manchester M239LT, UK
e-mail: angelika.wieck@manchester.ac.uk

J.C. Neill and J. Kulkarni (eds.), *Biological Basis of Sex Differences in Psychopharmacology*, 173
Current Topics in Behavioral Neurosciences 8, DOI 10.1007/7854_2011_127,

shown strong antimanic effects although further studies are necessary to test an effect in larger samples. Hormonal treatments should not yet be used in standard care but could be considered in women with treatment resistant psychoses.

Keywords Bipolar · Mechanism · Oestrogen · Oestradiol · Psychosis · SERM · Schizophrenia · Treatment

1 Introduction

Research into the biological mechanisms that mediate gender differences in psychosis and the influence of reproductive events on the course of psychosis in women has focussed on female sex steroids, and particularly oestradiol as the predominant ovarian steroid. In this chapter, the relationships between oestradiol, ovarian function and the course of the two major psychoses, namely schizophrenia and bipolar disorder, the biological mechanisms that may be involved and recent treatment trials that have used oestrogens and oestrogen receptor modulators will be discussed.

2 Clinical Studies That Link Oestradiol with Schizophrenia

There are several lines of evidence that link oestrogen with the course and severity of schizophrenia. In epidemiological studies, the age of illness onset has consistently been shown to be 4–6 years later than in men (Lewine 1988; Häfner et al. 1991; Castle and Murray 1991). This applies not only to the first sign, first negative symptom, first positive symptoms, and first full episode, but also to the first admission (Häfner 2003). Women have also been reported to have a second peak of onset in the perimenopause (Häfner 2003).

Several recent studies have reported that female inpatients and outpatients with schizophrenia tend to have hormonal profiles consistent with insufficient maturation of ovarian follicles and anovulatory cycles, even during treatment with prolactin-sparing antipsychotic medication (Riecher-Roessler et al. 1994; Huber et al. 2001; Canuso et al. 2002; Bergemann et al. 2005). This indicates a hypothalamic–pituitary dysfunction or a reduced ovarian sensitivity to pituitary peptide stimulation. However, alternative explanations for an impaired ovarian activity such as weight gain (Brewer and Balen 2010) induced by psychotropic medication and stress (Genazzani et al. 2010) on ovarian activity have not been ruled out. Despite the low oestradiol levels in schizophrenia, their fluctuations have nevertheless been shown to influence symptom severity. Riecher-Roessler et al. (1994) investigated 32 women, who were admitted for the treatment of an acute psychotic exacerbation and were on antipsychotic medication. In the weeks subsequent to admission, measures of overall illness severity, thought disturbance and paranoia inversely correlated with oestradiol levels. Furthermore, Seeman (1983) described that women ageing between 20 and 40 years require lower antipsychotic doses than older women or men and suggested that oestrogen may have anti-dopaminergic action. In line with this observation, Gattaz et al. (1994) showed a significant inverse association between oestradiol plasma concentrations and the required dose of antipsychotic medication.

Childbirth is the reproductive event that is associated with the largest changes in female sex steroid production. During pregnancy, oestradiol and progesterone serum concentrations gradually increase and reach values at term that are 100–200 times higher than in the early follicular phase of the menstrual cycle. Immediately following the expulsion of the placenta, levels begin to decline rapidly and reach follicular phase levels within 2–3 days. It has been known for some time that childbirth can trigger psychotic illnesses, but two recent epidemiological studies covering the whole population of Denmark over three decades have quantified this effect for different diagnoses more precisely (Munk-Olsen et al. 2006, 2009). In these studies, inpatient admissions for first onsets and recurrences of schizophrenia were 2–5.7 times more common in the 2 months after childbirth than later in the first postnatal year.

Based on these several strands of evidence, it has been suggested that oestrogen protects women from an early onset and severe course of illness, but that this advantage is lost in the perimenopause when oestrogen production declines (Hafner et al. 1989, Seeman and Lang 1990) These findings also suggest that women with schizophrenia may benefit from treatment with oestradiol to enhance this effect and that the dose required may not be large.

3 Clinical Studies That Link Oestradiol with Bipolar Disorder

There are no findings of a gender difference in the age of onset in bipolar disorder. However, one of the most remarkable findings in psychiatry is the powerful relationship of this illness with childbirth. In the studies by Munk-Olsen et al. (2006, 2009), the risk of a first or subsequent inpatient admission for bipolar disorder was increased more than 20 times in the first month after childbirth. Based on the dopamine hypothesis of bipolar disorder, Cookson (1982) suggested that bipolar episodes in the puerperium may be triggered by the rapid and massive decline of oestrogen or progesterone after childbirth via its effect on the dopaminergic system (Cookson 1982).

Whether the fall of sex steroid concentrations at the end of the menstrual cycle is associated with a worsening of bipolar symptoms is controversial. Two early retrospective studies found greater affective symptomatology in the premenstrual phase in women with bipolar disorder than healthy controls (Diamond et al. 1976; Price and DiMarzio 1986). However, in a cross-sectional interview study of women with recurrent major depressive disorder ($N = 509$) and bipolar 1 disorder ($N = 197$), Payne et al. (2007) found significant associations between histories of postpartum affective symptoms and premenstrual or perimenopausal mood changes in the major depression group (odds ratios 1.82 and 1.66) but not in the bipolar group. Karadag et al. (2004) prospectively followed 34 stable medicated patients with bipolar disorder and 35 healthy controls. Participants kept daily records of their mental and physical wellbeing over 2 months and completed an interview on their experiences across menstrual cycles. The bipolar group complained less often about mood changes, such as mood lability, depressive symptoms, and anger and irritability than the control group. In two prospective studies of rapid cycling bipolar patients, there were also no significant effects of the premenstrual cycle phase on bipolar mood changes (Wehr et al. 1988; Leibenluft et al. 1999).

There are two reasons why these studies are difficult to interpret. First, bipolar patients in these studies were on mood stabilizing medication, which may have suppressed premenstrual affective symptoms. Second, several authors commented on the high rate of menstrual cycle abnormalities in the bipolar subjects, and this may mean that their endocrine states across menstrual cycles were different from the controls. As in women with schizophrenia, several explanations need to be considered for such ovarian dysfunction. These include antipsychotic-induced hyperprolactinaemia (Wieck and Haddad 2003), valproate-induced polycystic ovarian syndrome (Joffe et al. 2006) or a dysregulation of the hypothalamic–pituitary– ovarian axis inherent in bipolar disorder. Rasgon et al. (2005) found that half of their 80 female study participants with bipolar disorder reported on interview that they had menstrual irregularities before commencing anti-bipolar medication. Similarly, in the Harvard Study of Moods and Cycles, Joffe et al. (2006) found that significantly more women with bipolar disorder reported early-onset menstrual cycle dysfunction (34.2%, before the onset of their illness) than healthy controls (21.7%).

Brockington (2005) reviewed the world literature on bipolar-like psychoses that repeatedly recur at the same phase of the menstrual cycle. However, this phenomenon has mostly been described in case reports or case series and does not appear to occur commonly.

4 Oestradiol in the Brain

About 1–3% of the total plasma oestrogen and progesterone circulate in serum unbound to proteins (Wu et al. 1976; Darne et al. 1987; Meulenberg and Hofman 1989) and are free to enter the brain by diffusion. The concentration of total oestradiol and progesterone in CSF correlates strongly and significantly with that in plasma (Bäckström et al. 1976), and the free fraction in plasma is of about the same magnitude as the total hormone in cerebrospinal fluid (Schwarz and Pohl 1992). On the other hand, CSF levels of the sex steroid-binding proteins (albumin, steroid hormone-binding globulin and cortisol-binding globulin) are several hundred times lower than in serum, suggesting that most of the total oestrogen and progesterone in CSF are free (Schwarz and Pohl 1992).

Oestrogens have a broad spectrum of actions in the central nervous system, which are mediated either via rapid alteration in signal transduction via membrane receptors or by slower modulation of gene transcription via receptors located intracellularly (Marshall 2011). Two oestrogen receptor subtypes, alpha and beta, have been identified, and their genes are located on different chromosomes. In the human forebrain, both subtypes are predominantly expressed in limbic-related areas although their distribution patterns differ (see review by Hughes et al. 2009). The mRNA expression of the alpha receptor appears to dominate in the hypothalamus and amygdala, areas concerned with autonomic and reproductive neuroendocrine functions as well as emotion interpretation and processing, whereas the beta isoform is dominant in the hippocampal formation, the entorhinal cortex and the thalamus, suggesting a possible role in cognition, non-emotional memory and motor functions (Hughes et al. 2009).

5 Mechanisms That May Mediate Oestradiol Effects in Schizophrenia

Oestrogen effects on the function of the mesolimbic and mesocortical dopaminergic and hippocampal glutamatergic systems are particularly relevant to psychoses. It is clear that dopaminergic neurones have oestrogen receptors (Creutz and Kritzer 2002), and that oestrogen interacts with dopaminergic systems (Sánchez et al. 2010). However, there is substantial variability in the direction of effects, reflecting differences in dose, duration of treatment, experimental protocols and outcomes studied (Chavez et al. 2010; Sánchez et al. 2010).

An experimental tool that has been widely used in the exploration of neurobiological mechanisms in schizophrenia is prepulse inhibition (PPI); see chapter by Veena Kumari (2011) for more details. This is the reduction of a startle response to a sudden loud noise if it is preceded by a weak prepulse within 30–500 ms. The inhibition of this reflex is modulated by brain circuits linking the limbic cortex, striatum, pallidum and pontine tegmentum and higher brain structures, such as the prefrontal cortex and hippocampus (Swerdlow et al. 1997), so that sensory information can be filtered and attention focussed. PPI is reduced in schizophrenia, some other psychiatric and several neurological disorders (Braff et al. 2001). A reduction in PPI during treatment with dopamine agonists and a reversal by antipsychotic agents in rodents suggest a contribution of the dopamine system in the modulation of PPI (Zhang et al. 2007; Mansbach et al. 1988) In schizophrenic patients, the PPI deficit may also be reversed by atypical antipsychotic treatment (Wynn et al. 2007; Aggernaes et al. 2010). Furthermore, recent studies suggest that dopaminergic modulation of PPI can be influenced by the hormonal milieu. For example, in female ovariectomized rats, high dose oestrogen treatment over 2 weeks has been shown to prevent apomorphine-induced disruptions of PPI, and that this may be mediated via an action on dopamine D2 receptors (Gogos et al. 2010). In healthy women, variations in sex steroid levels are also accompanied by changes in PPI. A greater PPI has been observed during the follicular phase relative to the luteal phase of the menstrual cycle (Swerdlow et al. 1997; Jovanovic et al. 2004). However, Kumari et al. (2010) did not find a correlation between changes in PPI and oestradiol levels between menstrual cycle phases but reported that a greater progesterone increase in the luteal phase was associated with a greater protection of PPI. The relationships between physiological changes in female steroid production, dopaminergic function and PPI have not yet been investigated in human females with or without schizophrenia.

However, the usefulness of PPI as a model for neural processes operating in schizophrenia is limited, since PPI decreases are neither specific for schizophrenia nor do they predict the pattern of symptoms, the course of illness or individual treatment responses (Swerdlow et al. 2008).

Although research of neurotransmitter function in schizophrenia focussed for many years on excess dopaminergic function in the striatum, more recently it has been suggested that this is secondary to dysfunctional glutamatergic neurotransmission (Carlsson et al. 2001), a hypothesis that has recently been confirmed by Stone et al. (2010). In this imaging study of drug-free subjects at very high risk of schizophrenia, the authors found a negative correlation between glutamate levels in the hippocampus and striatal dopaminergic activity. Preliminary evidence indicates that oestradiol can modulate glutamatergic function in the hippocampus and other brain regions (Smejkalova and Woolley 2010; Grove-Strawser et al. 2010). Whether this interaction is relevant to schizophrenia requires further research in preclinical and human studies.

6 Mechanisms That May Mediate Oestradiol Effects in Bipolar Disorder

Although effective treatments are available for bipolar disorder, a comprehensive pathophysiological model of the illness is still lacking (Cousins et al. 2009). In their review of the role of dopamine in bipolar disorder, Cousins et al. (2009) conclude that this neurotransmitter system is likely to play a central role in the understanding of the pathophysiology of this illness. They also suggested that it may be the second messenger systems and downstream pathways that are directly involved rather than presynaptic processes or postsynaptic dopamine receptors. Overactivity of one element in the dopamine signal transduction pathway, protein kinase C, has been associated with acute mania and several known antimanic agents, such as lithium, valproate, carbamazepine, aripiprazole and quetiapine, inhibit it (reviews by Zarate and Manji 2009; Cousins et al. 2009). Recent studies have also shown that the selective oestrogen receptor modulator tamoxifen is a centrally active protein kinase C inhibitor and has therefore been tested as an antimanic agent.

There are to date only two biological studies in humans that investigated biological mechanisms that might be involved in the triggering effect of childbirth on bipolar recurrences. To test the hypothesis that they are triggered by the effects of oestrogen withdrawal on the dopaminergic system, the apomorphine-induced growth hormone response was used as a measure of neurotransmission via hypothalamic D2 receptors. In the first study (Wieck et al. 1991), 15 drug-free women who had a history of bipolar illness but were currently well and 15 control women with no psychiatric history were recruited in late pregnancy. The neuroendocrine test was carried out on day 4 after delivery, and those who subsequently had a recurrence had a significantly higher response than those who remained well. However, the stimulation of hypothalamic growth hormone secretion is relatively blunted in the early postnatal period, and the study may have underestimated differences in the neuroendocrine response between the women who relapsed and those who did not. In the second study, hypothalamic D2 receptor sensitivity was tested across the menstrual cycle in eight medication-free well women with a history of puerperal bipolar illness and normal menstrual cycles and nine normally menstruating controls (Wieck et al. 2003). Subjects underwent the same apomorphine growth hormone test in the early follicular phase when oestrogen levels are low and in the midluteal phase when oestradiol levels are enhanced. The midluteal time point was defined as 7–11 days after the pre-ovulatory LH surge which was determined by ovulation test kits. Although the women with bipolar disorder were currently well, they had an enhanced response to apomorphine in the midluteal phase. This indicates that the hypothalamic dopaminergic system of women predisposed to puerperal bipolar illness responds more sensitively to changes in oestradiol levels. In view of recent findings, it is possible that this hypersensitivity is related to a dysfunction in the intracellular signalling system rather than the D2 receptors themselves. It is uncertain, however, whether such hypothalamic effects are representative of regulatory processes in the mesolimbic or mesocortical dopamine systems.

7 Treatment Studies in Schizophrenia

Chua et al. (2005) conducted a systematic review of oestrogen treatment trials in schizophrenia or related non-affective psychoses. They identified five randomized placebo-controlled studies including a total of 122 patients. Oestrogen treatments were adjunctive to antipsychotic therapy and lasted between 3 weeks and 6 months. No significant effect of active treatment was found on psychopathology. However, the authors concluded that the existing literature was difficult to interpret on account of the small sample sizes, randomization issues and the differences between studies in respect of menopausal status, phase of illness, type of oestrogen used, routes of administration and dose. A large double-blind randomized placebo-controlled trial ($N = 102$) of transdermal oestradiol treatment (100 μg over 24 h) over 28 days as adjunct to antipsychotic medication was published by Kulkarni et al. (2008). Participants were inpatients or outpatients, had a diagnosis of schizophrenia, schizophreniform disorder or schizoaffective depression and were in the acute or chronic phase of illness. At baseline, the mean score on the Positive and

Negative Symptoms Scale (PANSS) was similar to that reported by the authors of the scale for their sample of inpatients with schizophrenia (Kay et al. 1989). There were no differences between the groups in regard to adverse oestrogen-related or motor side effects. However, there were significant improvements in the active treatment group for the total PANSS score ($p < 0.002$), the positive symptom subscale ($p < 0.005$) and the general psychopathology subscale ($p < 0.01$), but not the negative symptom subscale. This pattern of response is often seen in trials of antipsychotic medication.

The transdermal application of 17-β oestradiol has the advantage of avoiding first pass liver metabolism. However, oestrogen treatment has several drawbacks. Its long-term effects in premenopausal women is not known, and in postmenopausal women it increases the risk of endometrial and breast cancer as well as myocardial infarct and stroke. Raloxifene is a promising alternative to 17-β oestradiol since it is an antagonist at the breast and does not promote endometrial cancer. However, it has been associated with a small increase in the risk of venous thromboembolism (Nelson et al. 2009). While the amounts entering the brain are reported to be small, pharmacological CNS effects have nevertheless been reported in animals (Littleton-Kearney et al. 2002) as well as in postmenopausal women (Neele et al. 2001).

Treatment with raloxifene was piloted in a dose-finding randomized controlled trial by Kulkarni et al. (2010) in 35 postmenopausal women in the acute phase of schizophreniform or schizoaffective disorder. Participants allocated to either 60 or 120 mg of raloxifene or placebo as adjuncts to antipsychotic medication over 12 weeks. Despite the small sample size, at the end of treatment there was a significant decrease in the total PANSS score ($p < 0.001$) and the general symptom score ($p < 0.02$) in the group treated with the larger dose.

8 Treatment Studies in Bipolar Disorder

Despite the high risk of bipolar recurrences in the immediate postpartum period, only few investigators have attempted testing the preventative or acute efficacy of pharmacological interventions. Due to ethical constraints, randomized controlled studies cannot be conducted in late pregnancy and large-scale observational naturalistic studies are logistically difficult, due to the relatively low number of patients at risk in the catchment areas of individual hospitals. Such studies require the collaboration of specialists in perinatal psychiatry from several centres.

Two open uncontrolled studies have prospectively tested the effect of oestrogen treatment commenced after delivery. The rationale was to soften the impact of the rapid decline of oestrogen levels following the expulsion of the placenta. Sichel et al. (1995) administered conjugated equine oestrogens to 11 drug-free women with histories of predominantly puerperal psychosis which is thought to be closely related to bipolar disorder. Oestrogens were administered immediately after childbirth at a high dose (eight times the maximum postmenopausal dose) with heparin protection, and the dose was gradually stepped down. Only one woman relapsed. This contrasts with the reported rates of postnatal recurrences of 26–57% in groups of women with a history of bipolar/schizoaffective disorder irrespective of their medication status (Reich and Winokur 1970; Dean et al. 1989; Robling et al. 2000; Jones and Craddock 2001; Robertson et al. 2005) and rates of 40% and 70%, respectively, in two medication-free samples (Marks et al. 1991; Viguera et al. 2000). In a dose-finding study of 29 drug-free women with histories of bipolar or schizoaffective disorder, transdermal patch treatment delivering 200, 400 and 800 μg oestradiol per 24 h was commenced within 48 h under heparin protection and stepped down over 12 days (Kumar et al. 2003). All women were admitted for at least the duration of the trial. The recurrence rate was not lower than expected (41%), but the highest starting dose (equivalent to 16 times the postmenopausal dose) was associated with a significantly shorter duration of admission than the two lower doses. It is possible that the difference between the two studies is due to the timing of the first dose. In the second study, treatment may have been initiated too late since the decline in oestradiol and progesterone is immediate and rapid after the expulsion of the placenta. However, due to the concern over postpartum thrombosis as well as postpartum bleeding during protective heparin treatment, the risks of commencing treatment immediately after childbirth could be too high to pursue this approach further.

In an open uncontrolled trial, ten women with established and severe puerperal psychosis were treated with oestradiol for 6 weeks (Ahokas et al. 2000). The term "puerperal psychosis" is often used in clinical practice to denote the proximity to childbirth without specifying the type of psychosis. Because a large proportion of these illnesses are on the bipolar spectrum, the study was included here. Participants had an illness onset on day 12 on average and entered the study at a mean of 12 weeks postpartum. Four patients had been treated unsuccessfully with antipsychotic medication before the trial. Oestradiol was administered sublingually to avoid first pass metabolism by the liver. None of the women had menstruated since delivery, and oestradiol levels were very low at baseline. The daily oestradiol dose was titrated according to serum concentrations with the aim of reaching concentrations of 400 pmol/L, i.e., about one-third of the peak level during the regular menstrual cycle. The score of the Brief Psychiatric Rating Scale fell dramatically within the first week from 78.3 to 18.8 ($p < 0.001$), and by week 2 the patients had become almost symptom-free. One woman discontinued oestradiol by week 5 and had a full recurrence of symptoms in week 6.

Recently, several studies have tested the antimanic properties of the selective oestrogen receptor modulator tamoxifen in male and female patients with bipolar disorder. In a single blind study, tamoxifen up to 80 mg daily lead to a rapid improvement in five of seven patients with mania (Bebchuk et al. 2000). In another small, 4-week, double-blind, placebo-controlled, add-on study, treatment with 40 mg tamoxifen ($n = 5$), medroxyprogesterone acetate ($n = 4$) and placebo ($n = 4$) was compared (Kulkarni et al. 2006). Subjects in the tamoxifen group had a significantly greater decrease in manic and positive psychotic symptoms compared to the placebo group ($p < 0.05$). All patients were receiving concomitant medication with either lithium or valproate. Similar results were obtained in two subsequent 3-week double-blind placebo-controlled, monotherapy studies. The study by Zarate et al. (2007) tested higher doses of up to 140 mg/day in a sample of 16 patients, and ratings were obtained daily during the first week. Significant improvements were seen as early as day 5, and the difference to placebo remained significant up to the end of treatment ($p < 0.001$) with a large effect size ($d = 1.08$). In the largest study ($N = 66$), Yildiz et al. (2008) used doses of up to 80 mg/day over 3 weeks. Significant improvement in the ratings of mania and clinical global impression (both $p < 0.001$) was reported at the end of treatment. In a fifth study (Amrollahi et al. 2011), tamoxifen or placebo were randomly added to treatment with lithium in 40 patients with acute mania. A significantly greater improvement was apparent at week 1 in the active treatment group, and this difference continued until the end of treatment in week 6.

The sample sizes in these trials are small, but the results are consistent and suggest that tamoxifen may be an effective antimanic agent that is well tolerated and acts rapidly. Drawbacks of the treatment are an increased risk of thromboembolic events and endometrial cancer. It may therefore be more suitable for short-term treatment of severe or treatment-resistant acute mania, particularly in men.

Although tamoxifen belongs to the group of selective oestradiol receptor modulators, it is thought that it involves a direct action on protein kinase C rather than the oestrogen receptor (O'Brian et al. 1986).

9 Conclusion

Insights into the various aspects of the relationship between oestradiol and psychotic illness are still limited. Although there is considerable evidence for a role of oestradiol, particularly for schizophrenia, not all available research supports this.

Nevertheless, in a recent well-designed randomized controlled treatment trial, women with acute schizophrenia-like illnesses responded well to the use of transdermal oestradiol (Kulkarni et al. 2008). The oestradiol dose of 100 μg/24 h is relatively low for a premenopausal patient group, since it only achieves levels that are similar to the early follicular phase of the menstrual cycle (Chetkowski et al. 1986). The mechanism by which the improvement was mediated is also not clear as yet. At the beginning of treatment, the patients were in different endocrine states. About one-third of the patients were taking antipsychotic medication with prolactin-elevating potential, and others entered the study in different phases of the menstrual cycle. The pattern of oestradiol levels across the 4 weeks of treatment that resulted from the interactions of endogenous hormone production, transdermally applied hormone and its potential to disrupt ovulation is difficult to predict. The mean oestradiol serum levels were low at baseline before treatment was begun (71.0 pg/ml). Potential mechanisms that could explain a therapeutic effect include that oestradiol levels stayed above a certain critical level or that the treatment led to less fluctuations over the treatment period than would have otherwise occurred. Further clinical trials by different research groups that examine pituitary gonadal peptide and female sex steroid secretion during treatment are required to test this promising treatment further. Preliminary data on the therapeutic efficacy of raloxifene are also encouraging, particularly because it has a less adverse side-effect profile. Further preclinical research is required as to its actions in the central nervous system.

In bipolar disorder, it is not clear whether there is a pre-existing immaturity in the hypothalamic–pituitary–ovarian axis, and future studies should clarify this important point. The strong relationship of bipolar disorder with childbirth is a much replicated finding, and the predominant hypothesis has been that the postpartum oestrogen withdrawal triggers a latent dopaminergic dysfunction. There is some indirect evidence to support this view. There is also uncontrolled evidence that oestradiol treatment may be useful in the treatment of puerperal bipolar or affective psychotic episodes although it may only be worth pursuing this treatment in established illness rather than in prevention due to the risks of oestradiol treatment immediately after childbirth. The potential of raloxifene or other new selective oestrogen receptor modulators in puerperal and nonpuerperal bipolar illness should be explored.

References

Aggernaes B, Glenthoj BY, Ebdrup BH, Rasmussen H, Lublin H, Oranje B (2010) Sensorimotor gating and habituation in antipsychotic-naive, first-episode schizophrenia patients before and after 6 months' treatment with quetiapine. Int J Neuropsychopharmacol 13(10):1383–1395

Ahokas A, Aito M, Rimón R (2000) Positive treatment effect of estradiol in postpartum psychosis: a pilot study. J Clin Psychiatry 61(3):166–169

Amrollahi Z, Rezaei F, Salehi B, Modabbernia AH, Maroufi A, Esfandiari GR, Naderi M, Ghebleh F, Ahmadi-Abhari SA, Sadeghi M, Tabrizi M, Akhondzadeh S (2011) Double-blind, randomized, placebo-controlled 6 week study on the efficacy and safety of the tamoxifen adjunctive to lithium in acute bipolar mania. J Affect Disord 129:327–331

Bäckström T, Carstensen H, Södergard R (1976) Concentration of estradiol, testosterone and progesterone in cerebrospinal fluid compared to plasma unbound and total concentrations. J Steroid Biochem 7(6–7):469–472

Bebchuk JM, Arfken CL, Dolan-Manji S, Murphy J, Hasanat K, Manji HK (2000) A preliminary investigation of a protein kinase C inhibitor in the treatment of acute mania. Arch Gen Psychiatry 57:95–97

Bergemann N, Mundt C, Parzer P, Jannakos I, Nagl I, Salbach B, Klinga K, Runnebaum B, Resch F (2005) Plasma concentrations of estradiol in women suffering from schizophrenia treated with conventional versus atypical antipsychotics. Schizophr Res 73:357–366

Braff DL, Geyer MA, Swerdlow NR (2001) Human studies of prepulse inhibition of startle: normal subjects, patient groups, and pharmacological studies. Psychopharmacology (Berl) 156(2–3):234–258. Review. PubMed PMID: 11549226

Brewer CJ, Balen AH (2010) The adverse effects of obesity on conception and implantation. Reproduction 140(3):347–364

Brockington I (2005) Menstrual psychosis. World Psychiatry 4(1):9–17

Canuso CM, Goldstein JM, Wojcik J et al (2002) Antipsychotic medication, prolactin elevation, and ovarian function in women with schizophrenia and schizoaffective disorder. Psychiatry Res 111:11–20

Carlsson A, Waters N, Holm-Waters S, Tedroff J, Nilsson M, Carlsson ML (2001) Interactions between monoamines, glutamate, and GABA in schizophrenia: new evidence. Annu Rev Pharmacol Toxicol 41:237–260

Castle DJ, Murray RM (1991) The neurodevelopmental basis of sex differences in schizophrenia. Psychol Med 20:285–304

Chavez C, Hollaus M, Scarr E, Pavey G, Gogos A, van den Buuse M (2010) The effect of estrogen on dopamine and serotonin receptor and transporter levels in the brain: an autoradiography study. Brain Res 1321:51–59

Chetkowski RJ, Meldrum DR, Steingold KA, Randle D, Lu JK, Eggena P, Hershman JM, Alkjaersig NK, Fletcher AP, Judd HL (1986) Biologic effects of transdermal estradiol. N Engl J Med 314(25):1615–1620

Chua WL, de Izquierdo SA, Kulkarni J, Mortimer A (2005) Estrogen for schizophrenia. Cochrane Database Syst Rev. (4):CD004719

Cookson JC (1982) Post-partum mania, dopamine, and estrogens. Lancet 2(8299):672

Cousins DA, Butts K, Young AH (2009) The role of dopamine in bipolar disorder. Bipolar Disord 11(8):787–806

Creutz LM, Kritzer MF (2002) Estrogen receptor-beta immunoreactivity in the midbrain of adult rats: regional, subregional, and cellular localization in the A10, A9, and A8 dopamine cell groups. J Comp Neurol 446(3):288–300

Darne J, McGarrigle HH, Lachelin GC (1987) Saliva oestriol, oestradiol, oestrone and progesterone levels in pregnancy: spontaneous labour at term is preceded by a rise in the saliva oestriol:progesterone ratio. Br J Obstet Gynaecol 94(3):227–235

Dean C, Williams RJ, Brockington IF (1989) Is puerperal psychosis the same as bipolar manic-depressive disorder? A family study. Psychol Med 19(3):637–647

Diamond SB, Rubinstein AA, Dunner DL, Fieve RR (1976) Menstrual problems in women with primary affective illness. Compr Psychiatry 17(4):541–548

Gattaz WF, Vogel P, Riecher-Rössler A, Soddu G (1994) Influence of the menstrual cycle phase on the therapeutic response in schizophrenia. Biol Psychiatry 36(2):137–139

Genazzani AD, Chierchia E, Santagni S, Rattighieri E, Farinetti A, Lanzoni C (2010) Hypothalamic amenorrhea: from diagnosis to therapeutical approach. Ann Endocrinol (Paris) 71 (3):163–169

Gogos A, Kwek P, Chavez C, van den Buuse M (2010) Estrogen treatment blocks 8-hydroxy-2-dipropylaminotetralin- and apomorphine-induced disruptions of prepulse inhibition: involvement of dopamine D1 or D2 or serotonin 5-HT1A, 5-HT2A, or 5-HT7 receptors. J Pharmacol Exp Ther 333:218–227

Grove-Strawser D, Boulware MI, Mermelstein PG (2010) Membrane estrogen receptors activate the metabotropic glutamate receptors mGluR5 and mGluR3 to bidirectionally regulate CREB phosphorylation in female rat striatal neurons. Neuroscience 170(4):1045–1055

Häfner H, Riecher A, Maurer K et al (1989) How does gender influence age at first hospitalization for schizophrenia? Psychol Med 19:903–918

Häfner H, Behrens S, De Vry J et al (1991) Estradiol enhances the vulnerability threshold for schizophrenia in women by an early effect on dopaminergic neurotransmission. Evidence from an epidemiological study and from animal experiments. Eur Arch Psychiatry Clin Neurosci 241:65–68

Häfner H (2003) Gender differences in schizophrenia. Psychoneuroendocrinology 2(Suppl):17–54

Huber TJ, Rollnik J, Wilhelms J, von zur Mühlen A, Emrich HM, Schneider U (2001) Estradiol levels in psychotic disorders. Psychoneuroendocrinology 26(1):27–35

Hughes ZA, Liu F, Marquis K, Muniz L, Pangalos MN, Ring RH, Whiteside GT, Brandon NJ (2009) Estrogen receptor neurobiology and its potential for translation into broad spectrum therapeutics for CNS disorders. Curr Mol Pharmacol 2(3):215–236

Joffe H, Cohen LS, Suppes T, McLaughlin WL, Lavori P, Adams JM, Hwang CH, Hall JE, Sachs GS (2006) Valproate is associated with new-onset oligoamenorrhea with hyperandrogenism in women with bipolar disorder. Biol Psychiatry 59(11):1078–1086

Jones I, Craddock N (2001) Familiality of the puerperal trigger in bipolar disorder: results of a family study. Am J Psychiatry 158(6):913–917

Jovanovic T, Szilagyi S, Chakravorty S, Fiallos AM, Lewison BJ, Parwani A, Schwartz MP, Gonzenbach S, Rotrosen JP, Duncan EJ (2004) Menstrual cycle phase effects on prepulse inhibition of acoustic startle. Psychophysiology 41(3):401–406

Karadag F, Akdeniz F, Erten E, Pirildar S, Yucel B, Polat A, Atmaca M (2004) Menstrually related symptom changes in women with treatment-responsive bipolar disorder. Bipolar Disord 6(3): 253–259

Kay SR, Opler LA, Lindenmeyer JP (1989) The Positive and Negative Syndrome Scale (Panss): rationale and standardisation. Br J Psychiatry Suppl 7:59–67

Kulkarni J, Garland KA, Scaffidi A, Headey B, Anderson R, de Castella A, Fitzgerald P, Davis SR (2006) A pilot study of hormone modulation as a new treatment for mania in women with bipolar affective disorder. Psychoneuroendocrinology 31:543–547

Kulkarni J, de Castella A, Fitzgerald PB, Gurvich CT, Bailey M, Bartholomeusz C, Burger H (2008) Estrogen in severe mental illness: a potential new treatment approach. Arch Gen Psychiatry 65(8):955–960

Kulkarni J, Gurvich C, Lee SJ, Gilbert H, Gavrilidis E, de Castella A, Berk M, Dodd S, Fitzgerald PB, Davis SR (2010) Piloting the effective therapeutic dose of adjunctive selective estrogen receptor modulator treatment in postmenopausal women with schizophrenia. Psychoneuroendocrinology 35(8):1142–1147

Kumar C, McIvor RJ, Davies T, Brown N, Papadopoulos A, Wieck A, Checkley SA, Campbell IC, Marks MN (2003) Estrogen administration does not reduce the rate of recurrence of affective psychosis after childbirth. J Clin Psychiatry 64(2):112–118

Kumari V, Konstantinou J, Papadopoulos A, Aasen I, Poon L, Halari R, Cleare AJ (2010) Evidence for a role of progesterone in menstrual cycle-related variability in prepulse inhibition in healthy young women. Neuropsychopharmacology 35:929–937

Kumari V (2011) Sex differences and hormonal influences in human sensorimotor gating: implications for schizophrenia. Curr Top Behav Neurosci. doi:10.1007/7854_2010_117

Leibenluft E, Ashman SB, Feldman-Naim S, Yonkers KA (1999) Lack of relationship between menstrual cycle phase and mood in a sample of women with rapid cycling bipolar disorder. Biol Psychiatry 46(4):577–580

Lewine RRJ (1988) Gender and schizophrenia. In: Nasrallah NH (ed) Handbook of schizophrenia, vol 3. Elsevier, Amsterdam, pp 121–136

Littleton-Kearney MT, Ostrowski NL, Cox DA, Rossberg MI, Hurn PD (2002) Selective estrogen receptor modulators: tissue actions and potential for CNS protection. CNS Drug Rev 8(3): 309–330

Mansbach RS, Geyer MA, Braff DL (1988) Dopaminergic stimulation disrupts sensorimotor gating in the rat. Psychopharmacology (Berl) 94(4):507–514

Marks MN, Wieck A, Checkley SA, Kumar R (1991) Life stress and post-partum psychosis: a preliminary report. Br J Psychiatry Suppl. (10):45–49

Marshall KM (2011) Introduction to the interaction between gonadal steroids and the central nervous system. Curr Topics Behav Neurosci. doi:10.1007/7854_2011_136

Meulenberg PM, Hofman JA (1989) Salivary progesterone excellently reflects free and total progesterone in plasma during pregnancy. Clin Chem 35(1):168–172

Munk-Olsen T, Laursen TM, Pedersen CB, Mors O, Mortensen PB (2006) New parents and mental disorders: a population-based register study. JAMA 296(21):2582–2589

Munk-Olsen T, Laursen TM, Mendelson T, Pedersen CB, Mors O, Mortensen PB (2009) Risks and predictors of readmission for a mental disorder during the postpartum period. Arch Gen Psychiatry 66(2):189–195

Neele SJ, Rombouts SA, Bierlaagh MA, Barkhof F, Scheltens P, Netelenbos JC (2001) Raloxifene affects brain activation patterns in postmenopausal women during visual encoding. J Clin Endocrinol Metab 86(3):1422–1424

Nelson HD, Fu R, Griffin JC, Nygren P, Smith ME, Humphrey L (2009) Systematic review: comparative effectiveness of medications to reduce risk for primary breast cancer. Ann Intern Med 151(10):703–715

O'Brian CA, Liskamp RM, Solomon DH, Weinstein IB (1986) Triphenylethylenes: a new class of protein kinase C inhibitors. J Natl Cancer Inst 76(6):1243–1246

Payne JL, Roy PS, Murphy-Eberenz K, Weismann MM, Swartz KL, McInnis MG, Nwulia E, Mondimore FM, MacKinnon DF, Miller EB, Nurnberger JI, Levinson DF, DePaulo JR Jr, Potash JB (2007) Reproductive cycle-associated mood symptoms in women with major depression and bipolar disorder. J Affect Disord 99(1–3):221–229

Price WA, DiMarzio L (1986) Premenstrual tension syndrome in rapid-cycling bipolar affective disorder. J Clin Psychiatry 47(8):415–417

Rasgon N, Altshuler L, Fairbanks L et al (2005) Reproductive function and risk for PCOS in women treated for bipolar disorder. Bipolar Disord 7:246–259

Reich T, Winokur G (1970) Postpartum psychoses in patients with manic depressive disease. J Nerv Ment Dis 151(1):60–68

Riecher-Roessler A, Häfner H, Dutsch-Strobel A et al (1994) Further evidence for a specfic role of estradiol in schizophrenia? Biol Psychiatry 36:492–494

Robertson E, Jones I, Haque S, Holder R, Craddock N (2005) Risk of puerperal and non-puerperal recurrence of illness following bipolar affective puerperal (post-partum) psychosis. Br J Psychiatry 186:258–259

Robling SA, Paykel ES, Dunn VJ, Abbott R, Katona C (2000) Long-term outcome of severe puerperal psychiatric illness: a 23 year follow-up study. Psychol Med 30(6):1263–1271

Sánchez MG, Bourque M, Morissette M, Di Paolo T (2010) Steroids-dopamine interactions in the pathophysiology and treatment of CNS disorders. CNS Neurosci Ther 16(3):e43–e71

Schwarz S, Pohl P (1992) Steroid hormones and steroid hormone binding globulins in cerebrospinal fluid studied in individuals with intact and with disturbed blood-cerebrospinal fluid barrier. Neuroendocrinology 55(2):174–182

Seeman M (1983) Interaction of sex, age and neuroleptic dose. Compr Psychiatry 24:125–128

Seeman MV, Lang M (1990) the role of estrogens in schizophrenia and gender difference. Schizophr Bull 16:185–194

Sichel DA, Cohen LS, Robertson LM, Ruttenberg A, Rosenbaum JF (1995) Prophylactic estrogen in recurrent postpartum affective disorder. Biol Psychiatry 38(12):814–818

Smejkalova T, Woolley CS (2010) Estradiol acutely potentiates hippocampal excitatory synaptic transmission through a presynaptic mechanism. J Neurosci 30(48):16137–16148

Stone JM, Howes OD, Egerton A, Kambeitz J, Allen P, Lythgoe DJ, O'Gorman RL, McLean MA, Barker GJ, McGuire P (2010) Altered relationship between hippocampal glutamate levels and striatal dopamine function in subjects at ultra high risk of psychosis. Biol Psychiatry 68(7):599–602

Swerdlow NR, Hartman PL, Auerbach PP (1997) Changes in sensorimotor inhibition across the menstrual cycle: implications for neuropsychiatric disorders. Biol Psychiatry 41(4):452–460

Swerdlow NR, Weber M, Qu Y, Light GA, Braff DL (2008) Realistic expectations of prepulse inhibition in translational models for schizophrenia research. Psychopharmacology (Berl) 199(3): 331–388

Viguera AC, Nonacs R, Cohen LS, Tondo L, Murray A, Baldessarini RJ (2000) Risk of recurrence of bipolar disorder in pregnant and nonpregnant women after discontinuing lithium maintenance. Am J Psychiatry 157(2):179–184

Wehr TA, Sack DA, Rosenthal NE, Cowdry RW (1988) Rapid cycling affective disorder: contributing factors and treatment responses in 51 patients. Am J Psychiatry 145(2):179–184

Wieck A, Haddad PM (2003) Antipsychotic-induced hyperprolactinaemia in women: pathophysiology, severity and consequences. Selective literature review. Br J Psychiatry 182:199–204

Wieck A, Kumar R, Hirst AD, Marks MN, Campbell IC, Checkley SA (1991) Increased sensitivity of dopamine receptors and recurrence of affective psychosis after childbirth. BMJ 303(6803): 613–616

Wieck A, Davies RA, Hirst AD, Brown N, Papadopoulos A, Marks MN, Checkley SA, Kumar RC, Campbell IC (2003) Menstrual cycle effects on hypothalamic dopamine receptor function in women with a history of puerperal bipolar disorder. J Psychopharmacol 17(2):204–209

Wu CH, Motohashi T, Abdel-Rahman HA, Flickinger GL, Mikhail G (1976) Free and protein-bound plasma estradiol-17 beta during the menstrual cycle. J Clin Endocrinol Metab 43(2): 436–445

Wynn JK, Green MF, Sprock J, Light GA, Widmark C, Reist C, Erhart S, Marder SR, Mintz J, Braff DL (2007) Effects of olanzapine, risperidone and haloperidol on prepulse inhibition in schizophrenia patients: a double-blind, randomized controlled trial. Schizophr Res 95(1–3): 134–142

Yildiz A, Guleryuz S, Ankerst DP, Ongur D, Renshaw PF (2008) Protein kinase C inhibition in the treatment of mania: a double-blind, placebo-controlled trial of tamoxifen. Arch Gen Psychiatry 65:255–263

Zarate CA, Manji HK (2009) Protein kinase C inhibitors: rationale for use and potential in the treatment of bipolar disorder. CNS Drugs 23(7):569–582

Zarate CA Jr, Singh JB, Carlson PJ, Quiroz J, Jolkovsky L, Luckenbaugh DA, Manji HK (2007) Efficacy of a protein kinase C inhibitor (tamoxifen) in the treatment of acute mania: a pilot study. Bipolar Disord 9:561–570

Zhang M, Ballard ME, Unger LV, Haupt A, Gross G, Decker MW, Drescher KU, Rueter LE (2007) Effects of antipsychotics and selective D3 antagonists on PPI deficits induced by PD 128907 and apomorphine. Behav Brain Res 182(1):1–11

Sex Differences Precipitating Anorexia Nervosa in Females: The Estrogen Paradox and a Novel Framework for Targeting Sex-Specific Neurocircuits and Behavior

Charlotte Keating

Contents

Abstract In anorexia nervosa (AN), *reward contamination* likely plays a significant role in maintenance of the illness. Reward contamination is a context in which patients' behaviors of self-starvation and excessive exercise, while initially

---

C. Keating
Monash Alfred Psychiatry Research Centre (MAPrc), The Alfred Hospital, 1st floor, Old Baker Building, Commercial Road, Prahran, VIC 3181, Australia
e-mail: charlottekeating1@gmail.com

J.C. Neill and J. Kulkarni (eds.), *Biological Basis of Sex Differences in Psychopharmacology*, 189
Current Topics in Behavioral Neurosciences 8, DOI 10.1007/7854_2010_99,

rewarding, become aversive, *even punishing*; but patients may not recognize the punishing and conflicted/contaminated behaviors. An emerging neurocircuit encompassing the anterior cingulate cortex (ACC) has been functionally linked to symptoms including reward contamination and body dysmorphic processing. Owing to the significantly greater prevalence of AN in females, evidence from clinical literature and preclinical models is spearheaded to provide a novel rationale for estrogen triggering sensitivity to the experience of stress and reward, precipitating AN disproportionately in females at the time of puberty. Paradoxically, however, estrogen may facilitate response to pharmacological interventions and (desensitization of the identified neurocircuits) via its contribution to serotonin modulation, hypothalamo-pituitary adrenal (HPA)-axis attenuation, and effects on dopamine.

Keywords ACC · Antipsychotics · DLPFC · Dopamine · DRN · Estrogen · HPA-axis · Neurocircuits · OFC · Reward contamination · Sensitivity · Serotonin · SSRI

1 Introduction

Anorexia nervosa (AN) is a relatively uncommon illness (affecting up to 4.3% of women across their lifetime, Wade, et al. 2001) but it is predominantly diagnosed in adolescent females, in up to 95% of cases, (DSM-IV-TR 2000) with a prevalence of 0.3–1.3% (Hoek and van Hoeken 2003; Bulik et al. 2006; van Kuyck et al. 2009). Several rationales for the sex-specific predominance of females diagnosed with AN have been proposed; however, this chapter will focus on a novel role for estrogen contributing to the disproportionate precipitation of the illness.

Self-starvation and excessive exercise are core illness features (Davis 1997) which have been linked to *anhedonia* [the reduced capacity to experience *reward* (Keating 2009)], which is consistent with reports that more excessive exercisers tend to be more anhedonic (Davis and Woodside 2002). The diagnostic relevance of reward to AN remains to be established, albeit the contribution of *sensitivity or capacity for reward* among patients has been considered an important factor influencing the clinical expression of AN (Davis and Woodside 2002; Keating 2009) and its development (Keating 2009) including the contribution to *reward contamination*, which likely plays a significant role in the disorders' maintenance (Keating 2009). Reward contamination is a context in which patients' behaviors of self-starvation and excessive exercise, while initially rewarding, become aversive *or punishing*, but patients may not *recognize* the punishing (and contaminated) features of their behaviors. Contaminated reward processing has been proposed to be mediated by abnormalities in the anterior cingulate cortex (ACC) among a broader network (Keating 2009).

Diagnostic criteria for AN does not encompass reward abnormalities, but includes weight relative to height (A), intense fear of gaining weight or becoming fat (B), body weight and shape distortions (C), and amenorrhea in postmenarcheal females (D) (DSM-IV-TR 2000), the latter demonstrating physiological dysfunction. With regard to amenorrhea, there are few differences in demographics, eating behaviors, body image perceptions, illness history, and psychiatric comorbidity distinguishing menstruating and nonmenstruating patients (Garfinkel et al. 1996; Abraham et al. 2005) with many questioning the relevance of this criterion to the illness (e.g., Roberto et al. 2008; Attia and Roberto 2009). Physiologically, elevated stress, hypothalamo-pituitary adrenal (HPA) axis activity is reported in the majority of patients with AN linked to their behaviors (Bergh and Sodersten 1996) and is logically consistent with hypothalamo-pituitary gonadal (HPG)-axis-suppression-induced amenorrhea.

This chapter will focus on a novel explanation for a role for estrogen (around puberty) facilitating illness onset disproportionately in females, via enhancing *sensitivity to stress and reward* associated with AN behaviors. In addition, a paradoxical role for estrogen in the neuropsychopharmacology of response to treatment in AN (Keating et al. 2010) will be developed, the latter owing to estrogen's relationship to serotonin modulation, HPA-axis modulation, and indirect evidence for an association with reward (dopamine) normalization, of which these systems have been putatively linked to AN.

1.1 Stress Induces Reward, and Both Are Linked to the Development of Anorexia Nervosa

Consistent with reports that AN behaviors are linked to anhedonia, the illness has been suggested to develop because it is initially rewarding to eat less food and AN behavior is maintained through conditioning to the situations that provide reward (Bergh and Sodersten 1996; Sodersten et al. 2006, 2008). At a pathophysiological level, the majority of patients present with HPA-axis hyperactivity (Bergh and Sodersten 1996; Keating et al. 2010; Licinio et al. 1996, among others), whereby elevated corticotropin-releasing hormone (CRH) levels, up to 170% of normal, have been linked to self-starvation (Bergh and Sodersten 1996). The physiological sequalae of CRH release is the stimulation and secretion of cortisol from the adrenal glands.

Adrenocortical hormones are rewarding because their secretion stimulates the release of dopamine (DA) from the ventral srtiatal terminals of the mesolimbic neurons in the brain via both adrenocortical hormone feedback on these terminals and activation of DA cell bodies in the ventral tegmentum of the mesencephalon (at least in an animal model of the illness, Bergh and Sodersten 1996). Importantly, adrenocortical secretions enhance the *reward value* of AN-linked behaviors by increasing the release of DA in the terminals of these neurons (Bergh and Sodersten 1996; Piazza et al. 1993), presenting a potently physiologically reinforcing mechanism for AN behaviors.

This explanation encompasses why patients experience their behaviors (excessive exercise and food restriction) rewarding in the first instance (Keating 2009) and has been extended psychologically and physiologically, to rationalize why patients continue engaging in these behaviors despite the fact that they are aversive, even punishing (Keating 2009).

1.2 In the Disorder Maintenance Phase, a Neurocircuit for Reward Contamination May Play a Significant Role in Persistent Poor Decision Making

There are several neurobiological explanations for development of AN; however, few consider the distinction between factors influencing onset relative to maintenance of the illness. This chapter will focus on stress-induced reward linked to the onset of AN, as well as the ultimate contamination of reward and punishment (Keating 2009). While behaviors linked to AN are initially rewarding and associated with illness onset, reward contamination likely plays a significant role in illness maintenance, due to the fact that patients may ultimately fail to recognize the punishing (or contaminated) features of their behaviors. Reward contamination has been proposed to be physiologically mediated by *overlapping* neural circuits responsible for processing *both* reward and punishment (Keating 2009) that become contaminated and reinforced, as a result of pathological and unrelenting engagement in disordered-linked behaviors.

A body of literature implicates dysfunction in reward processing in AN (reviewed in Keating 2009); however, no studies have systematically investigated reward contamination, that is, patients' perceptions of *punishment* associated with challenges where the *hedonistic* (reward) rating of tasks is assessed (i.e., testing whether a reward-based task is considered punishing, and vice versa, whether a punishing task is considered rewarding, Keating 2009). Recently, however, an investigation in individuals recovered from the illness may indirectly shed clinical light on reward contamination linked to AN.

During functional magnetic resonance imaging (fMRI), participants recovered from AN (and healthy controls) were required to process *reward* and *punishing* stimuli (e.g., wins and losses on a gambling task) with results demonstrating in recovered individuals that, neurally, the ventral striatum failed to distinguish reward from punishment, whereas healthy controls distinguished these stimuli in the same region (Wagner et al. 2007). While this study did not intend to investigate reward contamination per se, a failure to distinguish between rewarding and punishing stimuli in ventral striatum (the region of interest, ROI, in this study, Wagner et al. 2007) in recovered individuals supports the contention of reward contamination, given healthy controls were able to distinguish between these experiences in the same region.

Extrapolated to the reward contamination theory (Keating 2009), these results (from Wagner et al. 2007) support the ventral striatum as a reward-contaminated locus in recovered individuals. Moreover, the persistence of neural contamination within this region, following recovery from the illness, suggests that reward contamination in the ventral striatum (at least) reflects a vulnerability marker for the illness (Keating 2009), which is furthermore supported by data from an investigation in *ill* patients, similarly demonstrating dysfunction in this region, again linked to reward contamination, albeit not intending to test this.

For example, Fladung et al. (2010) investigated in ill AN patients, activity in the ventral striatum in relation to *self-referent* processing of pictures of women at different weights, including *underweight* and *normal weight*. Although Fladung et al. (2010) were interested in demonstrating reward-linked starvation dependence associated with the ventral striatum (Fladung et al. 2010; Keating 2010), their results may also infer important information regarding body dysmorphic self-perceptions and reward contamination linked to this region in AN (Keating 2010). An argument for interpreting the results by (Fladung et al. 2010) as an illustration of reward contamination is based on the fact that when processing pictures of women at different weights in a *self-referent manner*; patients demonstrated more *positive appraisals* correlated with activity in the ventral striatum to *underweight* relative to normal-weight pictures, reflecting an indirect illustration of reward-conflict given that it would *otherwise* be expected that an underweight picture may induce feelings of *aversion*, which is consistent with and supported by the fact that healthy control participants demonstrated the *opposite association*; e.g., more positive ratings of normal weight compared with underweight women also linked to activity in this region (Fladung et al. 2010).

The extent to which reward contamination may reflects a marker for AN in recovered individuals (e.g., interpretations based on data presented in Wagner et al. 2007) as well as symptoms of AN beyond body dysmorphic perceptions (Keating 2010) and neuroanatomically, whether the substrate for these psychological processes are diffuse within other brain regions, or the likelihood of a neurocircuit, remains to be directly tested.

1.2.1 An Emerging Neurocircuit for Anorexia Nervosa

Neuroanatomy

Regarding other regions involved in reward contamination, the ACC has been proposed as a key locus and nexus in reward contamination contributing to disorder maintenance in AN (Keating 2009). This rationale is consistent with both preclinical and clinical literature demonstrating its heterogeneous functions in reward (Petrovic et al. 2008; Kennerley et al. 2006) punishment (Wrase et al. 2007) and conflict (Pochon et al. 2008) processing (reviewed in detail in Keating 2009), as well as reward-punishment contamination demonstrated in another psychiatric illness, major depression (Knutson et al. 2008). Abnormalities in structure and function of the ACC region are some of the most frequently cited in AN literature (reviewed in Keating 2009); however, there remains to be a consensus on a disorder-specific explanation for a role of the ACC in AN.

Empirical evidence for the functional contribution of the ACC to AN, linked to reward, punishment, and the *contamination* of these experiences has been provided by an analyses of data (Kaye et al. 2009) from recovered individuals (Wagner et al. 2007). Kaye et al. (2009) revealed data that enables an interpretation to support the assertion that recovered individuals fail to neurally distinguish reward and loss (punishment) extending from the ACC (hypoactivated in recovered individuals) through to the ventral striatum, providing further support for a neurocircuit (at least the ACC and ventral striatum) involved in reward contamination.

Although a region of interest analysis was not extended to the OFC and dorsolateral prefrontal cortex (DLPFC) among other executive regions, it is proposed too, these areas, owing to their links to decision making based on reward (and reward history) (e.g., Kerns et al. 2004; Kennerley et al. 2006) as well as abnormalities in set-shifting in AN (Zastrow et al. 2009) may also contribute to or enable, reinforcement of disorder linked poor (or contaminated) decision making and action (behavior) selection, for which these regions are putatively involved (e.g., Kerns et al. 2004; Kennerley et al. 2006; Rushworth et al. 2007). Further analysis of more executive regions in recovered individuals (Wagner et al. 2007), but also in ill patients, would reveal the extent to which a neurocircuit may contribute to reward contamination. Furthermore, in the illness state, whether patients "self-referent processing" of pictures of body images (or processing of their own body images) and their experience of *reward or punishment* (theoretically) linked to these appraisals, extends beyond the ventral striatum, and whether these abnormalities persist in recovered individuals may further reveal the contribution of reward contamination to this pervasive disorder–symptom, as well as presenting another functional biomarker for the illness which may be linked to relapse–risk.

Neurochemistry

The neurochemistry of AN is complex. Abnormalities present in several systems including the HPA-axis (e.g., Bergh and Sodersten 1996; Licinio et al. 1996), the dopaminergic (e.g., Frank et al. 2005), and the serotonergic systems (which has been extensively reviewed, e.g., Kaye 2008), with persistent abnormalities demonstrated in the latter two systems (e.g., Bosanac et al. 2005) following weight gain. Dysfunction in the DA system is commonly reported, including greater D2 and D3 receptor binding in the ventral striatum during positron emission tomography (PET) (at baseline) in patients recovered from the illness (Frank et al. 2005) as well as abnormalities (overactivity) of the DA system in patients (e.g., Barbato et al. 2006). Functionally, DA has also been linked to processing both *reward and punishment* (Matsumoto and Hikosaka 2009) (in an animal model). Taken together with the fact that AN is linked to abnormalities in the DA system in the ventral striatum (Frank et al. 2005), and functionally, *reward contamination* overlaps the same region (based on data in Wagner et al. 2007) and DA has been linked to processing reward and punishment (e.g., Matsumoto and Hikosaka 2009), these findings converge to support the plausible role of DA linked to reward contamination in AN (Keating 2009). Moreover, in the context that upregulated HPA-axis activity leads to stimulation and secretion of DA, it is likely that reward contamination (linked to DA) is stimulated and reinforced, and therefore maintained by upregulated HPA-axis activity in the illness.

Although estrogen has been largely dismissed as a diagnostic or clinical indicator for AN (e.g., Roberto et al. 2008), at puberty estrogen is proposed to contribute to sensitivity to stress-linked reward at the onset of AN. Paradoxically, however, evidence supports the hypothesis that estrogen also contributes to treatment response (Keating et al. 2010) consistent with its interactions with serotonin, the HPA-axis, and DA, postulated to facilitate response to medication that downregulates the HPA-axis.

2 Sex-Specific Psychobiological and Pharmacological Explanation for the Disproportionate Development of Anorexia Nervosa in Females: Stress and Reward Sensitivity Triggered by Estrogen at Puberty

2.1 Evolutionary Conservation: Females Are More Sensitive to Stress and Reward than Males

A complex interaction of factors likely contributes to a biological rationale for the diagnostic predominance of AN (up to 95% of cases, DSM-IV-TR 2000) among females. The disproportionate precipitation of the illness is postulated to reflect distinct evolutionarily conserved differences in females relative to males in *sensitivity to stress* and *reward* psychobiology, given stress is involved in the onset and maintenance of AN (e.g., Lo Sauro et al. 2008) and the majority of patients with AN demonstrate HPA-axis hyperactivity (Bergh and Sodersten 1996; Keating 2009; Licinio et al. 1996, among others) linked to illness behaviors (Bergh and Sodersten 1996; Keating 2009).

Physiologically, stress-induced upregulation of the HPA-axis produces elevated cortisol (reviewed in Tilbrook and Clarke 2006), the consequences of which include increased synthesis and secretion of reward-linked DA in AN (e.g., Bergh and Sodersten 1996). Females in general (clinically and preclinically) respond with greater perturbations of the HPA-axis than males in the context of psychosocial stressors (for review, see Tilbrook and Clarke 2006) and show greater sensitivity to reward than do males (e.g., Kamarajan et al. 2008) of which the *sensitivity to* the experience of reward is likely to be physiologically mediated via estrogen (e.g., Zakharova et al. 2009).

Although AN behaviors have been linked to stress-induced reward (e.g., Bergh and Sodersten 1996), the disproportionate diagnosis of AN in females linked to stress, reward and estrogen has until now not been formally hypothesized (neither in other stress-linked illnesses favoring females, e.g., major depression). Indirect support for the relationship between these physiological variables validates the potential for a relationship between these factors contributing to AN (Monteleone et al. 2001). For example, increased cortisol concentrations, but reduced concentrations of estrogens (i.e., 17β-estradiol) (among other hormones measured including dehydroepiandrosterone, DHEA), have been reported in AN patients (Monteleone et al. 2001), which is logically consistent with the impact of hyperactive HPA-axis-induced suppression of the HPG-axis.

Regarding the molecular dissemination of sex differences in stress and reward potentially contributing to AN, data from at least one psychobiological study can be extrapolated to address this relationship. Recently, differences in the stress, CRH system response to psychosocial stressor challenge were investigated between male and female sheep (Rivalland et al. 2007), and serendipitously for the purposes of the current contention, the positive control was enkephalin-staining, a reward-linked opioid. Following psychosocial isolation restraint stress, among results, female sheep demonstrated a greater proportion of enkephalin (reward-linked) cells staining for Fos than males in both control and stressed animals, suggesting (of particular relevance) *a greater volume of reward-linked cells in females*, which was present *before* the stress-induced challenge (i.e., independent of the effects of stressor). Extrapolated to the clinical condition, females may present with a greater proportion of reward-linked neural substrates opposed to the stress system (including those diagnosed with AN) predisposing females to the disproportionate expression of reward and stress-linked physiology and pathophysiology. Data from Rivalland et al. (2007) may also infer (although speculative) that females (relative to males) may experience dose-dependent greater reward from stress, on the basis of greater enkephalin Fos staining in females. Owing to the greater volume of reward-linked cells found in females in this study, it is plausible (at least indirectly) that estrogen may be putatively involved in mediating the difference in reward between males and females.

2.2 Estrogen Mediates Increased Sensitivity to Reward in Females

Evidence suggests that the mechanism mediating greater sensitivity in reward and stress systems in females is estrogen. Evidence from preclinical and clinical research (yet to be tested in patients with AN) suggests a strong rationale for endogenous estrogen concentrations contributing to the sex difference in the experience of reward favoring females (Zakharova et al. 2009; Fattore et al. 2007). In a recent study, younger female mice demonstrated greater behavioral *sensitivity* to rewarding stimuli (cocaine induced place preference) than older females or male littermates of any age (Zakharova et al. 2009), and consistent with estrogen's role in mediating sensitivity to reward favoring females, when animals were ovariectomized, the sex difference in reward sensitivity was lost. Further support for estrogen involvement in reward sensitivity can be illustrated by a study demonstrating that certain female rat strains (both Long–Evans and Lister Hooded, but not Sprague–Dawley strains) experience greater reward than males, evidenced via greater self-administration of the cannabinoid (CB1 receptor) agonist (WIN 55, 212–2), in addition to which ovary-intact females experienced greater reward than estrogen-treated ovariectomized females (Fattore et al. 2007).

These preclinical findings are consistent with clinical research, demonstrating (though less directly) reward sensitivity favoring females. For example, in response to wins and losses on a gambling task, healthy females showed greater sensitivity to reward and loss (wins and losses, or punishment) than males (Kamarajan et al. 2008) in addition to which males were *equally sensitive* to both wins and losses. The neural correlates of these experiences were determined via brain oscillation, revealing that for females, greater reward sensitivity was linked to greater theta power and posterior maxima in the posterior region of the brain and activity in the anterior region when compared to loss conditions, linked to anterior involvement only. Males, however, demonstrated posterior maxima to both stimuli linked to activation in the frontal midline (Kamarajan et al. 2008). These results support an indirect role for estrogen in reward sensitivity, given females relative to males were more sensitive on relevant conditions of the gambling task. Future studies involving fMRI and PET imaging during ligand-specific binding (e.g., targeting DA) would resolve more specifically brain regions and circuits involved in the sex-specific experience of reward and punishment (or wins and losses), providing relevant insight regarding mechanisms that may lead to reward dysfunction (e.g., contamination) and neural substrates driving and maintaining behaviors linked to AN.

Although the impact of increasing levels of estrogen, characteristic of changes at puberty on the reward system, have yet to be investigated in AN, evidence for a modulatory impact of estrogen on reward systems can be extrapolated from a study disseminating the impact of concentrations of estrogens (e.g., estradiol) across the menstrual cycle in primates on the expression of the DA (reward-linked) system (Czoty et al. 2009). Specifically, during the luteal phase (high estradiol and progesterone concentrations) there was greater availability of D2 binding sites (via [18] FCP binding) in the caudate and putamen (e.g., striatum), suggesting greater unoccupied receptors in primates during physiological periods of higher estradiol (and progesterone) (Czoty et al. 2009) which has been suggested to reflect *reduced DA* release during the high estradiol phase (Young and Becker 2009).

Extrapolated to AN, it would be logically consistent that when estrogen levels are chronically low, DA levels are likely high. This can be illustrated by evidence that DA levels in the illness are elevated, as assessed via eye-blink rate (e.g., Barbato et al. 2006), a peripheral marker of central DA activity in patients relative to healthy controls. Furthermore, empirically, in weight-restored patients relative to healthy controls, D2 and D3 receptor binding has been shown to be increased (via [11C] raclopride binding) in the anteroventral striatum (Frank et al. 2005). Although menstrual cycling was not controlled for, it might be that increased D2 and D3 receptor binding reflects at least in part an impact of *estrogen normalization* on D2 binding [given three normal cycles were required for inclusion in the study by Frank et al. (2005) (estrogen normalization and treatment response discussed in Keating, et al. 2010)]. These interpretations are logically consistent with the physiological reports from Czoty et al. (2009) demonstrating that higher levels of estrogen are associated with lower concentrations of DA (Young and Becker 2009) and greater D2 and D3 receptor availability (Czoty et al. 2009).

2.3 Females Are More Vulnerable to Anorexia Nervosa Due to Estrogen-Mediated Stress and Reward Sensitivity

Support for sex differences in *susceptibility to development of AN behaviors* favoring females (Hancock and Grant 2009) linked to *stress and reward* (e.g., Bergh and Sodersten 1996) has been demonstrated in an activity-based rodent model of anorexia nervosa (ABA). Specifically, females were significantly more susceptible to restricted feeding and excessive wheel running, as demonstrated by more rapid and pronounced reductions in body weight and lower levels of food intake, when compared to males (Hancock and Grant 2009). Although endocrine and neurochemical measures were not taken, sex differences in the effects of food restriction and wheel running in ABA likely reflect differences in stress (among other endocrinological) factors, consistent with findings that HPA-axis reactivity, i.e., levels of adrenocorticotropin hormone (ACTH)/corticosterone (the rat homologue of cortisol) are greater in females relative to males, at rest and following exposure to a range of stressors, including forced swim (Panagiotaropoulos et al. 2004; Wigger and Neumann 1999), motorized wheel running, immobilization, footshock (Kant et al. 1983), and the chronic mild stress paradigm (Dalla et al. 2005, see chapter x).

Taken together, these outcomes (e.g., Hancock and Grant 2009; Panagiotaropoulos et al. 2004; Wigger and Neumann 1999) are consistent with the majority of literature concerning greater responsiveness of the HPA-axis in females than males (Tilbrook and Clarke 2006), and heightened HPA-axis reactivity may provide a risk factor for greater impact of the *stress* (Hancock and Grant 2009) of a wheel-running stressor (or clinically relevant excessive exercising), and its reward- (DA) linked reinforcement.

To test whether estrogen is crucial in mediating the greater vulnerability of females to the development of AN, and the contribution of greater sensitivity to stress and reward-linked systems in this context, a similar paradigm presented in Hancock and Grant (2009) may be tested. In addition to the utilization of continually assessed HPA-axis and DA levels in males and females compared on vulnerability to the ABA model, a study could involve females that are ovary intact, or ovariectomized or ovariectomized with estrogen replacement, at *puberty*, to determine the contribution of estrogen to the greater vulnerability of ABA in females.

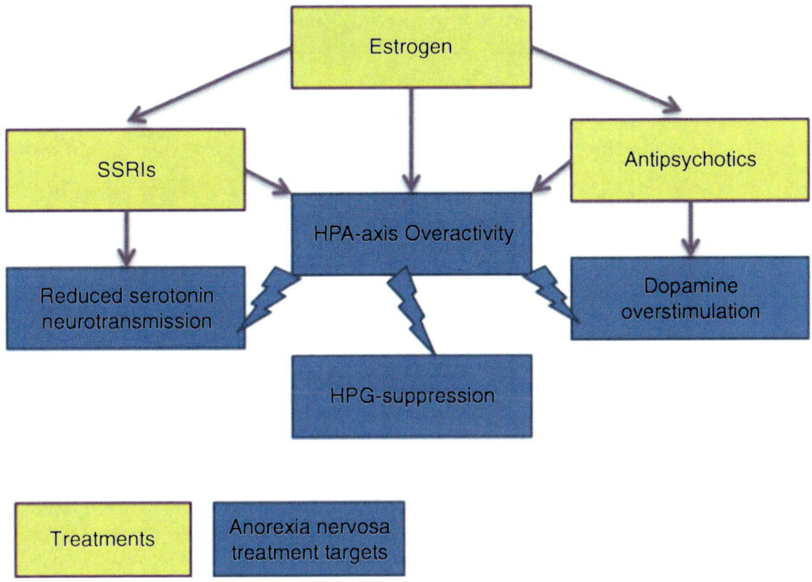

Fig. 1 A model for treatment targets in anorexia nervosa

2.3.1 Summary

In summary, in a model of ABA, females demonstrated greater sensitivity or vulnerability to the development of behaviors mimicking the AN clinical condition (excessive exercise and food restriction) that have been linked to elevated activity of the HPA-axis in clinical participants (e.g., Bergh and Sodersten 1996) and reflect elevated DA secondary to elevated HPA-axis activity in an animal model (Bergh and Sodersten 1996). These assertions are furthermore consistent with reports for the greater impact of *varied stressor tasks* on females relative to males described. Evidence supports the hypothesis that estrogen mechanistically contributes to the greater sensitivity to reward seen in females relative to males (e.g., Fattore et al. 2007; Zakharova et al. 2009), which may contribute to greater sensitivity around the time of puberty, in the development of stress-induced reward linked to behaviors, disproportionately precipitating AN in females.

3 Neuropsychopharmacology of Treatment Response in Anorexia Nervosa

3.1 The Estrogen Paradox

While estrogen has been linked to greater reward sensitivity in females relative to males, empirically rationalized to contribute to the disproportionate diagnosis of

AN in women, paradoxically, estrogen may facilitate response to pharmacological interventions including selective serotonin reuptake inhibitors (SSRIs) as well as antipsychotics via estrogen's contribution to HPA-axis attenuation, indirect influence on DA and contribution to serotonin modulation. The importance of estrogen in a clinical response to pharmacotherapy has been previously discussed (Keating et al. 2010). A schematic depicting neurobiological treatment targets for AN based on illness linked suppression of the HPG-axis (e.g., reduced estrogen levels), hyperactivity of the HPA-axis, elevated DA levels, and reduced serotonin levels is presented in Fig. 1.

Central to illness maintenance is HPA-axis overactivity. HPA-axis overactivity stimulates DA secretion, reinforcing (contaminated) reward-linked behaviors in AN. HPA-axis overactivity leads to a reduction in serotonin, the consequences of which facilitate HPA-axis overactivity. HPA-axis overactivity ultimately suppresses the HPG-axis, leading to amenorrhea. Normalizing this context, estrogen treatment can inhibit the HPA-axis (Young et al. 2001) estrogen is involved in the regulation of serotonin (Gundlah et al. 2005) and may facilitate the therapeutic effects of serotonin modulators (e.g., SSRIs) (Keating et al. 2010), ultimately facilitating attenuation of the HPA-axis, the consequence of which may normalize reward via reduced DA stimulation. Estrogen may also facilitate the actions of antipsychotics (e.g., antagonizing DA), reducing (contaminated) reward-linked reinforcement of disorder behaviors. According to this model, treatment approaches involving estrogen and SSRIs or estrogen and antipsychotics may reduce HPA-axis hyperactivity-induced reward linked to the reinforcement of AN behaviors.

3.1.1 Treatments in Anorexia Nervosa

The first line pharmacological approach to treating AN has involved the use of SSRIs which act to increase synaptic concentrations of 5-hydroxytryptamine (5-HT, serotonin). Albeit several small trials involving testing a large proportion of antidepressant treatments (75%) in patients with AN have demonstrated some improvement on measures including eating behavior and associated weight gain (and reduction in obsessive symptoms) (e.g., Rossi et al. 2007), there is an extensive body of literature demonstrating a lack of efficacy of antidepressant approaches in positively impacting desire for thinness or weight gain across a range of antidepressants including tricyclics, SSRIs, and mood stablizers (e.g., Attia et al. 1998; Biederman et al. 1985; Gross et al. 1981; Halmi et al. 1986; Kaye et al. 1998; Walsh et al. 2006). There are several explanations for their lack of efficacy, including the notion that abnormalities in the 5-HT system are not putatively linked to AN (e.g., Sodersten et al. 2008), and that (despite a lack of empirical support) enhancing 5-HT may lead to a dysphoric mood state (e.g., Kaye et al. 2009). Consistent with estrogen's regulation of 5-HT (e.g., Gundlah et al. 2005), a lack of estrogen in the illness state is proposed to mediate nonresponse to SSRIs (Keating et al. 2010).

3.2 A Rationale for the Lack of Efficacy of Serotonin Modulators in Anorexia Nervosa: An Overlooked Mediator of Estrogen in Treatment Response

The use of serotonergic modulators in AN is consistent with abnormalities in this system in patients, and dysfunction in this system has been extensively reviewed elsewhere (Barbarich-Marsteller 2007; Kaye et al. 2009; and others). For example, in recovered individuals with AN (restricting subtype), 5-HT transporter binding in the dorsoraphe nucleus (DRN) and anteroventral striatum is elevated (e.g., significantly increased [11C] McN562 binding potential assed via PET) relative to AN patients with bulimic characteristics (Bailer et al. unpublished, cited in Barbarich-Marsteller 2007). Reduced illness linked 5-hydroxyindolascetic acid (5-HTIAA) levels and elevated recovered state levels (reviewed in Barbarich-Marsteller 2007, logically consistent with elevated transporter binding in the illness state) would infer an association between recovery and augmentation of 5-HT levels; however, empirically medication enhancing 5-HT (e.g., SSRIs) show little benefit (Sodersten et al. 2008; Bergh et al. 1996; Adokat and Kutlesic 1995; Barbarich-Marsteller 2007) beyond facilitating relapse-prevention (e.g., Kaye et al. 2009). The neurobiological differences in recovered individuals relative to ill individuals, permitting some treatment efficacy of antidepressant medication, is likely linked to estrogen concentrations, but this has yet to be hypothesized or experimentally addressed.

The recovered state is a physiological status reflecting re-established estrogen concentrations, as opposed to the ammenorheic illness state. A difference in estrogen concentrations between these phases may explain the lack of efficacy of 5-HT modulators during the illness as opposed to recovery phase. Support for this contention can be derived from two lines of evidence: that 5-HT is regulated by estrogen (e.g., Gundlah et al. 2005) and that a therapeutic response to SSRIs demonstrated in other related illnesses suggests that treatment efficacy is linked to greater endogenous estrogen concentrations in females (e.g., Pae et al. 2009). For example, where amenorrhea is unlikely to be diagnosed (e.g., in depressed patients that are not severely underweight), females respond more favorably to SSRIs than males (Morishita and Kinoshita 2008), and in women taking hormone replacement therapy (HRT) during menopause, response to SSRIs is greater than in women not taking HRT (e.g., Zanardi et al. 2007).

These converging lines of evidence suggest that a lack of efficacy of SSRIs in AN is likely due to severely reduced estrogen in the illness state (Keating et al. 2010), which is consistent with and supported by the fact that in the recovered state, SSRIs are more effective in preventing relapse (e.g., Kaye et al. 2009) when menstrual function is restored. In addition to the observation that 5-HT is regulated by estrogen (Gundlah et al. 2005), antidepressants have been shown to attenuate activity of the HPA-axis (Schule 2006); hence, estrogen may also facilitate recovery in AN via its proposed effects in downregulating the glucocorticoid stress response (Solomon and Herman 2009), the consequence of which, in the context of AN, leads to reduced stimulation of DA.

3.2.1 Estrogen Reduces Activity of the HPA-Axis

Key to addressing the pathophysiology of AN is reducing or *hypoactivating* over-activity of the HPA-axis, the consequences of which may reduce DA synthesis and secretion, ultimately reducing reinforcement in AN (Keating 2010).

Recent evidence from a study involving stress-responsivity in female rats suggests definitively that estrogen (but not progesterone) is an important inhibitor of the HPA-axis (Young et al. 2001).

Consistent with HPA-axis inhibiting effects of estrogen (Young et al. 2001) and the corollary of reduced estrogen *activating* the HPA-axis, evidence shows that in females undergoing restraint stress, estrogen antagonists (tamoxifen and C1628) increase responsivity of the HPA-axis; that is, *C1628* increased both ACTH and corticosterone, and *tamoxifen* increased ACTH. Conversely, in ovariectomized female rats, low dose estradiol (over 7 days) *decreased HPA-axis responsiveness* (e.g., reduced ACTH) to psychosocial restraint stress, whereas progesterone had no effect (Young et al. 2001) (discussed elsewhere, Keating et al. 2010), which is consistent with the fact that both estradiol and progesterone reduced ACTH but that this magnitude did not differ when estradiol treatment was given alone (Young et al. 2001). These findings provide strong evidence for a modulatory impact of estrogen on stress-attenuation and warrant follow-up in AN.

3.2.2 Estrogen Regulates Serotonin and Facilitates Therapeutic Response to Serotonin Modulators

Females respond more favorably than males to SSRIs (Morishita and Kinoshita 2008), an effect likely mediated by estrogen. Data from primate and rodent investigations provide evidence that estrogen regulates 5-HT (e.g., Gundlah et al. 2005). For example, in females, 17β-estradiol (E2) likely induces tryptophan hydroxylase (TPH) (5-HT-precursor) expression in the DRN (the largest 5-HT containing nucleus in the brain) of guinea pigs and macaques (e.g., Bethea et al. 2000; Lu et al. 1999) and in murine DRN (Gundlah et al. 2005). In addition, estrogen regulation of TPH (subtype 1) may be specific to the estrogen receptor (ER) β-subtype, on the basis that estrogen increases TPH (subtype 1) expression (selectively) in the DRN of wild-type and ER-α knockout mice, where this effect on TPH (subtype 1) is not seen in ER β-knockouts (Gundlah et al. 2005).

At least in the context of major depression, which similarly to AN presents with significant abnormalities in 5-HT function and HPA-axis hyperactivity; endogenous concentrations of estrogen have been shown to influence response to 5-HT modulators. In a recent study, Zanardi et al. (2007) have demonstrated that menopausal women taking HRT (despite complex drug combinations between patients) reported a significant improvement in response to treatment (and reduction in depressive symptoms). These findings are consistent with other reports for greater endogenous concentrations of estrogen favoring response to 5-HT modulators (e.g., Pae et al. 2009; Morishita and Kinoshita 2008), consistent with the finding that females respond to antidepressants more favorably than males (Morishita and Kinoshita 2008 and see chapter by Sramek and Cutler).

Although yet to be experimentally addressed in patients with AN, a lack of empirical support for the efficacy of medication that enhances 5-HT (e.g., SSRIs) (Adokat and Kutlesic 1995; Barbarich-Marsteller 2007; Bergh et al. 1996; Sodersten et al. 2008) beyond facilitating relapse-prevention (e.g., Kaye et al. 2009) in AN may be due to a lack of estrogen. In sum, estrogen may facilitate SSRI (i.e., 5-HT) induced hypoactivation of the HPA-axis, the consequences of which reduce DA stimulation and reward-linked behavioral reinforcement.

3.3 A Rationale for the Greater Clinical Response to Antipsychotics Versus SSRIs in Anorexia Nervosa?

Antipsychotic medication has more recently begun being trialed in patients with AN, and evidence suggests that antipsychotics show promise in the treatment of AN albeit further treatment controlled trials are necessary to conclusively support their widespread clinical use (Court et al. 2008).

In the context of the current neurobiological explanation for AN, it is plausible that antipsychotics show enhanced efficacy over SSRIs on the basis that they antagonize DA receptors, which may theoretically reduce the reward-linked reinforcement associated with illness behaviors. DA blocking is a direct target of antipsychotics, whereas it is a downstream consequence of SSRIs, that is, achieved via a consequence of HPA-axis downregulation. Assuming reward-linked reinforcement is key to illness maintenance (via maintaining reward-conflict) in some patients, an approach directly targeting this system may be warranted, whereas in patients with lesser reward sensitivity (e.g., less excessive exercisers) SSRIs and estrogen may be appropriate (Keating et al. 2010).

The therapeutic value of antipsychotics in AN may furthermore be augmented by an estrogen adjunctive. This assertion is supported by evidence demonstrating, *physiologically*, a link between normalization of estrogen levels in recovered AN patients and DA, and is consistent with primate literature demonstrating that greater concentrations of estrogen (according to the luteal phase of the menstrual cycle) in association with greater availability of D2 binding sites (or unoccupied receptors during *relatively* elevated estradiol) (Czoty et al. 2009) and reduced DA release [consistent with findings from (Czoty et al. 2009) discussed in (Young and Becker 2009)]. So while in healthy primates, higher levels of estrogen are associated with lower concentrations of DA (Young and Becker 2009) based on the findings by Czoty et al. (2009) in AN, the opposite is seen. That is, concentrations of DA are high (Barbato et al. 2006) (e.g., overactivity of reward) and estrogen levels are low (Monteleone et al. 2001) reflecting amenorrhea.

It will be important in future to establish the relative efficacy of SSRIs and estrogen versus antipsychotics and estrogen in this patient group.

4 Conclusions and Future Directions: Estrogen Augmentation of SSRIs or Antipsychotics to Induce/Facilitate Response to Treatment via Normalizing Stress Hyperactivity and the Experience of Reward

Given HPA-axis activity is significantly elevated in AN (e.g., Bergh and Sodersten 1996; Licinio et al. 1996) consistent with elevated DA concentrations (Barbato et al. 2006) and estrogen has been shown to inhibit the HPA-axis during psychosocial stress (Young et al. 2001), as well as regulate 5-HT metabolism (Gundlah et al. 2005), it is logically consistent then that an AN-linked deficit in estrogen concentrations may contribute to elevated activity of the HPA-axis, and furthermore enable hyperactivity due to disadvantaging the capacity of 5-HT to attenuate the HPA-axis. Taken together, reduced estrogen may explain a lack of response to 5-HT modulation. The corollary of this explanation is that estrogen treatment in combination with SSRIs should be trailed in AN, in addition to which studies are required to assess the clinical utility of estrogen and antipsychotics in AN. These treatment directions are consistent with the theoretical proposal that stress-induced reward (albeit ultimately *contaminated* reward) underpins behaviors linked to the illness, and that estrogen-induced normalization of the HPA-axis may lead to the reduced reward-linked reinforcement of behaviors, in conjunction with SSRIs or antipsychotics.

References

Abraham SF, Pettigrew B, Boyd C, Russell J, Taylor A (2005) Usefulness of amenorrhoea in the diagnoses of eating disorder patients. J Psychosom Obstet Gynaecol 26:211–215

Adokat C, Kutlesic V (1995) Pharmacotherapy of the eating disorders: a commentary. Neurol Biobehav Rev 19:59–66

Attia E, Haiman C, Walsh BT, Flater SR (1998) Does fluoxetine augment the inpatient treatment of anorexia nervosa? Am J of Psychiatry 155:548–551

Attia E, Roberto CA (2009) Should amenorrhea be a diagnostic criterion for anorexia nervosa? Int J Eat Disord 42:581–589

Barbarich-Marsteller N (2007) Neurochemistry and pharmacological treatments: where is the field of anorexia nervosa heading? Cent Nerv Syst Agents Med Chem 7:35–43

Barbato G, Fichele M, Senatore I, Casiello M, Muscettola G (2006) Increased dopaminergic activity in restricting-type anorexia nervosa. Psychiatry Res 142:253–255

Bergh C, Sodersten P (1996) Anorexia nervosa, self-starvation and the reward of stress. Nat Med 2:21–22

Bergh C, Eriksson M, Lindberg G, Sodersten P (1996) Selective serotonin reuptake inhibitors in anorexia. Lancet 348:1459–1460

Bethea CL, Mirkes SJ, Shively CA, Adams MR (2000) Steroid regulation of tryptophan hydroxylase protein in the dorsal raphe of macaques. Biol Psychiatry 47:562–576

Biederman J, Herzog DB, Rivinus TM, Harper GP et al (1985) Amitriptyline in the treatment of anorexia nervosa: a double-blind, placebo-controlled study. J Clin Psychopharmacol 5:10–16

Bosanac P, Norman T, Burrows G, Beumont P (2005) Serotonergic and dopaminergic systems in anorexia nervosa: a role for atypical antipsychotics? Aust NZ J Psychiatry 39:146–153
Bulik CM, Sullivan PF, Tozzi F, Furberg H, Lichtenstein P, Pedersen NL (2006) Prevalence, heritability, and prospective risk factors for anorexia nervosa. Arch Gen Psychiatry 63:305–312
Court A, Mulder C, Hetrick SE, Purcell R, McGorry PD (2008) What is the scientific evidence for the use of antipsychotic medication in anorexia nervosa? Eat Disord 16:217–223
Czoty PW, Riddick NV, Gage HD, Sandridge M, Nader SH, Garg S, Bounds M, Garg PK, Nader MA (2009) Effect of menstrual cycle phase on dopamine D2 receptor availability in female cynomolgus monkeys. Neuropsychopharmacology 34:548–554
Dalla C, Antoniou K, Drossopoulou G, Xagoraris M, Kokras N, Sfikakis A, Papadopoulou-Daifoti Z (2005) Chronic mild stress impact: are females more vulnerable? Neuroscience 135:703–714
Davis C (1997) Eating disorders and hyperactivity: a psychobiological perspective. Can J Psychiatry 42:168–175
Davis C, Woodside DB (2002) Sensitivity to the rewarding effects of food and exercise in the eating disorders. Comp Psychiatry 43:189–194
DSM-IV-TR (2000) American psychiatric association. Diagnostic and statistical manual of mental disorders, 4th edn., text revision. American Psychiatric Association, Washington, DC
Fattore L, Spano MS, Altea S, Angius F, Fadda P, Fratta W (2007) Cannabinoid self-administration in rats: sex differences and the influence of ovarian function. Br J Pharmacol 152:795–804
Fladung AK, Gron G, Grammer K, Herrnberger B, Schilly E, Grasteit S, Wolf RC, Walter H, Wietersheim JV (2010) A neural signature of anorexia nervosa in the ventral striatal reward system. Am J Psychiatry 167(2):206–212
Frank GK, Bailer UF, Henry SE, Drevets W, Meltzer CC, Price JC, Mathis CA, Wagner A, Hoge J, Ziolko S, Barbarich-Marsteller N, Weissfeld L, Kaye WH (2005) Increased dopamine D2/D3 receptor binding after recovery from anorexia nervosa measured by positron emission tomography and [11c]raclopride. Biol Psychiatry 58:908–912
Garfinkel PE, Lin E, Goering P, Spegg C, Goldbloom D, Kennedy S, Kaplan AS, Woodside DB (1996) Should amenorrhoea be necessary for the diagnosis of anorexia nervosa? Evidence from a Canadian community sample. Br J Psychiatry 168:500–506
Gross HA, Ebert MH, Faden VB, Goldberg SC, et al (1981) A double-blind controlled trial of lithium carbonate primary anorexia nervosa. J Clin Psychopharmacol 1:376–381
Gundlah C, Alves SE, Clark JA, Pai LY, Schaeffer JM, Rohrer SP (2005) Estrogen receptor-beta regulates tryptophan hydroxylase-1 expression in the murine midbrain raphe. Biol Psychiatry 57:938–942
Halmi KA, Eckert E, LaDu TJ, Cohen J (1986) Anorexia nervosa. Treatment efficacy of cyproheptadine and amitriptyline. Arch Gen Psychiatry 43:177–181
Hancock S, Grant V (2009) Early maternal separation increases symptoms of activity-based anorexia in male and female rats. J Exp Psychol Anim Behav Process 35:394–406
Hoek HW, van Hoeken D (2003) Review of the prevalence and incidence of eating disorders. Int J Eat Disord 34:383–396
Kamarajan C, Rangaswamy M, Chorlian DB, Manz N, Tang Y, Pandey AK, Roopesh BN, Stimus AT, Porjesz B (2008) Theta oscillations during the processing of monetary loss and gain: a perspective on gender and impulsivity. Brain Res 1235:45–62
Kant GJ, Lenox RH, Bunnell BN, Mougey EH, Pennington LL, Meyerhoff JL (1983) Comparison of stress response in male and female rats: pituitary cyclic AMP and plasma prolactin, growth hormone and corticosterone. Psychoneuroendocrinology 8:421–428
Kaye W, Gendall K, Strober M (1998) Serotonin neuronal function and selective serotonin reuptake inhibitor treatment in anorexia and bulimia nervosa. Bio Psychiatry 44:825–838
Kaye W (2008) Neurobiology of anorexia and bulimia nervosa. Physiol Behav 94:121–135
Kaye WH, Fudge JL, Paulus M (2009) New insights into symptoms and neurocircuit function of anorexia nervosa. Nat Rev Neurosci 10:573–584

Keating C (2009) Theoretical perspective on anorexia nervosa: the conflict of reward. Neurosci Biobehav Rev 34:73–79

Keating C (2010) Anorexia Nervosa: a single ROI for self-referent processing of pictures of body images? doi: 10.1176/appi.ajp.2010.09121712.

Keating C, Tilbrook A, Kulkarni J (2010) Oestrogen: an overlooked mediator in the neuropsychopharmacology of treatment response? The International Journal of Neuropsychopharmacology. doi:10.1017/S1461145710000982. Available on CJO 22 September 2010

Kennerley SW, Walton ME, Behrens TE, Buckley MJ, Rushworth MF (2006) Optimal decision making and the anterior cingulate cortex. Nat Neurosci 9:940–947

Kerns JG, Cohen JD, MacDonald AW 3rd, Cho RY, Stenger VA, Carter CS (2004) Anterior cingulate conflict monitoring and adjustments in control. Science 303:1023–1026

Knutson B, Bhanji JP, Cooney RE, Atlas LY, Gotlib IH (2008) Neural responses to monetary incentives in major depression. Biol Psychiatry 63:686–692

Licinio J, Wong ML, Gold PW (1996) The hypothalamic-pituitary-adrenal axis in anorexia nervosa. Psychiatry Res 62:75–83

Lo Sauro C, Ravaldi C, Cabras PL, Faravelli C, Ricca V (2008) Stress, hypothalamic-pituitary-adrenal axis and eating disorders. Neuropsychobiology 57:95–115

Lu NZ, Shlaes TA, Gundlah C, Dziennis SE, Lyle RE, Bethea CL (1999) Ovarian steroid action on tryptophan hydroxylase protein and serotonin compared to localization of ovarian steroid receptors in midbrain of guinea pigs. Endocrine 11:257–267

Matsumoto M, Hikosaka O (2009) Two types of dopamine neuron distinctly convey positive and negative motivational signals. Nature 459:837–841

Monteleone P, Luisi M, Colurcio B, Casarosa E, Ioime R, Genazzani AR, Maj M (2001) Plasma levels of neuroactive steroids are increased in untreated women with anorexia nervosa or bulimia nervosa. Psychosom Med 63:62–68

Morishita S, Kinoshita T (2008) Predictors of response to sertraline in patients with major depression. Hum Psychopharmacol 23:647–651

Pae CU, Mandelli L, Kim TS, Han C, Masand PS, Marks DM, Patkar AA, Steffens DC, De Ronchi D, Serretti A (2009) Effectiveness of antidepressant treatments in pre-menopausal versus post-menopausal women: a pilot study on differential effects of sex hormones on antidepressant effects. Biomed Pharmacother 63:228–235

Panagiotaropoulos T, Papaioannou A, Pondiki S, Prokopiou A, Stylianopoulou F, Gerozissis K (2004) Effect of neonatal handling and sex on basal and chronic stress-induced corticosterone and leptin secretion. Neuroendocrinology 79:109–118

Petrovic P, Pleger B, Seymour B, Kloppel S, De Martino B, Critchley H, Dolan RJ (2008) Blocking central opiate function modulates hedonic impact and anterior cingulate response to rewards and losses. J Neurosci 28:10509–10516

Piazza PV, Deroche V, Deminiere JM, Maccari S, Le Moal M, Simon H (1993) Corticosterone in the range of stress-induced levels possesses reinforcing properties: implications for sensation-seeking behaviors. Proc Natl Acad Sci USA 90:11738–11742

Pochon JB, Riis J, Sanfey AG, Nystrom LE, Cohen JD (2008) Functional imaging of decision conflict. J Neurosci 28:3468–3473

Rivalland ET, Clarke IJ, Turner AI, Pompolo S, Tilbrook AJ (2007) Isolation and restraint stress results in differential activation of corticotrophin-releasing hormone and arginine vasopressin neurons in sheep. Neuroscience 145:1048–1058

Roberto CA, Steinglass J, Mayer LE, Attia E, Walsh BT (2008) The clinical significance of amenorrhea as a diagnostic criterion for anorexia nervosa. Int J Eat Disord 41:559–563

Rossi G, Balottin U, Rossi M, Chiappedi M, Fazzi E, Lanzi G (2007) Pharmacological treatment of anorexia nervosa: a retrospective study in preadolescents and adolescents. Clin Pediatr 46:806–811

Rushworth MF, Behrens TE, Rudebeck PH, Walton ME (2007) Contrasting roles for cingulate and orbitofrontal cortex in decisions and social behaviour. Trends Cogn Sci 11:168–176

Schule C (2006) Neuroendocrinological mechanisms of actions of antidepressant drugs. J Neuroendocrinol 19:213–226

Sodersten P, Bergh C, Zandian M (2006) Understanding eating disorders. Horm Behav 50:572–578

Sodersten P, Nergardh R, Bergh C, Zandian M, Scheurink A (2008) Behavioral neuroendocrinology and treatment of anorexia nervosa. Front Neuroendocrinol 29:445–462

Solomon MB, Herman JP (2009) Sex differences in psychopathology: of gonads, adrenals and mental illness. Physiol Behav 97:250–258

Tilbrook AJ, Clarke IJ (2006) Neuroendocrine mechanisms of innate states of attenuated responsiveness of the hypothalamo-pituitary adrenal axis to stress. Front Neuroendocrinol 27:285–307

van Kuyck K, Gerard N, Van Laere K, Casteels C, Pieters G, Gabriels L, Nuttin B (2009) Towards a neurocircuitry in anorexia nervosa: evidence from functional neuroimaging studies. J Psychiatr Res 43:1133–1145

Wade T, Bergin M, Tiggemann M, Bulik CM, Fairburn CG (2001) Assessing the prevalence of eating disorders in an Australian twin population. Aust NZ J Psychiatry 30:845–851

Wagner A, Aizenstein H, Venkatraman VK, Fudge J, May JC, Mazurkewicz L, Frank GK, Bailer UF, Fischer L, Nguyen V, Carter C, Putnam K, Kaye WH (2007) Altered reward processing in women recovered from anorexia nervosa. Am J Psychiatry 164:1842–1849

Walsh BT, Kaplan AS, Attia E, Olmsted M, Parides M, et al (2006) Fluoxetine after weight restoration in anorexia nervosa: a randomized controlled trial. J Am Med Assoiation 295:2605–2612

Wigger A, Neumann ID (1999) Periodic maternal deprivation induces gender-dependent alterations in behavioral and neuroendocrine responses to emotional stress in adult rats. Physiol Behav 66:293–302

Wrase J, Kahnt T, Schlagenhauf F, Beck A, Cohen MX, Knutson B, Heinz A (2007) Different neural systems adjust motor behavior in response to reward and punishment. Neuroimage 36:1253–1262

Young EA, Becker JB (2009) Perspective: sex matters: gonadal steroids and the brain. Neuropsychopharmacology 34:537–538

Young EA, Altemus M, Parkison V, Shastry S (2001) Effects of estrogen antagonists and agonists on the ACTH response to restraint stress in female rats. Neuropsychopharmacology 25:881–891

Zakharova E, Wade D, Izenwasser S (2009) Sensitivity to cocaine conditioned reward depends on sex and age. Pharmacol Biochem Behav 92:131–134

Zanardi R, Rossini D, Magri L, Malaguti A, Colombo C, Smeraldi E (2007) Response to SSRIs and role of the hormonal therapy in post-menopausal depression. Eur Neuropsychopharmacol 17:400–405

Zastrow A, Kaiser S, Stippich C, Walther S, Herzog W, Tchanturia K, Belger A, Weisbrod M, Treasure J, Friederich HC (2009) Neural correlates of impaired cognitive-behavioral flexibility in anorexia nervosa. Am J Psychiatry 166:608–616

Gender Differences in Neurodevelopmental Disorders: Autism and Fragile X Syndrome

Nicole J. Rinehart, Kim M. Cornish, and Bruce J. Tonge

Contents

Abstract Gender is an important factor to consider in understanding the clinical presentation, management, and developmental trajectory of children with neuropsychiatric disorders. While much is known about the clinical and neurobehavioural profiles of boys with neuropsychiatric disorders, surprisingly little is known about girls. The aim of this chapter was to review our understanding of gender by considering the most prevalent childhood onset neuropsychiatric disorders, autism and Fragile X syndrome. This chapter highlights findings which suggest that girls with autism and Fragile X syndrome show some unique differences in cognitive and clinical profiles when compared to boys with these conditions; this may indicate the need for innovative assessment and management approaches which take gender into consideration. Our understanding of how differences emerge in boys and girls with neuropsychiatric disorders is unclear, future

N.J. Rinehart, K.M. Cornish, and B.J. Tonge (✉)
Centre for Developmental Psychiatry and Psychology, School of Psychology and Psychology, Faculty of Medicine, Nursing and Health Sciences, Monash University, Melbourne, Australia
e-mail: bruce.tonge@monash.edu

J.C. Neill and J. Kulkarni (eds.), *Biological Basis of Sex Differences in Psychopharmacology*, 209
Current Topics in Behavioral Neurosciences 8, DOI 10.1007/7854_2010_96,

research needs to focus on the role of biological maturation rates, sex hormones, and psychosocial factors in order to progress this field.

Keywords Autism · Cognition · Fragile X syndrome (FXS) · Genetic Disorder · Gender · Intellectual Disability · Mental Health · Neurodevelopmental Disorder · Social Functioning

Sex differences in the prevalence of psychiatric disorders in children and adolescents have consistently been identified in epidemiological studies beginning with the landmark Isle of Wight study (Rutter 1989). Approximately twice as many prepubertal boys have psychopathology compared to girls, but the rates become more equivalent in adolescence. Changes in rates are also influenced by the type of disorder. Males are more vulnerable to suffer from attention-deficit hyperactivity disorder (ADHD) and autism in childhood, but as boys mature ADHD may become less prevalent and conduct disorder and substance abuse more prevalent (Rutter et al. 2003). A meta-analysis of 26 population studies found that the prevalence of depression doubles from childhood to approximately 5.6% in adolescence and is higher in females (5.9%) than in males (4.6%) (Costello et al. 2006). More young girls than boys are likely to suffer from depression, and following puberty this female preponderance for anxiety increases to around twice that for males (Lewinsohn et al. 1998). The causes of these gender differences are unclear, but an interaction between biological, psychological, and social factors are likely (Rutter et al. 2003; Nolen-Hoeksema et al. 1999). Lewinsohn et al. (1998) suggest that the female vulnerability to anxiety is based on genetic factors, but others claim that vulnerability to depression and emotional problems in females is strongly influenced by the effects of socialization, sensitivity to interpersonal relationships, the sense of mastery, self regulation, and response to stress (Nolen-Hoeksema et al. 1999; Kennan and Shaw 1997). Epidemiological studies of populations of twins and more recently the application of molecular genetics in longitudinal studies provide a method to describe gene–environment interactions contributing to psychopathology (Foley et al. 2004). A particularly useful way to investigate the "nature versus nurture" effects of gender on psychopathology is through studies of disorders that have a strong genetic component (henceforth referred to as "neurodevelopmental disorders"). Approximately, 3% of children will be born with a neurodevelopmental disorder equating to approximately 650,000 individuals in Australia alone. The impact of "nature" versus "nurture" is infinitely more complex in neurodevelopmental disorders where the interaction between prenatal hormones, specific susceptibility genes, and risk or protective genes on the X chromosome operates (Rutter 2005). The genetic mechanisms that result in a higher prevalence of males with child-onset neurodevelopmental disorders, including those for which genetic etiology, are still not determined, such as autism, as well as those disorders for which the genetic origins are known, such as fragile X syndrome (FXS), allow investigation of gender-related differences in prevalence, clinical manifestation, and cognitive and neuropsychiatric profiles. However, across the neurodevelopmental disorder literature, little attention has been paid to the way in which downstream biological maturation rates, sex hormones, and psychosocial environments differentially impact on the developmental trajectory for male and females.

The focus of this chapter will be to overview the clinical, cognitive, and neurobiological literature on gender differences in the most prevalent childhood onset neuropsychiatric disorders, autism and FXS, the latter being the most common cause of hereditary intellectual disability in males, resulting from the silencing of a single gene on the X chromosome. Thompson et al. (2003) have extensively reviewed the literature on gender differences in neurodevelopmental disorders and concluded that gender has yet to matter sufficiently in neurodevelopmental disorder research, and by extension clinical practice. The almost exclusive focus on males with neurodevelopmental disorders in the biological and clinical research literature has led to an imbalance in gender-specific information, which can inform the clinical assessment and management of females with neurodevelopmental disorders. In this chapter, we will highlight the important gender differences that necessitate careful investigation of male and female profiles across development to facilitate targeted gender-specific clinical and educational interventions and treatments.

1 Autism

1.1 Diagnosis

Autism is a generic term referring to a group of related conditions defined in the DSM-IV-TR (American Psychiatric Association 2000) and the ICD-10 (World Health Organisation 1992) as pervasive developmental disorders (PDD). These disorders have their onset within the first 3 years of life, but the clinical picture may change with development. In DSM IV TR (American Psychiatric Association 2000), the PDDs comprise the categories of autistic disorder, Asperger disorder, Rett's disorder, childhood disintegrative disorder, and PDD – not otherwise specified (PDD-NOS). Autism was first described by Leo Kanner in 1943 in a group of 11 children who had the distinctive core features of social, language, and communication disturbance, and an obsessive desire for sameness (Kanner 1943). In the following year, Hans Asperger described a group of 16 children and adolescents in Vienna who had deficits in communication and social skills together with obsessional interest, intolerance of change, and motor clumsiness (Asperger 1944). Unlike the children described by Kanner, these young people were of normal intellectual ability and did not have any delay or abnormality in their language development. This has become the differentiating feature of Asperger disorder from autistic disorder.

All children with autism have impaired social interactions, which may change as they develop. Infants with autism do not anticipate social interactions, such as being picked up, or seek physical comfort or parental attention. Preschool children with autism usually avoid eye contact and do not engage in social imitation such as waving goodbye. They are unresponsive to the feelings and emotions of others. They are aloof and unable to engage effectively with other children or understand reciprocal social interactions. As the child grows older, there may be an increased interest in other people, but social skills are often stilted and learned in an inflexible manner, leading the child to appear odd and socially clumsy. Parents usually first seek help because their child has language delay and a lack of nonverbal communication, and easily becomes frustrated. About 50% of children with autism fail to develop functional speech and only learn slowly to compensate with gesture. Language development is often abnormal in the remainder, with echolalia, self-directed jargon, and the repetition of irrelevant phrases, for example from a TV show. The correct use of pronouns and the related development of a sense of self and others are delayed. Poor comprehension, problems expressing needs by words and gesture, and difficulty in social understanding are frequently the causes of frustration and disturbed behavior. Children who do develop functional language usually have difficulty in using language socially and in initiating or sustaining a reciprocal conversation. For example, the child may talk at others in a socially inappropriate manner. In contrast to children with autistic disorder, young people with Asperger disorder have no delay in the development of normal expressive and receptive language, including the use of communicative phrases by the age of three. However, children with Asperger disorder have problems in their social use of language, for example being verbose and preoccupied with a favorite topic. Their speech may appear odd due to the use of an unusual accent, or the presence of abnormalities in pitch and volume, leading, for example, to a flat and monotonous delivery.

The play, behavior, and daily life of children with autism are usually rigid and repetitive. Younger children may line up toys or objects, or be preoccupied with special objects such as stones, and become distressed if these activities are interrupted. Their ritualistic play lacks imagination and social imitation. With development, play may become more complex, such as re-enacting scenes from a favorite video story, but is usually still repetitive. The older child may develop preoccupations with themes such as train timetables or dinosaurs, and this will be the focus of their play, drawing, and conversation. They may have a number of rituals associated with daily life, such as a fixed order for bathing and dressing, or an insistence on wearing the same clothes or taking the same route to a familiar place. Change or unexpected events can be distressing: for example, the arrival of a new student in the classroom. There may be a number of perceptual or sensory abnormalities such as hyperacusis or tactile sensitivity, manifesting, for example, as an aversion to having their hair brushed. Some children with autism have a remarkable lack of sensitivity to pain. Children with autism are usually visually attentive: for example, they may study the detail in a picture book or closely observe spinning wheels, the edges of objects, or reflections in water. There are usually some motor mannerisms, such as hand flapping, or tiptoe walking and gait abnormality.

Approximately 80% of children with autism also have intellectual disability, and a range of other emotional and behavioral disturbances is common. Children with autistic disorder who have intellectual abilities within the normal range are referred to as high functioning. The individual cognitive profiles of children with autism usually show a wide scatter of abilities, with deficits in verbal and social comprehension tasks, and more ability with visuomotor performance skills. In contrast, children with Asperger disorder have overall normal intellectual abilities, but usually have relative deficits in visuomotor tasks and motor skills compared with their verbal performance. For the sake of brevity, autism and Asperger disorder will henceforth be referred to as "autism."

2 Fragile X Syndrome

FXS was first described in 1943 by Martin and Bell (originally labeled the "Martin-Bell" syndrome) and is the world's most common hereditary cause of developmental delay in males. The disorder is caused by the silencing of a single gene on the long arm of the X chromosome at q27.3. The gene, named the Fragile X Mental Retardation Gene-1 (FMR1), was identified in 1991 and is "turned off" in affected individuals. When this occurs, there is an expansion of a trinucleotide (CGG) in the repeat region. In individuals unaffected by FXS, there are between 7 and 55 CGG repeats, with 30 repeats the most common number. In clinically affected individuals (known as the full-mutation status), the CGG region expands to over 200 repeats, resulting in silencing of the gene and loss of the fragile mental retardation protein (FMRP). The extent to which these molecular discoveries explain the behavioral phenotypic outcomes in FXS is beginning to be revealed (see also Cornish et al. 2008, for a recent review of the genetic-cognitive/behavioral correlates that comprise the fragile X continuum).

The clinical features that can characterize FXS include an elongated face, large prominent ears, and forehead, and in males, postpubertal macroorchidism (Cornish et al. 2007a; Lachiewicz et al. 2000). More subtle features can include narrow inter-eye distance, a highly arched palate of the mouth, and hyperextensible joints. However, the wide variability in manifestation in both boys and girls makes a diagnosis based on physical features alone almost impossible. It is precisely because of their relatively "normal" appearance that many affected children are not diagnosed with FXS until relatively late in their development. Undoubtedly, the most defining feature, especially in boys with the disorder, is intellectual disability and the resulting cognitive-behavioral phenotype, most notably the attentional control difficulties, language, and spatial impairments that can accompany the syndrome from very early in development (Abbeduto et al. 2007; Cornish and Wilding 2010; Cornish et al. 2007b; Scerif et al. 2004, 2007).

Undoubtedly, behavioral problems often link with cognitive impairment, and two disorders in particular that co-exist in many individuals with FXS are autism and ADHD. Due to the pervasiveness of these symptoms, a diagnosis of FXS may occur later in childhood or never at all if clinicians are unable to dissociate the FXS phenotype from a diagnosis of autism or ADHD. To address this concern and provide much needed information to clinicians and educators, the findings from a series of recent studies have begun to elucidate the FXS "signature" associated with both autism and ADHD.

Among the most distinctive and pervasive behavioral features of young boys with FXS are attentional and hyperactivity problems (Cornish et al. 2001a; Hatton et al. 2002; Sullivan et al. 2007), the severity of which often leads to a clinical diagnosis of ADHD. At a finer-tuned level, however, the FXS profile appears to be characterized by unexpectedly extreme levels of inattentiveness, restlessness, fidgetiness, impulsive tendencies, and distractibility even when their level of general development is taken into account. In one of the most comprehensive study's to date, Turk (1998) compared the ADHD profiles of 49 FXS boys (aged 4–16 years) to that of 45 boys with Down syndrome (aged 4–16 years), and 42 boys with mental retardation of an unknown cause (aged 4–16 years). Although both groups of boys showed similar levels of motor activity, the boys with FXS show significantly more inattentiveness, restlessness, fidgetiness, distractibility, and impulsive tendencies suggestive of DSM-IV ADHD predominantly inattentive type. Moreover, there is some evidence that these features do not necessarily improve with age (in contrast to most children with these traits), emphasizing the need for early diagnosis and multidisciplinary intervention. In girls with FXS, the research is not as extensive as that undertaken in affected boys but nonetheless points to a substantive minority of girls presenting with ADHD-inattentive type symptoms (Mazzocco et al. 1998). Together, these findings highlight a distinctive ADHD profile in the FXS full mutation that is not solely the artifact of intellectual disability.

One of the most intriguing and complex of relationships in FXS is its association with autism. There are currently very few single-gene disorders for which there is a certainty of the involvement of autism; FXS is one. As a single-gene disorder, FXS offers an interesting genetic model to explore the functions of FMRP regulation and the repercussions of its loss in early brain development. Commonalities across core social and language domains define the link between FXS and autism, and it therefore seems highly plausible that similar neurobiological mechanisms are affected in both disorders. A recent study by Loesch and colleagues found that a common impairment in verbal skills best described the comorbidity of FXS and autism at the cognitive level (Loesch et al. 2007). In an earlier study, Philofsky et al. (2004) reported a similar link in children with FXS between exceptionally low verbal ability, in this case receptive language, and a dual diagnosis of autism, compared to children with FXS alone in which verbal skills appeared to be a relative strength. Overall, children with a dual diagnosis tend to display more impaired cognitive performance than children with either autism alone or FXS alone.

However, although commonalities between FXS and autism appear quite striking at one level, subtle differences are notable. For example, in terms of eye gaze, in children with autism atypical eye gaze is most acute in social interactions and appears to be motivated both by a lack of understanding of the social situation itself and by the absence of a desire to communicate. In contrast, eye gaze behavior in FXS does not appear to be guided by a lack of social awareness or communication. The majority of individuals with FXS, although tending to avoid social interactions, will offer what is now classically termed the "fragile X handshake," whereby an initial wish to communicate socially, with a "handshake," a socially acceptable remark or even brief initial eye contact, is coupled with active and even persistent gaze avoidance. Subsequent interactions with familiar persons may be marked by the same active gaze avoidance despite the growing relationship. The gaze avoidance persists even when attempts are made to extinguish it; it may, in fact, increase in intensity. It has been suggested that FXS is associated with a unique pattern of hyperarousal and social anxiety that can cause them to avert their eyes in a social situation (to avoid the sensory stimulation of eye contact) but may still wish to communicate socially (Cornish et al. 2004a; Wolff et al. 1989). Thus, individuals with FXS are more likely to exhibit autistic-like behaviors such as eye gaze aversion, which are more symptomatic of their hyperarousal and social anxiety rather than from an inherent lack of understanding of the social situation. See Cornish et al. (2007a) for more detailed descriptions of the commonalities and differences between FXS and autism across cognitive domains.

The frequency of autism among FXS individuals is still controversial, but approximately 2–6% of children with autism will have the FMR1 mutation (Reddy 2005; Wassink et al. 2001), and between 33% and 67% of children with FXS will fulfill the diagnostic criteria for autism (Clifford et al. 2007; Rogers et al. 2001) with more males than females reportedly meeting the autism cut-off.

3 Autism: Prevalence and Gender Differences

It is well established that there is a greater preponderance of males than females with autism. The magnitude of gender ratio differs with intellectual ability. For children with autism who have associated intellectual disability (referred to in the literature as low-functioning autism, LFA), the gender ratio is approximately 2:1. For children with autism who have above-average intellectual abilities (referred to as high-functioning autism or Asperger disorder), the ratio goes up to 10:1 (Baron-Cohen et al. 2005).

There is evidence that the prevalence of autism is increasing. Fombonne, DuMazaubrunc, and Grandjean (Fombonne 2003) reported rates of 10–12 per 10,000 using rigorous diagnostic criteria and standardized diagnostic assessment. Prevalence estimates increase substantially when the broader autism spectrum is included, and a broader range of assessment techniques are used. In Australia, the estimate of around 1 in 160 was given for children between the ages of 6 and 12 years (Australian Advisory Board on Autism Spectrum disorders). Similar figures have been reported by the US Centre for Disease Control (CDC 2007), estimating a prevalence figure of 1 in 150, for 8-year-old children diagnosed with an "autism spectrum disorder." Liu et al. (2010) recently reported prevalence clusters of children with autism which related to groups of parents who were more informed about the condition through knowing a family or affected child with autism within the community. Liu et al. (2010) noted that these parents were more likely to seek a formal diagnosis, than those parents who had not had community contact with the diagnosis of autism.

4 Fragile X Prevalence and Gender Differences

FXS affects approximately 1 in 2,500 males (Hagerman 2008). The focus initially in females with FXS has been on those with significant cognitive deficits. However, the spectrum of involvement includes those with learning difficulties and/or emotional problems with an IQ in the broad range of normal. Since most females with the full mutation fall in this category, the prevalence of affected females including this milder involvement will be greater than 1 in 8,000. In the FXS full mutation (>200 repeats), X linkage means that males are especially vulnerable to the full effects of the condition at the brain, behavioral, and cognitive levels. Almost all boys will present with moderate intellectual disability with profiles emerging as young as 3 years of age (Skinner et al. 2005). In females, the phenotypic variation is such that some girls only show subclinical learning disabilities (Bennetto and Pennington 2002), while approximately 25% display more significant cognitive impairment (most with mild ID and rare individuals with moderate ID) similar in profile to boys with FXS. Genetic variation in the form of X-inactivation (when one of the two X chromosomes remains inactive and the other active) is seen as the major contributor to the heterogeneity of intellectual disability and the broad range of cognitive deficits seen in females with FXS. For obvious reasons, this is not an issue of concern in FXS males whose impairment, without the protection of X-inactivation, shows greater severity. In males, FMRP level accounts for about 75% of the variance in IQ, but in females the proportion is much smaller (Lightbody et al. 2006).

5 Autism: Clinical Presentation and Gender Differences

There is no evidence that the core symptoms of autism differ for males and females. Volkmar et al. (1993) compared males and females with autism on the Autism Behavior Checklist, age of onset, and ratings of ICD-10 (World Health Organisation 1992) items. They found no consistent gender differences between males and females that were not related to intelligence (IQ). These findings were consistent with those of Pilowsky et al. (1998) who found no gender differences in a matched sample of 18 males and 18 females on two commonly used autism assessment tools, the ADI-R (Le Couteur et al. 1989) and the Childhood Autism Rating Scale (Schopler et al. 1988).

While the core symptoms of autism are the same in males and females according to DSM-IV-TR criteria, the clinical picture may differ in a way which is consistent with gender-related differences in the normal population of male and female children. For example, differences in emotional expression, play interests, language, cognitive profile, and social relationships (Thompson et al. 2003) may give rise to differences in the way males and females with autism present clinically. While boys may have an intense preoccupation with vehicles, such as trains and cars, girls may develop their special interests in areas such as, teddy-bear or doll collections, and spend unusual amounts of time on craft, and art, to the exclusion of other types of play (Gillberg and Coleman 2000). Gender-neutral activities such as "drawing" may form the "special interest" for either males or females with autism. Gillberg and Coleman (2000) reported that females tend to have fewer special interests than males with autism, although this may relate to male interests being more circumscribed (e.g., trains, cars) than female interests (e.g., craft, doll play), and thus more clinically salient.

Normally intelligent females with autism may present as more passive, socially competent, and with better communication skills than boys. This may again link to what we know about gender differences in males and females in the typically developing population. For example, Skuse (2009) reported that among children who have higher verbal intellectual abilities in the general population, males are more vulnerable than females to deficits in social-communicative functioning. If we extend these typical gender differences to the autism population, we would expect that females would present more competently than IQ-matched males with autism who may be more obviously socially awkward. The explanation for females with autism presenting as more socially competent may be more simply explained in terms of gender differences in biological maturation rates which affords developmental advantages for females, but not males with autism.

While the clinical presentation of females with autism may appear superficially different to that of males, females experience the same, if not greater, levels of social isolation and disadvantage as males with autism. Using the Autism Diagnostic Interview (ADI) items, McLennan et al. (2005) reported that females were described as having more severe social deficits than males with autism. These social deficits were particularly evident during adolescence in peer relationships with none of the females in the study having any sort of reciprocal friendship after age 10. Holtmann et al. (2007) reported that females have more social, thought, and attentional problems than matched boys with autism as rated by their parents using the Child Behavior Checklist. However, Holtmann et al. (2007) suggest that these parent reports may themselves represent a bias in behavioral expectations for males and females, rather than a "true" gender difference in psycholopathology and behavioral functioning (Rutter 2005).

The superficially more competent social functioning of females with autism may result in a late diagnosis or misdiagnosis of the disorder. Gillberg and Coleman

(2000) reported that normally intelligent females with autism are referred to clinical services later than males with high-functioning autism. This finding may fit with the common referral pattern to psychiatric/psychological services to preference males who tend to have higher levels of disruptive behavioral disorders which impact more broadly on the family and school setting, than females who tend to have higher levels of internalizing symptoms. Children with autism who present with a passive and often "dreamy" clinical presentation, and also experience learning difficulties at school, may initially be given a diagnosis of attention-deficit disorder (ADD). It may be that this referral pathway is more common for females than males with autism, given the gender-mediated differences in clinical presentation, although there is no data to specifically make this distinction in diagnostic pathway. Other potential misdiagnoses for females with autism include a primary anxiety disorder, selective mutism, and anorexia nervosa (Thompson et al. 2003). Another potential clinical pathway is for females with autism to be initially assessed as having an insecure attachment disorder; this presentation together with the often overprotective parenting response it elicits (Soppitt 2006) may erroneously lead to psychological interventions focused on parenting or family functioning, rather than appropriate autism-focused intervention. The focus on "over" parenting to account for the insecure attachment presentation of girls with autism has many parallels with the "refrigerator," "under" parenting hypothesis used to account for the aloof and detached presentation of boys with autism in the 1950s.

6 Fragile X: Clinical Presentation and Gender Differences

Severity of clinical/behavioral difficulties in FXS males and females will vary across and within gender. For example, hypersensitivity and hyperarousal are recognized as early prominent behavioral features of children with FXS, with and without autism (Belser and Sudhalter 1995; Miller et al. 1999). For example, in girls with FXS, Keysor et al. (2002) demonstrated higher arousal levels at baseline relative to girls with Turner syndrome and typical developing control children, with a small rise on anxiety-provoking cognitive tasks compared to that seen in girls with Turner syndrome, resulting in eventual comparable anxiety in both groups on these tasks. In terms of ADHD symptomology, there is a well-documented profile in males (Asperger 1944), but in contrast in FXS girls, the profile is less well documented with more variability. Unlike affected boys, only about one-third of girls appear to meet the DSM-IV diagnostic criteria for ADHD although many will present with some ADHD symptomology, notably inattentiveness rather than hyperactivity (Mazzocco et al. 1998). For recent reviews, see Cornish et al. (2004b) and Hatton et al. (2009). However, it is also likely that gender differences in clinical presentation will be related to severity of IQ and the presence or absence of comorbid autism.

7 Autism: Cognition and Gender Differences

In addition to environmental (e.g., parental expectation) and biological explanations (e.g., biological maturation rates) which may account for gender differences in autism, there may be a neuropsychological explanation for differences in clinical presentation between males and females.

Early studies comparing cognitive functioning in males and females with autism reported no noteworthy gender differences other than the level of intelligence (Volkmar et al. 1993; Tsai and Beisler 1983; Pilowsky et al. 1988). In one of the earliest studies of gender differences in autism, Lord et al. (1982) compared male and female children (3–8 years) on a range of nonverbal measures including intelligence, adaptive functioning, receptive vocabulary, perception, and eye–hand integration, as well as on affect, play, and relating human interest. While males were found to exhibit more unusual visual responses and inappropriate stereotypical play than females, the females in the study were found to be more impaired on every measure relating to cognitive functioning than boys including IQ, the Vineland social quotient, receptive vocabulary, eye–hand integration tasks, and perceptual skills.

More recent studies focused on normally intelligent children with autism (i.e., high-functioning autism or Asperger disorder) have indicated that gender differences may exist, laying suggestion for a gender-related neuropsychological phenotype (Kopp and Gillberg 1992). A recent study by Carter et al. (2007) comparing the cognitive profiles of toddlers with autism showed that males displayed stronger verbal and motor skills than girls, whereas girls exhibited better nonverbal problem-solving ability than boys. In the most recent study on gender differences in autism, Koyama et al. (2008) compared the cognitive profiles of children with high-functioning autism using a Japanese version of the Wechsler Intelligence Scale for Children – Third Edition. While no differences were observed in overall IQ scores, gender differences were observed on particular subtests: females outperformed males on tests of processing speed, coding, and symbol search, whereas boys were significantly better than girls on the block design subtest.

Nyden et al. (2000) have shown that females with autism have more impaired executive functioning when compared to males. Lemon et al. (2010) recently showed that females with autism have significantly more impaired executive functioning, in particular, response inhibition, when compared to typically developing females. No differences were found between males with and without autism on this test of executive function.

The subtle profile and intellectual differences between males and females are largely consistent with the only brain imaging study to look at gender differences in autism. Using magnetic resonance imaging, Bloss and Courchesne (2007) found that when males and females are matched on verbal and nonverbal intellectual ability, there was no difference in the pattern of brain abnormalities; however, females were associated with additional abnormalities in cerebral white and grey matter, and temporal grey matter (Bloss and Courchesne 2007). It was suggested that these neurological gender differences may index etiological and downstream biological maturational differences in males and females with autism. Such neurobiological differences may also account for some of the suggested phenotypical gender-related differences between males and females with autism. More significant white and grey matter abnormalities may account for Gillberg and Steffenberg's (1987) earlier finding that being female puts an individual with autism at greater risk for behavioral and or cognitive deterioration postpuberty, with 12% of males and 50% of females showing deterioration in one or both domains. Minshew (1991) refers to puberty as marking a "second-wave" of frontal deficits for young people with autism: Future research should focus on different trajectories of fronto-striatal executive impairment for females and males with autism pre- and postpuberty. Put in a clinical context, mapping of executive dysfunction for females across this critical developmental period may provide an important clinical indicator of risk for postpuberty clinical deterioration. This type of research would help clarify whether clinical reports of high levels of frontally mediated psychopathology [e.g., social, thought, and attention problems (Holtmann et al. 2007)] are best accounted for by *"nature"* (e.g., parental expectations for females) or *"nurture"* [e.g., increased frontal-susceptibility in females; see Lemon et al. (2010)].

It will also be important for future research to feed-in what we know from the typically developing literature about differences in cognitive processes which characterize males and females. For example, higher levels of psychopathology in females with autism may relate to Canli and Amin's (2002) finding that females are better than males at remembering emotional issues and events; this possibly gender-mediated cognitive predisposition, accompanied with the obsessional thought patterns which are core to autism, may result in a tendency for females to ruminate more and become increasingly mentally unwell over time.

8 Fragile X: Cognition and Gender Differences

At the intellectual level, the majority of FXS males (>95%) will display IQs within the moderate-to-severe range of impairment, with profiles emerging as young as 3 years of age. In females, the phenotypic variation is such that some girls only show subclinical learning disabilities, while approximately 50% display more moderate-to-severe mental retardation similar in profile to boys with FXS. As already noted, the X-inactivation status (ratio of normal alleles on the X chromosome) in FXS females is seen as the major contributor to the heterogeneity of intellectual disability and cognitive deficits. Surprisingly, however, few studies have addressed the issue of whether degree of intellectual impairment changes with age, albeit progressing at a slower pace compared to typically developing children, or whether intellectual level actually decreases with increasing age, or even remains static across developmental time. Although published studies have consistently observed a decline in cognitive abilities from middle-late childhood onwards (Fisch et al. 1999, 2002), Cornish et al. (2004b) have previously argued that this decline might possibly be due to the FXS child's increasing problems in maintaining and developing successful cognitive strategies that keep pace with their age-normed cohort rather than an actual regression in intellectual level or failure in neural development. However, few, if any, studies have examined intellectual decline and its relation to cognition in females, and fewer still have examined any profile beyond late childhood.

At the cognitive level, at least in males by late adulthood, relative strengths in vocabulary (Roberts et al. 2007), long-term memory for meaningful and learned information (Munir et al. 2000), and visual-perceptual skills (Cornish et al. 1999; Kogan et al. 2004) are accompanied by relative weaknesses in the storage and manipulation of complex information in working memory (Munir et al. 2000; Lanfranchi et al. 2008), linguistic processing (Abbeduto et al. 2007; Belser and Sudhalter 2001), visuo-spatial cognition (Cornish et al. 1998, 1999), and inhibition (Cornish et al. 1998, 1999). Disappointingly, there is an imbalance in the ratio of male to female studies with a greater proportion of research focused on the male phenotype as cited above. There is also a paucity of studies that have investigated cognitive *signatures across* both genders in the same research design.

In the domains of *attention* and *working memory,* toddlers and children with FXS when compared to typically developing children have significant impairments in attention and memory functioning. An early study by Hooper et al. (2000) showed that children with FXS from 4 years of age display striking difficulties in attention and memory subscales of the Leiter International Performance Scale-Revised, a nonverbal assessment tool. In a series of more recent studies, Scerif et al. (2004) have demonstrated even earlier difficulties in the control of attention in children with FXS as young as 12 months of age (Scerif et al. 2004, 2007). These studies aimed at tracing developmental trajectories of attentional control in both children with FXS and in typically developing groups. While typically developing toddlers displayed gradual improvements in the accuracy with which they searched their visual environment, toddlers with FXS tended to perseverate and were unable to shift attention away from previously correct responses, regardless of their overall developmental level. These findings replicate the pattern of difficulties seen in older boys (7–12 years) (Wilding et al. 2002) and in young adult men (18–30 years) (Cornish et al. 2001b). Similar difficulties in attention switching in FXS girls (8–16 years) have also been reported (Kovar 1993). Thus, difficulties in perseveration and in shifting attentional focus are core deficits in FXS and appear to remain constant with age. In contrast, the ability to select relevant from irrelevant information (selective attention) is a relative strength at least in FXS males that continues to develop linearly with increasing chronological age (Cornish et al. 2007b). Comparable developmental studies in females are needed to understand the range of attention dysfunction and its relation with age and IQ. To date, however, the current findings underscore the importance of recognizing and treating early attention deficits that if left untreated will impact significantly across development and learning in both males and females. It is of note that a pervasive deficit in attentional control may account for the prevalence of specific impairments across other cognitive domains that also involve inhibition as a core component, for example, the high incidence of repetitive speech seen in boys and girls with FXS (Cornish et al. 2004b).

In the domain of memory, in boys there is evidence clearly pointing to *relative* strengths in long-term and short-term recall for meaningful information including memory for faces (Turk and Cornish 1998) and story recall (Munir et al. 2000) with performance at a level equivalent to typical children matched on *developmental* level (but not chronological age level). No equivalent published studies have been conducted in FXS girls. In terms of *working memory* (the ability to retain and manipulate information "online" over short periods which is crucial in guiding attention and behavior during the course of an activity), accumulating evidence points to a relative weakness in visuo-spatial working memory compared to verbal working memory (Munir et al. 2000; Cornish et al. 1999). For example, Munir et al. (2000) examined working memory performance in 25 boys with FXS aged 8–15 years, 25 boys with Down syndrome (trisomy 21) aged 7–15 years, and two groups (25 in each) of typically developing children matched to the syndrome groups on developmental level (mental age) and on chronological age. At first glance, the findings indicated general weaknesses across both verbal and visual memory skills that were not syndrome specific but suggestive of developmental delay. However, closer inspection revealed that the impairment of the FXS group relative to that of the DS group was significantly larger on tasks that tapped visuo-spatial memory skills than for tasks that tapped verbal memory skills. In comparison to FXS boys, few studies have addressed working memory in affected girls and those that have, focused almost exclusively on adult women. However, two recent studies by Mazzocco et al. (1998) and Kirk et al. (2005) highlight difficulties in working memory thresholds that also include a specific deficit in visual memory. Less than 53% of affected girls compared to 96% of typically developing females were able to recreate the gestalt of a design by memory even though they could correctly identify the object. This finding lends some support to a tentative hypothesis that visuo-spatial impairment may be a defining feature of the phenotype in both boys and girls irrespective of degree of intellectual impairment. However, one must show some caution here in giving the impression that visual memory is a global weakness in FXS. Variability especially in the female phenotype is inevitable, and studies of adult women have reported visual memory skills that are within the normal range (Mazzocco et al. 1993).

In one of the few FXS studies to date to assess both males and females on the same cognitive measures, Cornish and Wilding (2006) found evidence of gender-specific FXS "*signatures*" whereby males and females of similar verbal mental age display different cognitive profiles in the domains of spatial cognition and language. FXS males as a group performed worse overall, followed by FXS females and then typically developing controls. When gender differences emerged, they were only within the FXS groups. Females were superior to males on tasks that tapped visuo-spatial skill, visuo-constructive skill, and articulation, but both genders displayed comparable performance on tasks that involved visual perception such as gestalt integration and language comprehension. The pattern of *gender differences* in FXS represents a unique and important finding that highlights a specific male weakness for skills that depend on visual and spatial construction of abstract, meaningless designs to form a whole, and for recall and construction of a familiar design. The fact that both tasks require relatively good visuomotor integration for successful completion may be the key to understanding the fragile X male deficit. Similar gender-driven studies across other core domains (e.g., attention and working memory) would be of tremendous benefit in adding to our knowledge about the impact of gender in X-linked genetic disorders.

9 Autism and Fragile X Gender Difference Theories

In FXS, genetic variation in the form of X-inactivation (when one of the two X chromosomes remains inactive and the other active) is seen as the major contributor to the broad range of cognitive deficits seen in females with FXS. This may also account for why in some females there is a significantly less impaired phenotype compared to their male counterparts who do not benefit from the protection of X-inactivation, and hence show greater severity.

Explanations for the gender differences in autism are less clear. Skuse (2000, 2009) and Baron-Cohen (Baron-Cohen 2002; Baron-Cohen et al. 2005) have proposed the two main theories to account for the preponderance of males versus females diagnosed with autism. Skuse (2009) proposed that males are more vulnerable to autism by virtue of a single X chromosome. Based on parallels between the high male preponderance of autism and social skill deficits in monosomy in females (XO), known as Turner syndrome, Skuse's theory suggests that a single X chromosome could reduce the threshold at which symptoms of autism manifest. Females with a second X chromosome have protection from the impact of autosomally mediated genetic vulnerability by "its influence on the development, structure, and function of the social brain" (Skuse 2009). Skuse (2009) proposed that X-imprinting would lead to male preponderance irrespective of gender hormones. This theory is proposed to account for the overrepresentation of males diagnosed with neuropsychiatric disorders, including ADHD, a disorder which like autism involves in social communication impairment (see a discussion of ADHD gender issues below) (Skuse 2000).

The evidence about whether X-linked genes contribute to individual risk factors for autism is controversial (Gong et al. 2008). One of the main limitations of Skuse's theory is that it cannot explain father-to-son transmission (Rutter 2005). Rutter (2005) noted that if an imprinted gene of the X chromosome is responsible for an increased male liability, it is likely that it operates in all males rather than through variable allelic transmission. Rutter (2005) suggests that "epigenetic effects remain a possibility and it is possible that prenatal hormones interact with susceptibility genes, but this possibility can be investigated more satisfactorily once susceptibility genes have been indentified" (Rutter 2008).

While Skuse's theory discounts a role for male sex hormones in the etiology of autism, Baron-Cohen's extreme male brain (EMB) theory (Baron-Cohen et al.

2005; Baron-Cohen 2002) proposes that exposure to higher levels of fetal testosterone (fT) may be an important causal factor in the etiology of autism, and could explain the male preponderance. The cognitive sequelae to increased testosterone in utero is hypothesized to be a "systemizing" cognitive style in males, characterized by focused learning of facts and rules of systems. Reduced testosterone in females is thought to result in a more "empathetic" cognitive style. Earlier empirical evidence to support the EMB comes from the finding that individuals with autism have a lower second to fourth finger length ratio (2D:4D); this finger length ratio is a gender dimorphic finding which has been attributed to high levels of prenatal testosterone in the male brain. In other studies to support the EMB, Baron-Cohen has reported that fT is inversely related to eye contact, vocabulary, empathy, and embedded figures performance. The most recent data to support the EMB theory were reported by Aueyung et al. (2009) who followed up a cohort of 235 children, where amniotic measures of fT were sampled in the course of clinical screening to detect genetic risk factors. When the children reached 6–10 years of age, autistic traits were assessed using parent report on standardized instruments. A positive relationship between high "autistic symptom" scores and fT was reported.

While the EMB has been referred to as the only theory of autism to provide a link between etiology, neuropsychology, and the neural basis of autism (Klin 2009), there have been several criticisms of its empirical evidence base. Skuse (2009) notes that data supporting the EMB do not make it clear whether fT actually increases the risk for a clinical diagnosis of autism, or whether fT merely reduces the threshold for detection of autistic-like symptoms. Barbeau et al. (2009) argue that the link between autism and an exaggerated male behavioral profile is problematic because individuals with autism actually do better than non-autistic individuals on some tasks considered to be measures of female traits, and which are not associated with high testosterone levels. For example, Barbeau et al. point out that the performance of a group of individuals with autism on a lexical knowledge task was more similar to the performance typically seen in females than males (Walenski et al. 2008). Barbeau et al. (2009) also argues that testosterone favors right hemisphere development, which should result in superior global processing ability; however, it is well documented that individuals with autism show a preference for processing "local" details. Barbeau et al.'s (2009) main critique of Aueyung et al.'s (2009) data linking autistic traits to fT is that the traits being measured cannot be considered "autistic since the children are not autistic, they are typical traits found in typical individuals at different levels, together with all the normal behavioral traits" (p. 27). More broadly, perhaps the most significant criticism of the EMB is the issue of testosterone levels affecting cognitive, but not physical traits in individuals with autism (Barbeau et al. 2009). There are some indirect links between testosterone-related medical conditions and females with autism; for example, Baron-Cohen et al. reported that, compared to controls, significantly more women with autism reported via a medical screening instrument a range of testosterone-linked conditions, including hirsutism, irregular menstrual cycle, polycystic ovary syndrome, severe acne, epilepsy, and a family history of ovarian, uterine, and prostate cancers, tumors, or growths. Furthermore, significantly more mothers of females with autism reported severe acne, breast and uterine cancers, tumors, or growths, and a family history of ovarian and uterine cancers, tumors, or growths. Further research is needed to investigate whether these elevated physical and medical risks are directly related to serum levels of testosterone, and not some other hormonal risk factor.

10 Summary

This chapter highlights a new focus in the neurodevelopmental literature: gender-specific research. To paraphrase Thompson et al. (2003), "sex" is beginning to "matter" in our understanding of children with neurodevelopmental disorders, the two most common being autism and FXS. There is evidence that the clinical presentation, cognitive profile, and neurobiological trajectory which characterizes males and females with neurodevelopmental disorders are different. Our understanding of how these differences emerge is not yet clear, although biological maturation rates, sex hormones, and psychosocial factors all play a role. The clinical diagnosis, educational and mental health management plans for girls with neurodevelopmental disorders will necessarily differ from that which is currently prescribed for males, but to date there is no empirical evidence available to indicate what gender-specific management plans should entail. The findings which suggest that girls with autism have more significant fronto-striatal executive impairment and may show greater deterioration in their mental health postpuberty may flag a need for increased psychiatric/psychological review and psychoeducation for parents, leading up to this critical developmental period. The lack of gender-specific research in the neurodevelopmental disability field is perhaps in, and of itself, a contributing environmental "risk" factor for girls. To illustrate, while psychoeducation is the front-line treatment for children with neurodevelopmental disorders, current information available to parents is based on research conducted with predominately male samples, some of which is not (according to parent reports) directly applicable or salient to girls with autism. This may delay the assessment process by invoking the "wait and see" method. The lack of salient information which informs the different ways autism may manifest in girls may complicate the subsequent grieving and adjustment periods for families. Autism is often referred to as a "silent disability"; for girls, this silence equates to a greater risk for under-diagnosis, misdiagnosis, and reduced opportunities for considering gender-specific interventions.

The EMB theory is currently suggested to be the most comprehensive explanation available for the male preponderance of autism; while there is a growing body of research to support the links between fT and "autistic symptoms" in males with autism, it remains to be seen whether the clinical picture for girls with autism is really that of an "extreme male brain" or whether it is that of a "less extreme female brain." The complex pattern of gender differences for males and females affected by autism would call for future theoretical accounts to perhaps be gender-neutral, or based on the chromosomes involved, as is the case with FXS and Skuse's X-imprinting theory. Future new discoveries about the "nature versus nurture" effects of gender in the context of neurodevelopmental disorders will have broader implications for how we understand the causes of gender differences in childhood and adolescent psychiatry.

References

Abbeduto L, Brady N, Kover ST (2007) Language development and fragile X syndrome: profiles, syndrome-specificity, and within-syndrome differences. Ment Retard Dev Disabil Res Rev 13(1):36–46

American Psychiatric Association (2000) Diagnostic and Statistical Manual of Mental Disorders IV-TR, 4th edn. American Psychiatric Association Press, Washington, DC

Asperger H (1944) Die autistichen Psychopathen im Kindersalter. Arch Psychiatr Nervenkr 117:76–136

Auyeung B et al (2009) Fetal testosterone and autistic traits. Br J Psychiatry 100(1):1–22

Barbeau E, Mendrek A, Mottron L (2009) Are autistic traits autistic? Br J Psychiatry 100(1):23–28

Baron-Cohen S (2002) The extreme male brain theory of autism. Trends Cogn Sci 6(6): 248–254

Baron-Cohen S, Knickmeyer RC, Belmonte M (2005) Sex differences in the brain: implications for explaining autism. Science 310(5749):819–823

Belser RC, Sudhalter V (1995) Arousal difficulties in males with fragile X syndrome: a preliminary report. Dev Brain Dysfunct 8:270–279

Belser RC, Sudhalter V (2001) Conversational characteristics of children with fragile X syndrome: repetitive speech. Am J Ment Retard 106(1):28–38

Bennetto L, Pennington BF (2002) Neuropsychology. In: Hagerman RJ, Hagerman PJ (eds) Fragile X syndrome: diagnosis, treatment and research, 3rd edn. The John Hopkins University Press, Baltimore

Bloss C, Courchesne E (2007) MRI neuroanatomy in young girls with autism: a preliminary study. J Am Acad Child Adolesc Psychiatry 46(4):515–523

Canli T, Amin Z (2002) Neuroimaging of emotion and personality: scientific evidence and ethical considerations. Brain cogn 50(3):414–431

Carter A et al (2007) Sex differences in toddlers with autism spectrum disorders. J Autism Dev Disord 37(1):86–97

Clifford S et al (2007) Autism spectrum phenotype in males and females with fragile X full mutation and premutation. J Autism Dev Disord 37(4):738–747

Cornish KM, Wilding J (2006) Does cognitive performance increase or decrease across development in Fragile X syndrome: impact of verbal mental age and gender. In: Mallard J (ed) Advances in fragile X research. Nova Publishing, New York, pp 23–36

Cornish KM, Wilding J (2010) Attention, genes & developmental disorders. Oxford University Press, New York

Cornish KM, Munir F, Cross G (1998) The nature of the spatial deficit in young females with Fragile-X syndrome: a neuropsychological and molecular perspective. Neuropsychologia 36(11):1239–1246

Cornish KM, Munir F, Cross G (1999) Spatial cognition in males with Fragile-X syndrome: evidence for a neuropsychological phenotype. Cortex 35(2):263–271

Cornish KM, Munir F, Wilding J (2001a) A neuropsychological and behavioural profile of attention deficits in fragile X syndrome. Rev Neurol 1:S24–S29

Cornish KM, Munir F, Cross G (2001b) Differential impact of the FMR-1 full mutation on memory and attention functioning: a neuropsychological perspective. J Cogn Neurosci 13(1):144–150
Cornish KM et al (2004a) Annotation: deconstructing the attention deficit in fragile X syndrome: a developmental neuropsychological approach. J Child Psychol Psychiatry 45(6): 1042–1053
Cornish K, Sudhalter V, Turk J (2004b) Attention and language in fragile X. Ment Retard Dev Disabil Res Rev 10(1):11–16
Cornish KM, Levitas A, Sudhalter V (2007a) Fragile X syndrome: the journey from genes to behavior. In: Mazzocco MMM, Ross JL (eds) Neurogenetic developmental disorders: manifestation and identification in childhood. MIT Press, Cambridge, MA, pp 73–103
Cornish KM, Scerif G, Karmiloff-Smith A (2007b) Tracing syndrome-specific trajectories of attention across the lifespan. Cortex 46(3):672–685
Cornish KM, Turk J, Hagerman R (2008) The fragile X continuum: new advances and perspectives. J Intellect Disabil Res 52(6):469–482
Costello JE, Erkanli A, Angold A (2006) Is there an epidemic of child or adolescent depression? J Child Psychol Psychiatry 47(12):1263–1271
Fisch GS et al (1999) Longitudinal assessment of adaptive and maladaptive behaviors in fragile X males: growth, development, and profiles. Am J Med Genet 83(4):257–263
Fisch GS, Simensen RJ, Schroer RJ (2002) Longitudinal changes in cognitive and adaptive behavior scores in children and adolescents with the fragile X mutation or autism. J Autism Dev Disord 32(2):107–114
Foley DL et al (2004) Childhood adversity, monoamine oxidase. A genotype, and risk for conduct disorder. Arch Gen Psychiatry 16:738–744
Fombonne E (2003) Epidemiological surveys of autism and other pervasive developmental disorders. J Autism Dev Disord 33:365–382
Gillberg C, Coleman M (2000) The biology of the autistic syndromes, 3rd edn. Mac Keith Press, London UK
Gillberg C, Steffenburg S (1987) Outcome and prognostic factors in infantile autism and similar conditions. A population based study of 46 cases followed through puberty. J Autism Dev Disord 17:273–287
Gong X et al (2008) Analysis of X chromosome inactivation in autism spectrum disorders. Am J Med Genet B Neuropsychiatr Genet 147B:830–835
Hagerman R (2008) The fragile X prevalence paradox. Br Med J 45(8):498–499
Hatton DD et al (2002) Problem behavior in boys with fragile X syndrome. Am J Med Genet 108(2):105–116
Hatton D et al (2009) Developmental trajectories of young girls with fragile x syndrome. Am J Intellect Dev Disabil 114(3):161–171
Holtmann M, Bölte S, Poustka F (2007) Autism spectrum disorders: sex differences in autistic behaviour domains and coexisting psychopathology. Dev Med Child Neurol 49(5):361–366
Hooper SR et al (2000) Nonverbal assessment of cognitive abilities in children with fragile X syndrome: the utility of the Leiter International Performance Scale-Revised. J Psychoeduc Assess 18:255–267
Kanner L (1943) Autistic disturbances of affective contact. Nerv Child 2:217–250
Kennan K, Shaw D (1997) Developmental and social influences on young girls early problem behaviour. Psychol Bull 121:95–113
Keysor CS et al (2002) Physiological arousal in females with fragile X or Turner syndrome. Dev Psychobiol 41(2):133–146
Kirk JW, Mazzocco MM, Kover ST (2005) Assessing executive dysfunction in girls with fragile X or Turner syndrome using the contingency naming test (CNT). Dev Neuropsychol 28:755–777
Klin A (2009) Embracing the challenge of bold theories of autism. Br J Psychol 100(1):29–32

Kogan CS et al (2004) Differential impact of the FMR1 gene on visual processing in fragile X syndrome. Brain cogn 127(3):591–601

Kopp S, Gillberg C (1992) Girls with social deficits and learning problems: autism, atypical Asperger syndrome or a variant of these conditions. Eur Child Adolesc Psychiatry 1(2): 89–99

Kovar C (1993) The neurocognitive phenotype of Fragile X girls. Unpublished master's thesis, University of Denver, Denver, Colorado

Koyama T et al (2008) Cognitive and symptom profiles in Asperger's disorder and high-functioning autism. Seishin Shinkeigaku Zasshi 110(6):469–474

Lachiewicz AM, Dawson DV, Spiridigliozzi GA (2000) Physical characteristics of young boys with fragile X syndrome: reasons for difficulties in making a diagnosis in young males. Am J Med Genet 92(4):229–236

Lanfranchi S et al (2008) Working memory in individuals with fragile X syndrome. Child Neuropsychol 15(2):105–119

Le Couteur A et al (1989) Autism diagnostic interview: a standardized investigator-based instrument. J Autism Dev Disord 19(3):363–387

Lemon J et al (2010) Brief Report: Executive functioning in Autism Spectrum Disorders: A gender comparison of response inhibition. J Autism Dev Disord (in press)

Lewinsohn PM, Rohde P, Seeley JR (1998) Major depressive disorder in older adolescents: prevalence, risk factors and clinical implications. Clin Psychol Rev 18(7):765–794

Lightbody AA, Hall SS, Reiss AL (2006) Chronological age, but not FMRP levels, predicts neuropsychological performance in girls with fragile X syndrome. Am J Med Genet B Neuropsychiatr Genet 141B(5):468–472

Liu K, Zerubavel N, Bearman P (2010) Social demographic change for autism. Demography 47(2):327–343

Loesch DZ et al (2007) Molecular and cognitive predictors of the continuum of autistic behaviours in fragile X. Neurosci Biobehav Rev 31(3):315–326

Lord C, Schopler E, Revicki D (1982) Sex differences in autism. J Autism Dev Disord 12(4): 317–330

Mazzocco MMM, Pennington BF, Hagerman RJ (1993) The neurocognitive phenotype of female carriers of fragile X: additional evidence for specificity. J Dev Behav Pediatr 14(5):328–335

Mazzocco M et al (1998) Social functioning among girls with fragile X or Turner syndrome and their sisters. J Autism Dev Disord 28(6):509–517

McLennan J, Lord C, Schopler E (2005) Sex differences in higher functioning people with autism. J Autism Dev Disord 23(2):217–227

Miller LJ et al (1999) Electrodermal responses to sensory stimuli in individuals with fragile X syndrome: a preliminary report. Am J Med Genet 83:268–279

Minshew N (1991) Indices of neural function in autism: clinical and biologic implications. Pediatrics 87(5):774–780

Munir F, Cornish KM, Wilding J (2000) Nature of the working memory deficit in Fragile-X syndrome. Brain Cogn 44(3):387–401

Nolen-Hoeksema S, Larson J, Grayson C (1999) Exploring the gender differences in depressive symptoms. J Per Soc Psychol 77:1061–1072

Nyden A, Hjelmquist E, Gillberg C (2000) Autism spectrum and attention-deficit disorders in girls. Some neuropsychological aspects. Eur Child Adolesc Psychiatry 9(3):180–185

Philofsky A et al (2004) Linguistic and cognitive functioning and autism symptoms in young children with fragile X syndrome. Am J Ment Retard 109(3):208–218

Pilowsky T et al (1988) The Autism Diagnostic Interview-Revised and the Childhood Autism Rating Scale: differences between diagnostic systems and comparison between genders. J Autism Dev Disord 28(2):143–151

Pilowsky T et al (1998) The Autism Diagnostic Interview-Revised and the Childhood Autism Rating Scale: differences between diagnostic systems and comparison between genders. J Autism Dev Disord 28(2):143–151

Reddy KS (2005) Cytogenetic abnormalities and fragile-X syndrome in autism spectrum disorder. BMC Med Genet **6**(3). doi:10.1186/1471-2350-6-3
Roberts J et al (2007) Receptive vocabulary, expressive vocabulary, and speech production of boys with fragile X syndrome in comparison to boys with Down syndrome. Am J Ment Retard 112(3):177–193
Rogers SJ, Wehner EA, Hagerman R (2001) The behavioral phenotype in fragile X: symptoms of autism in very young children with fragile X syndrome, idiopathic autism, and other developmental disorders. J Dev Behav Pediatr 22(6):409–417
Rutter M (1989) Isle of Wight revised: twenty five years of child psychiatry epidemiology. J Am Acad Child Adolesc Psychiatry 28:633–643
Rutter M (2005) Aetiology of autism: findings and questions. J Intellect Disabil Res 49(4): 231–238
Rutter M (2008) Introduction: whither gene-environment interactions. Novartis Found Symp 293: 1–12; 68–70
Rutter M, Caspi A, Moffitt T (2003) Using sex differences in psychopathology to study causal mechanisms: unifying issues and research strategies. J Child Psychol Psychiatry 44(8): 1092–1115
Scerif G et al (2004) Visual search in typically developing toddlers and toddlers with Fragile X or Williams syndrome. Dev Sci 7(1):116–130
Scerif G et al (2007) Delineation of early attentional control difficulties in fragile X syndrome: focus on neurocomputational changes. Neuropsychologia 45(8):1889–1898
Schopler E, Reichter RJ, Renner A (1988) The childhood autism rating scale (CARS). Western Psychological Services, Los Angeles
Skinner M et al (2005) Mapping nonverbal IQ in young boys with fragile X syndrome. Am J Med Genet A 132A:25–32
Skuse D (2000) Imprinting, the X-chromosome, and the male brain: explaining sex differences in the liability to autism. Pediatr Res 47(1):9
Skuse D (2009) Is autism really a coherent syndrome in boys, or girls? Br J Psychol 100(Pt 1): 39–47
Soppitt R (2006) Clinical observations in children with autistic spectrum disorders. Autism 10:429
Sullivan K et al (2007) Sustained attention and response inhibition in boys with fragile X syndrome: measures of continuous performance. Am J Med Genet B Neuropsychiatr Genet 144B(4):517–532
Thompson T, Caruso M, Ellerbeck K (2003) Sex matters in autism and other developmental disabilities. J Intellect Disabil 7(4):345–362
Tsai LY, Beisler JM (1983) The development of sex differences in infantile autism. Br J Psychiatry 142:373–378
Turk J (1998) Fragile X syndrome and attentional deficits. J Appl Res Intellect Disabil 11:175–191
Turk J, Cornish K (1998) Face recognition and emotion perception in boys with fragile X syndrome. J Intellect Disabil Res 42(6):490–499
Volkmar FR, Szatmari P, Sparrow SS (1993) Sex differences in pervasive developmental disorders. J Autism Dev Disord 26(4):579–591
Walenski M et al (2008) Brief report: enhanced picture naming in autism. J Autism Dev Disord 37(8):1395–1399
Wassink TH, Piven J, Patil SR (2001) Chromosomal abnormalities in a clinic sample of individuals with autistic disorder. Psychiatr Genet 11(2):57–63
Wilding J, Cornish K, Munir F (2002) Further delineation of the executive deficit in males with fragile-X syndrome. Neuropsychologia 40(8):1343–1349
Wolff P et al (1989) The greeting behavior of fragile X males. Am J Ment Retard 93(4):406–411
World Health Organisation (1992) The ICD-10 classification of mental and behavioural disorders: Clinical descriptions and diagnostic guidelines. World Health Organisation, Geneva

The Impact of Gender on Antidepressants

John J. Sramek and Neal R. Cutler

Contents

Abstract There is a large body of literature debating whether and how gender affects the metabolism, side-effect profile, and efficacy of antidepressants. Gender differences in antidepressant pharmacokinetics and efficacy profiles have been attributed to not only anatomic and physiological differences between the sexes, but also behavioral factors, comorbid disorders, and gender-specific conditions, such as pregnancy and menopause. Despite the large body of research on this topic, few definitive conclusions regarding effects of gender on antidepressant treatment exist, and much of this research is incomplete, contradictory, or not fully used to optimize the administration of antidepressants and the response to treatment. This chapter will review the latest research on gender-specific effects of antidepressant treatment, focusing on the overall, gender-related differences in efficacy, metabolism, and side-effect profile of antidepressants, and how these differences can be used to better optimize treatment of depression in a clinical setting.

Keywords Antidepressant · Depression · Efficacy · Estrogen · Female · Gender · Hormones · Male · Menopause · Metabolism · Pathogenesis · Phamacokinetics · Sex · SSRI · TCA

J.J. Sramek (✉) and N.R. Cutler
Worldwide Clinical Trials, Inc., 401 N. Maple Drive, Beverly Hills, CA 90210, USA
e-mail: jsramek@wwctrials.com, neal.cutler@wwctrials.com

J.C. Neill and J. Kulkarni (eds.), *Biological Basis of Sex Differences in Psychopharmacology*, 231
Current Topics in Behavioral Neurosciences 8, DOI 10.1007/7854_2010_118,

1 Introduction

There is a large body of literature debating whether and how gender affects the metabolism, side-effect profile, and efficacy of antidepressants. Gender differences in antidepressant pharmacokinetic and efficacy profiles have been attributed to not only anatomic and physiological differences between the sexes, but also behavioral factors, comorbid disorders, and gender-specific conditions, such as pregnancy and menopause. For example, gender-related variance in body fat, hormone levels, liver metabolism, gastric properties, and blood flow has been shown to affect the absorption, distribution, metabolism, and elimination of a drug. In addition, behaviors such as smoking and alcohol intake, typically more prevalent among men, may also influence the pharmacokinetics of antidepressants. Pregnancy and menopause, as well as concomitant, female-specific medications, such as birth-control treatment and hormone therapy, can have an effect on the disposition and dose requirement of antidepressants. Finally, the clinical presentation, prevalence, and resiliency of the depression itself, as well as the existence of comorbid disorders such as anxiety disorder, often have gender-specific components that may influence the treatment of the disorder and the response of patients to medication.

Despite the large body of research on this topic, few definitive conclusions regarding effects of gender on antidepressant treatment exist, and much of this research is incomplete, contradictory, or not fully used to optimize the administration of antidepressants and the response to treatment. This chapter will review the latest research and literature on gender-specific effects of antidepressant treatment, focusing on the overall, gender-related differences in efficacy, metabolism, and side-effect profile of antidepressants, and how these differences can be used to better optimize treatment of depression in a clinical setting.

2 Gender Differences in the Pathogenesis of Depression

Women are at least twice as likely as men to suffer from depression (Weissman and Klerman 1977; Weissman et al. 1993). This holds true across the spectrum of depressive disorders, including atypical depression, unipolar depression, dysthymia, and seasonal affective disorder (Rapaport et al. 1995; Lucht et al. 2003; Leibenluft et al. 1995). These differences have been found in every country surveyed (Weissman et al. 1996) and across all age groups (Kessler et al. 1993). In the USA, the national lifetime risk of major depression is 21% in women versus 13% in men (Kessler et al. 1994). Greater incidence of depression in females appears to be linked to a higher first onset rate (Kessler 2003). Symptom presentation is also typically more severe in females. Depressed women are more likely to experience chronic or recurrent depression than depressed men, with an earlier age of onset and poorer quality of life (Kornstein et al. 2000b). In addition, depressed women tend to have more weight gain, anxiety, and physical symptoms than depressed men (Frank et al. 1988; Young et al. 1990; Angst and Dobler-Mikola 1984; Williams et al. 1995), although a study of depressed adolescents found younger girls tended to lose weight (Baron and Joly 1988). Depressed women also experience more crying, guilt, and body image dissatisfaction (Wilhelm et al. 2002), while depressed men tend to experience more work inhibition, health concerns, and social withdrawal (Vredenburg et al. 1986).

The reasons for these dramatic differences between the sexes are still unknown, but many theories exist. Since gender differences in depression prevalence do not typically emerge until adolescence, studies have suggested that female's greater susceptibility may be tied to increases in the female reproductive hormones estrogen and progesterone at puberty (Nolen-Hoeksema 1990). Fluctuations in these female hormones have been suggested to make females more susceptible to both depression and anxiety (Seeman 1997), and women may experience hormonal triggers of depressive episodes tied to reproductive events (Parry 1989). In fact, numerous studies have found variations in hormonal levels tied to increases in depressive symptoms in women undergoing puberty, menstruation, and menopause (Schmidt et al. 2004; Freeman et al. 2004). In support of this theory, the postmenopausal incidence of depression in women (when reproductive hormones stabilize) was found to be similar to that of men (Bebbington et al. 2003). There is even evidence that circadian rhythms interact with reproductive hormones and confer susceptibility to depression in some women during premenstrual dysphoric disorder (PMDD), pregnancy, and the postpartum period (Parry and Newton 2001).

Differences in monamine functioning and processing may also contribute to the gender differences in the prevalence of depression. A recent study investigated tryptophan depletion in women and men who were currently in remission from a DSM-IV-defined major depressive episode. Depletion of the monoamine tryptophan produces a transient reduction in serotonin transmission and can be used for inducing depressive episodes in formerly depressed patients. This procedure triggered significantly greater depressive symptoms in women versus men, suggesting depressive vulnerability may be related to gender differences in monoaminergic function (Moreno et al. 2006). In support of this theory, positron emission tomography (PET) showed serotonin synthesis was 48% lower in females compared with males following tryptophan depletion (Nishizawa et al. 1997).

Several studies suggest that depression manifests itself with fundamentally different characteristics in males versus females. For example, young women have been found to experience more negative self-evaluation and rumination than young men, making them more susceptible to depression, and causing them to experience deeper and more prolonged depressive episodes than men (Hankin and Abramson 2001; Nolen-Hoeksema et al. 1999). In addition, changes in brain function induced by depression and antidepressants have been found to be different between the sexes. Women with a history of childhood depression exhibit mid-frontal alpha wave suppression on the opposite side of the brain than men (Miller et al. 2002), and depressed women who respond to the antidepressant fluoxetine have lower purine levels compared to non-responders than men (Renshaw et al. 2001).

In addition, several neurotransmitter systems associated with mood and depression show altered levels in males and females. Here we remind the reader of the large body of evidence correlating dysfunction of the norepinephrine (Delgado and Moreno 2000; Anand and Charney 2000; Lambert et al. 2000) and serotonergic systems (Owens and Nemeroff 1994) with the pathogenesis of depression, particularly the depletion of both of these neurotransmitters. The prolongation of serotonin (5-HT) and norepinephrine (NE) in the synaptic cleft is the mechanism of action for two major classes of antidepressants, appropriately named selective serotonin reuptake inhibitors (SSRIs) and serotonin-norepinephrine reuptake inhibitors (SNRIs). Different concentrations of serotonin and its metabolites have been found in the brains of same-aged males and females (Legato 1997). Baseline serotonin production was found to be 52% higher in the brains of young adult men compared to young adult women when measured with a PET radioligand (Nishizawa et al. 1997), and this was supported by another recent PET study (Sakai et al. 2006). Older, postmenopausal women were found to have lower levels of serotonin and cortisol than older men (Young 1995), and a percentage of depressed women showed lower-than-normal norepinephrine activity for their age (Halbreich and Lumley 1993). Serotonin metabolism and the serotonin metabolite 5HIAA were found to be elevated in females versus males (Young et al. 1980), possibly related to higher availability of the serotonin transporter 5-HTT in women (Staley et al. 2001). In addition, the serotonin 5-HT_{1A} receptor subtype, which is implicated in the pathogenesis of depression, shows a significant decrease in binding potential with age in males but not females (Meltzer et al. 2001). Brain levels of serotonin and norepinephrine show greater age-related changes in females compared to males (Legato 1997). Platelet 5-HT and levels of the serotonin metabolite 5-HIAA were significantly higher in older women compared to younger women (89.41 ng/108 platelet for younger vs. 112.9 for older; 1.20 for younger vs. 2.19 for older, respectively) (Kumar et al. 1998).

The dopaminergic system has also been found to play a role in the pathogenesis of depression, and a large body of recent literature has focused on altering brain levels of dopamine (DA) as a therapeutic intervention in depression. DA turnover has been shown to be reduced in depressed patients (Brown and Gershon 1993), and chronic antidepressant treatment can alter DA transmission (Dhir and Kulkarni 2007). Homovanilic acid (HVA), a major metabolite of DA, has been found to be reduced in the CSF (Roy et al. 1985; Hamner and Diamond 1996) and plasma (Lambert et al. 2000) of depressed patients. Neuroimaging and histopathological studies have found D2/D3-receptor binding to be increased, and dopamine transporter (DAT) binding to be decreased, in patients suffering from major depressive disorder (MDD) in comparison to healthy controls (Klimek et al. 2002; D'Haenen and Bossuyt 1994; Shah et al. 1997; Meyer et al. 2001). DA may also promote neurotrophic processes in the adult hippocampus (Guiard et al. 2009), and is the target for new classes of antidepressants, such as triple monoamine reuptake inhibitors and dopamine agonists (Rakofsky et al. 2009).

Gender effects appear to play a role in the dopaminergic system as well. Neuroimaging studies have suggested that women have a higher synaptic concentration of dopamine in the striatum than men. They also show that women have a higher striatal dopamine synthesis capacity than men, and that age decreases this capacity in men more than in women (Laakso et al. 2002). Preclinical studies have suggested that female sex hormones enhance presynaptic dopamine turnover (Shimizu and Bray 1993; Xiao and Becker 1994). Women experience reduced striatal dopamine release in response to amphetamines than men (Munro et al. 2006). In addition, women in the luteal phase of the menstrual cycle, which is associated with high levels of progesterone, display reduced responses to amphetamine and cocaine (inhibitors of DA reuptake) compared to men (Sofuoglu et al. 2004; White et al. 2002). Females have been shown to exhibit higher levels of DAT activity than males (Dluzen and McDermott 2008), supported by preclinical studies suggesting higher DAT density in female rats versus male rats (Rivest et al. 1995). In contrast, male rats exhibit higher production and turnover of the D1 and D2 dopamine receptors (although both genders show comparable receptor density) (Andersen et al. 1997).

While numerous and varied gender differences have been reported in the serotonergic, dopaminergic, and norepinephrine systems, the significance of these changes, and how they interact to increase or decrease the susceptibility of each gender to depression, requires further study.

Depression in females may also correlate with reductions in reproductive hormones due to menopause. Surgical menopause has also been found to increase depressive symptoms (Shifren and Avis 2007). In addition, oophorectomized females who received estrogen replacement therapy (ERT) reported diminished anxiety and depression compared to oophorectomized females not receiving ERT (Nathorst-Boos et al. 1993). While suggestive of a palliative role for estrogen in the treatment of depression, the evidence for this remains ambiguous, as other studies have not shown postmenopausal women to be at an increased risk for depression (Winokur 1973).

3 Gender Differences in Response to Antidepressants

Despite decades of research on sex-related differences in antidepressant treatment, the consensus is still out on whether such a difference exists, and if so, whether it is dependent on the class of antidepressant, menopausal state, BMI index, hormonal interactions, or PD/PK characteristics, among other variables. There are numerous studies that have shown significant gender differences in regards to the efficacy of certain antidepressants (see Table 1 for a list of studies showing gender effects). For example, several studies have found a significantly greater therapeutic response among men compared to women after taking the tricyclic antidepressant (TCA) imipramine (Dawkins and Potter 1991; Hamilton et al. 1996; Kornstein et al. 2000a; Frank et al. 1988). Another study found that women had more therapeutic benefit with the SSRI sertraline versus the TCA imipramine, whereas men responded similarly to both antidepressants

Table 1 Studies finding gender-based efficacy differences with antidepressants

Reference	Drug type	Study type	Subjects	Results
Haykal and Akiskal (1999)	SSRIs, TCAs	TCA-type antidepressants or fluoxetine	25 male and 17 female primary dysthymic patients	Females responded better than males to SSRIs
Kornstein et al. (2000a, b)	SSRIs, TCAs	12-week double-blind trial with sertraline or imipramine	235 male and 400 female outpatients with chronic major depression or double depression	Females responded better to SSRI sertraline, males responded better to TCA imipramine; differences observed primarily in premenopausal females
Martenyi et al. (2001)	SSRIs, TCAs	6-week, double-blind trial of SSRI (fluoxetine) and a norepinephrinergic TCA (maprotiline)	105 male and female depressed patients	Females in their reproductive period were more responsive to SSRIs than norepinephrinergic TCAs
Quitkin et al. (2001)	TCAs, MAOIs, SSRIs	20-year review of 8 placebo-controlled antidepressant trials and 1 open-label study	1,746 depressed patients between 18 and 65	Older females had superior response to TCAs than younger females; females had statistically superior response to MAOIs than males
Khan et al. (2005)	SSRIs, SNRIS	Review of 15 randomized, placebo-controlled trials for sex differences in antidepressant efficacy	323 depressed patients	Females had a significantly greater response than males to SSRI and (to a lesser extent) SNRI treatment
Berlanga and Flores-Ramos (2006)	SSRIs, SNRIs	8-week, double-blind clinical trial for gender differences in SSRI citalopram and SNRI reboxetine	86 depressed patients (48 females, 38 males) 18–40 years old	Premenopausal females respond better than males to serotonergic antidepressants
Young et al. (2009)	SSRI	12–14-week study of citalopram	1,043 male and 1,833 female patients with single or recurrent nonpsychotic MDD	Females have a better response to the SSRI citalopram than males

(Baca et al. 2004). In addition, a study of atypical depression found monoamine oxidase inhibitors (MAOIs) were superior to TCAs in women, while TCAs were superior to MAOIs in men (Davidson and Pelton 1986). Several studies also suggest women may respond better to SSRIs than men (Kornstein et al. 2000a; Thase et al. 2000; Martenyi et al. 2001; Haykal and Akiskal 1999; Khan et al. 2005; Sagud et al. 2002; Young et al. 2009), although the results of one of these studies has been disputed (Quitkin et al. 2001). A recent study found that younger women exposed to the SSRI fluvoxamine not only had a better response than men, but also a better response than older women (>44 years old) (Naito et al. 2007).

In contrast, there are a large number of studies that have detected no gender differences in regards to the efficacy of various antidepressants, even when examining the same class of antidepressants as the studies which found gender differences (see Table 2 for a list of studies showing no gender effects). For example, a study found the serotonin-norepinephrine reuptake inhibitor (SNRI) venlafaxine and SSRIs produced a comparable response in both men and women (Entsuah et al. 2001). A large, retrospective study of TCAs, MAOIs, and the SSRI fluoxetine found no difference based on sex or menopausal status on drug efficacy (Quitkin et al. 2002), and additional studies found women did not respond preferentially to SSRIs (Thiels et al. 2005), nor did men respond preferentially to TCAs (Parker et al. 2003). Adding fuel to this argument, Hildebrandt et al. (2003) found no effect of gender on therapeutic outcome in patients treated with the TCA clomipramine, nor with the SSRIs citalopram and paroxetine, nor with the MAOI moclobemide. A meta-analysis of 30 randomized placebo-controlled trials of imipramine or amitriptyline also found no effect of gender on TCA efficacy(Wohlfarth et al. 2004).

While it is unclear why such contradictory data exist, there are numerous methodological issues which could explain such differences. For one thing, these studies did not all use the same criteria for determining what constituted a significant response to treatment. For example, one study used a paired *t*-test to compare total HAMD17 baseline to posttreatment scores (Martenyi et al. 2001), another study required a 50% or greater decrease in HAMD21 scores to indicate a significant response to treatment (Entsuah et al. 2001). In addition, the varied ages of the female patients in these studies could also play a role, as older, postmenopausal women have very different levels of female sex hormones than younger, premenopausal women, and these hormones may affect antidepressant efficacy and metabolism. Other potential sources of variability include the clinical presentation of the depressed patients (i.e., typical versus atypical), whether the patient was treated with antidepressants previously, the various drugs and doses used across studies, and how well the patient adhered to the treatment regimen. In addition, some studies were prospective, while others were a meta-analysis of large data pools. Better-controlled studies are needed to account for these numerous variables and tease out any influence of gender on antidepressant efficacy.

Table 2 Studies finding no gender-based efficacy differences with antidepressants

Reference	Drug type	Study type	Subjects	Results
Entsuah et al. (2001)	SSRIs, SNRIs	Meta-analysis of 8 comparable double-blind, active-controlled, randomized SSRI or venlafaxine clinical trials	2,045 patients with major depression or major depressive disorder, aged 18–83 years	Males and females have comparable responses to SSRIs and SNRIs across various age groups
Hildebrandt et al. (2003)	SSRIs, TCAs, MAOIs	Review of 3 Danish double-blind randomized, controlled trials	292 inpatients (96 males, 196 females) with major and predominantly melancholic depression	No relationship between plasma concentrations, gender, and therapeutic outcome
Wohlfarth et al. (2004)	TCAs	Review of 30 randomized, placebo-controlled trials of antidepressant efficacy	3,886 patients (1,555 males and 2,331 females) with depression	Tricyclic antidepressant response is independent of gender
Thiels et al. (2005)	SSRIs	Review of data from a 6-month prospective sertraline utilization observation study	1,594 male and 3,858 female depressed patients	No gender difference in side-effects, treatment termination, or treatment response to SSRI
Pinto-Meza et al. (2006)	SSRIs	6-month follow-up study of antidepressant treatment with a SSRI (citalopram, fluoxetine, paroxetine, or sertraline)	242 females (95 in their menopause) and 59 males with major depression	No gender differences were observed in treatment response, depression severity, and symptomatology

3.1 Pharmacokinetic Differences

Women and men have been found to exhibit different pharmacokinetic (PK) profiles of antidepressants. Different PK profiles stem from many different sources, including gender-related variance in body weight, plasma volume, gastric acid production, gastric emptying time, plasma protein levels, enzyme activity, drug transporter function, and drug clearance rates, among others. Women have been shown to have higher plasma levels (Hamilton et al. 1996; Preskorn and Mac 1985) and lower clearance of TCAs (Gex-Fabry et al. 1990). This may partially be attributed to the fact that women have a higher percentage of adipose tissue and body fat compared to men. Antidepressants are lipophilic and have a strong affinity for adipose tissue, typically resulting in a greater volume of drug distribution in women (Yonkers and Brawman-Mintzer 2002; Yonkers et al. 1992). Lipophilic drugs tend to have a greater distribution in women, while water soluble drugs tend to have a greater distribution in men. In addition, women tend to secrete less gastric acid and empty the contents of their stomach more slowly than men, and have slower gastric motility in the presence of female sex hormones, resulting in reduced breakdown and clearance of antidepressants (Hutson et al. 1989; Young et al. 2009).

A review study summarized data supporting sex differences in the activity of various antidepressant-metabolizing enzymes, although little research exists into how these differences might translate into differences in clinical efficacy (Yonkers and Brawman-Mintzer 2002). Several studies found lower levels of the drug transport protein P-glycoprotein in females, which could result in altered bioavailability of certain drugs in women (Lan et al. 2000; Schuetz et al. 1995). These differences were not observed in a subsequent study, suggesting that further investigation of this transport protein is necessary (Kim et al. 2001). Studies have also found that estrogen affects the binding of metabolic enzymes to certain glycoproteins, which could potentially affect drug metabolism if those drugs were composed of glycoproteins or glycoprotein homologues (Succari et al. 1990).

The cytochrome P450 (CYP450) enzyme superfamily is one of the major drug metabolizing systems in humans, and several of its constituent parts exhibit gender differences which could contribute to observed gender differences in the PK profile of antidepressants. The metabolic enzyme CYP3A4 is highly expressed in the liver and has the largest range of substrates of all CYP enzymes, making it one of the most important enzymes for breaking down xenobiotics. It plays a major role in the metabolism of many of the drugs taken into the body, including numerous SSRIs (including citalopram, escitalopram, fluoxetine, and sertraline) and TCAs (including amitriptyline, imipramine, and clomipramine). Drugs that are substrates of CYP3A4 have often been found to clear faster in women than in men (Meibohm et al. 2002), which may be due to higher observed CYP3A4 enzymatic activity in women than in men (Hunt et al. 1992; Schmidt et al. 2001). On the other hand, drugs that are substrates of CYP2D6, another major metabolizer of xenobiotics, including the TCAs desipramine (Abernethy et al. 1985) and mirtazapine (Timmer et al. 2000), have often been shown to clear faster in men than in women (Labbe et al. 2000). Substrates of CYP1A2 have generally been found to clear faster in men than in women (Ou-Yang et al. 2000; Ereshefsky et al. 1991; Bruno et al. 1997), although there are studies that have shown the exact opposite (Nafziger and Bertino 1989). In addition, studies have shown that substrates of CYP2E1 also clear faster in men than in women (Lucas et al. 1995; Kim and O'Shea 1995).

In addition to the CYP450 system, there are many other metabolic enzymes that display gender differences in function, and these enzymes also contribute to the PK profile of xenobiotics. For example, substrates of the enzymes thiopurine methyl transferase, glucuronidation, dihydropyrimidine dehydrogenase, UDP-gluronosyl transferase, and catechol-*O*-methyl transferase all show greater clearance in men than in women (Franconi et al. 2007). On the other hand, substrates of xanthine oxidase show greater clearance in women than in men (Bock et al. 1994).

Determining how these numerous (and often contradictory) gender-based differences in metabolic enzymes and functions combine to affect the breakdown and distribution of antidepressants is an extremely complex task. It remains to be determined exactly what the clinical significance these differences are in regards to antidepressant efficacy. Most likely this effect will vary not only by the class of antidepressant, but also by the individual structure of each unique antidepressant compound.

3.2 Female Reproductive Hormones

Levels of estrogen and other female reproductive hormones have been suggested to play a role in the pathogenesis of depression and the efficacy of antidepressants. This has been evidenced, in part, by the difference in susceptibility to depression, and the difference in response to antidepressants, in women before and after menopause – an event associated with a dramatic drop in reproductive hormone levels. Premenopausal women respond better to certain classes of antidepressants than postmenopausal women. Likewise, depressed, postmenopausal women receiving ERT in combination with a SSRI showed better clinical response than depressed, postmenopausal women receiving the SSRI alone (Schneider et al. 1997). A 6-week, open-label, naturalistic study looked at the response of premenopausal and postmenopausal women with MDD to various antidepressants, including SSRIs, SNRIs, and TCAs. The study found that postmenopausal status predicted a poor response to antidepressants, and this also correlated with high levels of follicle-stimulating hormone (FSH) (Pae et al. 2009). Several studies have found both FSH and luteinizing hormone (LH) to be elevated in postmenopausal depression (Harlow et al. 2003; Abe et al. 1985; Freeman et al. 2006). Supporting this finding, improvements in depressive symptoms in peri-menopausal women have been correlated with decreases in FSH levels (Daly et al. 2003). Levels of estradiol (E2) have also been reported lowered in depressed premenopausal and peri-menopausal women (Harlow et al. 2003; Young et al. 2000).

Female reproductive hormones have effects that may interfere with or enhance the efficacy of antidepressants. For example, progesterone and estrogen have been found to modulate neurotransmitter synthesis, release, and reuptake – characteristics that many antidepressants also modulate (Frackiewicz et al. 2000). Estrogen has also been found to increase new dendritic spine formation and regulate neurotropic factors (Bryant et al. 2006). In addition, progesterone in the luteal phase has been found to slow gastric emptying, potentially modifying an antidepressant's pharmacokinetic properties (Yonkers et al. 1992). Estrogen has been found to interact with the serotonergic system, which is the target of many antidepressants, including SSRIs and SNRIs. Serotonergic agents have been found to be more potent in the presence of estrogen (Halbreich et al. 1995). In addition, the serotonin releaser, fenfluramine, had a greater effect on postmenopausal women taking ERT than ERT-naive, postmenopausal women (van Amelsvoort et al. 2001).

3.3 The Role of Menopause

Studies looking at menopause and antidepressants tend to support the theory that estrogen and other reproductive hormones enhance antidepressant efficacy. For example, a recent study looked at 242 depressed women (95 in menopause), and 59 depressed men beginning antidepressant treatment with an SSRI (citalopram, fluoxetine, paroxetine, or sertraline) from 16 primary care centers over the course of 6 months. The study found that menopause, characterized by significantly reduced female reproductive hormones (including estrogen), seems to negatively affect the SSRI treatment response of depressed women (Pinto-Meza et al. 2006). Another study examined 115 female outpatients with MDD (divided into premenopausal and postmenopausal status), as well as 86 age-matched male outpatients, before and after 8 weeks of treatment with the SSRI nefazodone or venlafaxine. The study found that women are more responsive to SSRIs during their reproductive period than during their menopausal period (Grigoriadis et al. 2003). Another study looked at the response of 86 depressed male and premenopausal female patients (18–40 years old) to the SSRI citalopram and the serotonin-norepinephrine reuptake inhibitors (SNRI) reboxetine, in an 8-week, double-blind clinical trial. The study concluded that premenopausal women respond better than men to serotonergic antidepressants, further suggesting that female gonadal hormones such as estrogen may have an enhancing effect on antidepressant efficacy (Berlanga and Flores-Ramos 2006).

These results are far from conclusive, as estrogen may function as a general mood-enhancer, independent of its hypothesized role as an antidepressants-enhancer. Estrogen given to peri-menopausal women not on antidepressants was found to be an effective treatment for depression (Soares et al. 2001). In addition, abrupt interruption of ERT in women aged over 40 with recurrent depression can quickly induce a new depressive episode (Stewart et al. 2004). On the other hand, numerous studies have not found a higher risk for depression in women during their postmenopause phase, when reproductive hormones drop significantly (McKinlay et al. 1987; Avis et al. 1994). In support of this, another study found that estrogen alone was ineffective at relieving depression in most postmenopausal women (Morrison et al. 2004).

Low LH levels also correlate and may be predictive of better antidepressant response in postmenopausal women (Zanardi et al. 2007). Serotonin levels have been found to be inversely correlated with LH levels (Carretti et al. 2005). Since low LH levels imply higher serotonin levels, this may provide a more potent substrate for antidepressants whose mechanism of efficacy is dependent upon the availability of serotonin (such as SSRIs and SNRIs). Low LH levels also correlate to hyperactivity of the hypothalamic–pituitary–adrenal (HPA) axis (Vadakkadath Meethal and Atwood 2005; Swaab et al. 2005), which represents the complex interactions between these organs that is essential in mood regulation, and is disrupted in MDD (Pariante 2003). As antidepressants typically work by reducing HPA activity, patients with low LH levels may have a hyperactive HPA axis, and are therefore good candidates for experiencing symptom improvement via antidepressant treatment (Holsboer and Barden 1996; Holsboer 2000).

The menstrual cycle also may play a role in attenuating antidepressant efficacy. The menstrual cycle can affect gastric motility, as fluid retention may result in diluted plasma levels of a drug (Yonkers et al. 1992). In addition, fluctuating levels of female sex hormones during the menstrual cycle may interact in complex ways with antidepressant metabolism. All of these factors contribute to the complex pharmacokinetic profile of antidepressants in women. More research is required to understand how each of these factors work together to produce observed gender differences in the efficacy of many antidepressants.

4 Conclusion

Despite decades of research into the effect of gender on depression pathogenesis and antidepressant efficacy, the consensus is still out on what these effects are, how these effects are produced, and in some cases, whether these effects exist at all. While many studies have found a higher incidence of depression in females, with a different array of symptoms and severity of symptom presentation, the reasons for these differences remain unknown. Likely, female sex hormones are at least partially involved, as periods of fluctuating levels of estrogen and progesterone have been linked to increased incidence of depression, such as during puberty and the transition to menopause. Differences in monoamine function, neurotransmitter metabolism, and even innate brain structures may also play a role in depression susceptibility, as numerous studies have identified an array of gender-specific differences in structures implicated in the pathogenesis of depression.

These systems are thought to play a role in producing gender-specific differences in antidepressant therapy, although their discrete role is far from clear, and even whether such gender differences exist remains controversial. There are almost an equal number of published studies showing gender differences, and no gender differences, in the efficacy profile of identical classes of antidepressants. This contradiction may be due to numerous factors, including methodological differences between studies in the measurement of a statistically significant treatment response.

A better consensus exists regarding gender differences in antidepressant metabolism. Sex-specific variance has been identified in numerous pharmacokinetic components, including plasma volume, gastric acid production, gastric emptying time, plasma protein levels, metabolic enzyme activity, drug transporter function, and drug clearance rates, among others. In addition, a large body of literature exists examining the effects of female hormones on the efficacy of antidepressants. For example, numerous studies suggest that estrogen has an enhancing effect on antidepressant efficacy, as evidenced by the poorer response of postmenopausal women to a variety of antidepressants when compared with premenopausal controls. This finding has direct therapeutic implications for treating depressed, postmenopausal women. On the other hand, lower levels of LH have been correlated with improved antidepressant efficacy.

Most likely it is the delicate interplay between all of these complex, multivariate systems that produces the observed gender differences in depression susceptibility and treatment response. More studies are required to tease apart each of these factors to understand how they contribute to the disease phenotype and the efficacy of antidepressants.

Acknowledgment The authors acknowledge Andrew Kuhlman for his assistance in compiling, writing, and editing this chapter.

References

Abe T, Suzuki M, Wada Y, Yamaya Y, Moritsuka T (1985) Clinical and endocrinological features of statistical clusters of women with climacteric symptoms. Tohoku J Exp Med 146:59–68

Abernethy DR, Greenblatt DJ, Shader RI (1985) Imipramine and desipramine disposition in the elderly. J Pharmacol Exp Ther 232:183–188

Anand A, Charney DS (2000) Norepinephrine dysfunction in depression. J Clin Psychiatry 61(Suppl 10):16–24

Andersen SL, Rutstein M, Benzo JM, Hostetter JC, Teicher MH (1997) Sex differences in dopamine receptor overproduction and elimination. Neuroreport 8:1495–1498

Angst J, Dobler-Mikola A (1984) Do the diagnostic criteria determine the sex ratio in depression? J Affect Disord 7:189–198

Avis NE, Brambilla D, McKinlay SM, Vass K (1994) A longitudinal analysis of the association between menopause and depression. Results from the Massachusetts Women's Health Study. Ann Epidemiol 4:214–220

Baca E, Garcia-Garcia M, Porras-Chavarino A (2004) Gender differences in treatment response to sertraline versus imipramine in patients with nonmelancholic depressive disorders. Prog Neuropsychopharmacol Biol Psychiatry 28:57–65

Baron P, Joly E (1988) Sex differences in the expression of depression in adolescents. Sex Roles 18:1–7

Bebbington P, Dunn G, Jenkins R, Lewis G, Brugha T, Farrell M, Meltzer H (2003) The influence of age and sex on the prevalence of depressive conditions: report from the National Survey of Psychiatric Morbidity. Int Rev Psychiatry 15:74–83

Berlanga C, Flores-Ramos M (2006) Different gender response to serotonergic and noradrenergic antidepressants. A comparative study of the efficacy of citalopram and reboxetine. J Affect Disord 95:119–123

Bock KW, Schrenk D, Forster A, Griese EU, Morike K, Brockmeier D, Eichelbaum M (1994) The influence of environmental and genetic factors on CYP2D6, CYP1A2 and UDP-glucuronosyltransferases in man using sparteine, caffeine, and paracetamol as probes. Pharmacogenetics 4:209–218

Brown AS, Gershon S (1993) Dopamine and depression. J Neural Transm Gen Sect 91:75–109

Bruno R, Vivier N, Montay G, Le Liboux A, Powe LK, Delumeau JC, Rhodes GR (1997) Population pharmacokinetics of riluzole in patients with amyotrophic lateral sclerosis. Clin Pharmacol Ther 62:518–526

Bryant DN, Sheldahl LC, Marriott LK, Shapiro RA, Dorsa DM (2006) Multiple pathways transmit neuroprotective effects of gonadal steroids. Endocrine 29:199–207

Carretti N, Florio P, Bertolin A, Costa CV, Allegri G, Zilli G (2005) Serum fluctuations of total and free tryptophan levels during the menstrual cycle are related to gonadotrophins and reflect brain serotonin utilization. Hum Reprod 20:1548–1553

Daly RC, Danaceau MA, Rubinow DR, Schmidt PJ (2003) Concordant restoration of ovarian function and mood in perimenopausal depression. Am J Psychiatry 160:1842–1846

Davidson J, Pelton S (1986) Forms of atypical depression and their response to antidepressant drugs. Psychiatry Res 17:87–95

Dawkins K, Potter WZ (1991) Gender differences in pharmacokinetics and pharmacodynamics of psychotropics: focus on women. Psychopharmacol Bull 27:417–426

Delgado PL, Moreno FA (2000) Role of norepinephrine in depression. J Clin Psychiatry 61(Suppl 1):5–12

D'Haenen HA, Bossuyt A (1994) Dopamine D2 receptors in depression measured with single photon emission computed tomography. Biol Psychiatry 35:128–132

Dhir A, Kulkarni SK (2007) Involvement of dopamine (DA)/serotonin (5-HT)/sigma (sigma) receptor modulation in mediating the antidepressant action of ropinirole hydrochloride, a D2/D3 dopamine receptor agonist. Brain Res Bull 74:58–65

Dluzen DE, McDermott JL (2008) Sex differences in dopamine- and vesicular monoamine-transporter functions. Ann N Y Acad Sci 1139:140–150

Entsuah AR, Huang H, Thase ME (2001) Response and remission rates in different subpopulations with major depressive disorder administered venlafaxine, selective serotonin reuptake inhibitors, or placebo. J Clin Psychiatry 62:869–877

Ereshefsky L, Saklad SR, Watanabe MD, Davis CM, Jann MW (1991) Thiothixene pharmacokinetic interactions: a study of hepatic enzyme inducers, clearance inhibitors, and demographic variables. J Clin Psychopharmacol 11:296–301

Frackiewicz EJ, Sramek JJ, Cutler NR (2000) Gender differences in depression and antidepressant pharmacokinetics and adverse events. Ann Pharmacother 34:80–88

Franconi F, Brunelleschi S, Steardo L, Cuomo V (2007) Gender differences in drug responses. Pharmacol Res 55:81–95

Frank E, Carpenter LL, Kupfer DJ (1988) Sex differences in recurrent depression: are there any that are significant? Am J Psychiatry 145:41–45

Freeman EW, Sammel MD, Liu L, Gracia CR, Nelson DB, Hollander L (2004) Hormones and menopausal status as predictors of depression in women in transition to menopause. Arch Gen Psychiatry 61:62–70

Freeman EW, Sammel MD, Lin H, Nelson DB (2006) Associations of hormones and menopausal status with depressed mood in women with no history of depression. Arch Gen Psychiatry 63:375–382

Gex-Fabry M, Balant-Gorgia AE, Balant LP, Garrone G (1990) Clomipramine metabolism. Model-based analysis of variability factors from drug monitoring data. Clin Pharmacokinet 19:241–255

Grigoriadis S, Kennedy SH, Bagby RM (2003) A comparison of antidepressant response in younger and older women. J Clin Psychopharmacol 23:405–407

Guiard BP, El Mansari M, Blier P (2009) Prospect of a dopamine contribution in the next generation of antidepressant drugs: the triple reuptake inhibitors. Curr Drug Targets 10: 1069–1084

Halbreich U, Lumley LA (1993) The multiple interactional biological processes that might lead to depression and gender differences in its appearance. J Affect Disord 29:159–173

Halbreich U, Rojansky N, Palter S, Tworek H, Hissin P, Wang K (1995) Estrogen augments serotonergic activity in postmenopausal women. Biol Psychiatry 37:434–441

Hamilton JA, Grant M, Jensvold MF (1996) Sex and treatment of depression: when does it matter? In: Jensvold MF (ed) Psychopharmacology and women: sex, gender, and hormones. American Psychiatric Press, Inc, Washington, DC, pp 241–257

Hamner MB, Diamond BI (1996) Plasma dopamine and norepinephrine correlations with psychomotor retardation, anxiety, and depression in non-psychotic depressed patients: a pilot study. Psychiatry Res 64:209–211

Hankin BL, Abramson LY (2001) Development of gender differences in depression: an elaborated cognitive vulnerability-transactional stress theory. Psychol Bull 127:773–796

Harlow BL, Wise LA, Otto MW, Soares CN, Cohen LS (2003) Depression and its influence on reproductive endocrine and menstrual cycle markers associated with perimenopause: the Harvard Study of Moods and Cycles. Arch Gen Psychiatry 60:29–36

Haykal RF, Akiskal HS (1999) The long-term outcome of dysthymia in private practice: clinical features, temperament, and the art of management. J Clin Psychiatry 60:508–518

Hildebrandt MG, Steyerberg EW, Stage KB, Passchier J, Kragh-Soerensen P (2003) Are gender differences important for the clinical effects of antidepressants? Am J Psychiatry 160: 1643–1650

Holsboer F (2000) The corticosteroid receptor hypothesis of depression. Neuropsychopharmacology 23:477–501

Holsboer F, Barden N (1996) Antidepressants and hypothalamic-pituitary-adrenocortical regulation. Endocr Rev 17:187–205

Hunt CM, Westerkam WR, Stave GM (1992) Effect of age and gender on the activity of human hepatic CYP3A. Biochem Pharmacol 44:275–283

Hutson WR, Roehrkasse RL, Wald A (1989) Influence of gender and menopause on gastric emptying and motility. Gastroenterology 96:11–17

Kessler RC (2003) Epidemiology of women and depression. J Affect Disord 74:5–13

Kessler RC, McGonagle KA, Swartz M, Blazer DG, Nelson CB (1993) Sex and depression in the National Comorbidity Survey. I: Lifetime prevalence, chronicity and recurrence. J Affect Disord 29:85–96

Kessler RC, McGonagle KA, Zhao S, Nelson CB, Hughes M, Eshleman S, Wittchen HU, Kendler KS (1994) Lifetime and 12-month prevalence of DSM-III-R psychiatric disorders in the United States. Results from the National Comorbidity Survey. Arch Gen Psychiatry 51:8–19

Khan A, Brodhead AE, Schwartz KA, Kolts RL, Brown WA (2005) Sex differences in antidepressant response in recent antidepressant clinical trials. J Clin Psychopharmacol 25: 318–324

Kim RB, O'Shea D (1995) Interindividual variability of chlorzoxazone 6-hydroxylation in men and women and its relationship to CYP2E1 genetic polymorphisms. Clin Pharmacol Ther 57:645–655

Kim RB, Leake BF, Choo EF, Dresser GK, Kubba SV, Schwarz UI, Taylor A, Xie HG, McKinsey J, Zhou S, Lan LB, Schuetz JD, Schuetz EG, Wilkinson GR (2001) Identification of functionally variant MDR1 alleles among European Americans and African Americans. Clin Pharmacol Ther 70:189–199

Klimek V, Schenck JE, Han H, Stockmeier CA, Ordway GA (2002) Dopaminergic abnormalities in amygdaloid nuclei in major depression: a postmortem study. Biol Psychiatry 52:740–748

Kornstein SG, Schatzberg AF, Thase ME, Yonkers KA, McCullough JP, Keitner GI, Gelenberg AJ, Davis SM, Harrison WM, Keller MB (2000a) Gender differences in treatment response to sertraline versus imipramine in chronic depression. Am J Psychiatry 157:1445–1452

Kornstein SG, Schatzberg AF, Thase ME, Yonkers KA, McCullough JP, Keitner GI, Gelenberg AJ, Ryan CE, Hess AL, Harrison W, Davis SM, Keller MB (2000b) Gender differences in chronic major and double depression. J Affect Disord 60:1–11

Kumar AM, Weiss S, Fernandez JB, Cruess D, Eisdorfer C (1998) Peripheral serotonin levels in women: role of aging and ethnicity. Gerontology 44:211–216

Laakso A, Vilkman H, Bergman J, Haaparanta M, Solin O, Syvalahti E, Salokangas RK, Hietala J (2002) Sex differences in striatal presynaptic dopamine synthesis capacity in healthy subjects. Biol Psychiatry 52:759–763

Labbe L, Sirois C, Pilote S, Arseneault M, Robitaille NM, Turgeon J, Hamelin BA (2000) Effect of gender, sex hormones, time variables and physiological urinary pH on apparent CYP2D6 activity as assessed by metabolic ratios of marker substrates. Pharmacogenetics 10:425–438

Lambert G, Johansson M, Agren H, Friberg P (2000) Reduced brain norepinephrine and dopamine release in treatment-refractory depressive illness: evidence in support of the catecholamine hypothesis of mood disorders. Arch Gen Psychiatry 57:787–793

Lan LB, Dalton JT, Schuetz EG (2000) Mdr1 limits CYP3A metabolism in vivo. Mol Pharmacol 58:863–869

Legato MJ (1997) Gender-specific physiology: how real is it? How important is it? Int J Fertil Womens Med 42:19–29

Leibenluft E, Hardin TA, Rosenthal NE (1995) Gender differences in seasonal affective disorder. Depression 3:13–19

Lucas D, Menez C, Girre C, Berthou F, Bodenez P, Joannet I, Hispard E, Bardou LG, Menez JF (1995) Cytochrome P450 2E1 genotype and chlorzoxazone metabolism in healthy and alcoholic Caucasian subjects. Pharmacogenetics 5:298–304

Lucht M, Schaub RT, Meyer C, Hapke U, Rumpf HJ, Bartels T, von Houwald J, Barnow S, Freyberger HJ, Dilling H, John U (2003) Gender differences in unipolar depression: a general population survey of adults between age 18 to 64 of German nationality. J Affect Disord 77:203–211

Martenyi F, Dossenbach M, Mraz K, Metcalfe S (2001) Gender differences in the efficacy of fluoxetine and maprotiline in depressed patients: a double-blind trial of antidepressants with serotonergic or norepinephrinergic reuptake inhibition profile. Eur Neuropsychopharmacol 11:227–232

McKinlay JB, McKinlay SM, Brambilla DJ (1987) Health status and utilization behavior associated with menopause. Am J Epidemiol 125:110–121

Meibohm B, Beierle I, Derendorf H (2002) How important are gender differences in pharmacokinetics? Clin Pharmacokinet 41:329–342

Meltzer CC, Drevets WC, Price JC, Mathis CA, Lopresti B, Greer PJ, Villemagne VL, Holt D, Mason NS, Houck PR, Reynolds CF III, DeKosky ST (2001) Gender-specific aging effects on the serotonin 1A receptor. Brain Res 895:9–17

Meyer JH, Kruger S, Wilson AA, Christensen BK, Goulding VS, Schaffer A, Minifie C, Houle S, Hussey D, Kennedy SH (2001) Lower dopamine transporter binding potential in striatum during depression. Neuroreport 12:4121–4125

Miller A, Fox NA, Cohn JF, Forbes EE, Sherrill JT, Kovacs M (2002) Regional patterns of brain activity in adults with a history of childhood-onset depression: gender differences and clinical variability. Am J Psychiatry 159:934–940

Moreno FA, McGahuey CA, Freeman MP, Delgado PL (2006) Sex differences in depressive response during monoamine depletions in remitted depressive subjects. J Clin Psychiatry 67:1618–1623

Morrison MF, Kallan MJ, Ten Have T, Katz I, Tweedy K, Battistini M (2004) Lack of efficacy of estradiol for depression in postmenopausal women: a randomized, controlled trial. Biol Psychiatry 55:406–412

Munro CA, McCaul ME, Wong DF, Oswald LM, Zhou Y, Brasic J, Kuwabara H, Kumar A, Alexander M, Ye W, Wand GS (2006) Sex differences in striatal dopamine release in healthy adults. Biol Psychiatry 59:966–974

Nafziger AN, Bertino JS Jr (1989) Sex-related differences in theophylline pharmacokinetics. Eur J Clin Pharmacol 37:97–100

Naito S, Sato K, Yoshida K, Higuchi H, Takahashi H, Kamata M, Ito K, Ohkubo T, Shimizu T (2007) Gender differences in the clinical effects of fluvoxamine and milnacipran in Japanese major depressive patients. Psychiatry Clin Neurosci 61:421–427

Nathorst-Boos J, von Schoultz B, Carlstrom K (1993) Elective ovarian removal and estrogen replacement therapy–effects on sexual life, psychological well-being and androgen status. J Psychosom Obstet Gynaecol 14:283–293

Nishizawa S, Benkelfat C, Young SN, Leyton M, Mzengeza S, de Montigny C, Blier P, Diksic M (1997) Differences between males and females in rates of serotonin synthesis in human brain. Proc Natl Acad Sci U S A 94:5308–5313

Nolen-Hoeksema S (1990) Sex differences in depression. Stanford University Press, Stanford CA

Nolen-Hoeksema S, Larson J, Grayson C (1999) Explaining the gender difference in depressive symptoms. J Pers Soc Psychol 77:1061–1072

Ou-Yang DS, Huang SL, Wang W, Xie HG, Xu ZH, Shu Y, Zhou HH (2000) Phenotypic polymorphism and gender-related differences of CYP1A2 activity in a Chinese population. Br J Clin Pharmacol 49:145–151

Owens MJ, Nemeroff CB (1994) Role of serotonin in the pathophysiology of depression: focus on the serotonin transporter. Clin Chem 40:288–295

Pae CU, Mandelli L, Kim TS, Han C, Masand PS, Marks DM, Patkar AA, Steffens DC, De Ronchi D, Serretti A (2009) Effectiveness of antidepressant treatments in pre-menopausal versus post-menopausal women: a pilot study on differential effects of sex hormones on antidepressant effects. Biomed Pharmacother 63:228–235

Pariante CM (2003) Depression, stress and the adrenal axis. J Neuroendocrinol 15:811–812

Parker G, Parker K, Austin MP, Mitchell P, Brotchie H (2003) Gender differences in response to differing antidepressant drug classes: two negative studies. Psychol Med 33:1473–1477

Parry BL (1989) Reproductive factors affecting the course of affective illness in women. Psychiatr Clin North Am 12:207–220

Parry BL, Newton RP (2001) Chronobiological basis of female-specific mood disorders. Neuropsychopharmacology 25:S102–S108

Pinto-Meza A, Usall J, Serrano-Blanco A, Suarez D, Haro JM (2006) Gender differences in response to antidepressant treatment prescribed in primary care. Does menopause make a difference? J Affect Disord 93:53–60

Preskorn SH, Mac DS (1985) Plasma levels of amitriptyline: effect of age and sex. J Clin Psychiatry 46:276–277

Quitkin FM, Stewart JW, McGrath PJ (2001) Gender differences in treatment response. Am J Psychiatry 158:1531–1533

Quitkin FM, Stewart JW, McGrath PJ, Taylor BP, Tisminetzky MS, Petkova E, Chen Y, Ma G, Klein DF (2002) Are there differences between women's and men's antidepressant responses? Am J Psychiatry 159:1848–1854

Rakofsky JJ, Holtzheimer PE, Nemeroff CB (2009) Emerging targets for antidepressant therapies. Curr Opin Chem Biol 13:291–302

Rapaport MH, Thompson PM, Kelsoe JR Jr, Golshan S, Judd LL, Gillin JC (1995) Gender differences in outpatient research subjects with affective disorders: a comparison of descriptive variables. J Clin Psychiatry 56:67–72

Renshaw PF, Parow AM, Hirashima F, Ke Y, Moore CM, Frederick Bde B, Fava M, Hennen J, Cohen BM (2001) Multinuclear magnetic resonance spectroscopy studies of brain purines in major depression. Am J Psychiatry 158:2048–2055

Rivest R, Falardeau P, Di Paolo T (1995) Brain dopamine transporter: gender differences and effect of chronic haloperidol. Brain Res 692:269–272

Roy A, Pickar D, Linnoila M, Potter WZ (1985) Plasma norepinephrine level in affective disorders. Relationship to melancholia. Arch Gen Psychiatry 42:1181–1185

Sagud M, Hotujac L, Mihaljevic-Peles A, Jakovljevic M (2002) Gender differences in depression. Coll Antropol 26:149–157

Sakai Y, Nishikawa M, Leyton M, Benkelfat C, Young SN, Diksic M (2006) Cortical trapping of alpha-[(11)C]methyl-l-tryptophan, an index of serotonin synthesis, is lower in females than males. Neuroimage 33:815–824

Schmidt R, Baumann F, Hanschmann H, Geissler F, Preiss R (2001) Gender difference in ifosfamide metabolism by human liver microsomes. Eur J Drug Metab Pharmacokinet 26:193–200

Schmidt PJ, Haq N, Rubinow DR (2004) A longitudinal evaluation of the relationship between reproductive status and mood in perimenopausal women. Am J Psychiatry 161:2238–2244

Schneider LS, Small GW, Hamilton SH, Bystritsky A, Nemeroff CB, Meyers BS (1997) Estrogen replacement and response to fluoxetine in a multicenter geriatric depression trial. Fluoxetine Collaborative Study Group. Am J Geriatr Psychiatry 5:97–106

Schuetz EG, Furuya KN, Schuetz JD (1995) Interindividual variation in expression of P-glycoprotein in normal human liver and secondary hepatic neoplasms. J Pharmacol Exp Ther 275:1011–1018

Seeman MV (1997) Psychopathology in women and men: focus on female hormones. Am J Psychiatry 154:1641–1647

Shah PJ, Ogilvie AD, Goodwin GM, Ebmeier KP (1997) Clinical and psychometric correlates of dopamine D2 binding in depression. Psychol Med 27:1247–1256

Shifren JL, Avis NE (2007) Surgical menopause: effects on psychological well-being and sexuality. Menopause 14:586–591

Shimizu H, Bray GA (1993) Effects of castration, estrogen replacement and estrus cycle on monoamine metabolism in the nucleus accumbens, measured by microdialysis. Brain Res 621:200–206

Soares CN, Almeida OP, Joffe H, Cohen LS (2001) Efficacy of estradiol for the treatment of depressive disorders in perimenopausal women: a double-blind, randomized, placebo-controlled trial. Arch Gen Psychiatry 58:529–534

Sofuoglu M, Mitchell E, Kosten TR (2004) Effects of progesterone treatment on cocaine responses in male and female cocaine users. Pharmacol Biochem Behav 78:699–705

Staley JK, Krishnan-Sarin S, Zoghbi S, Tamagnan G, Fujita M, Seibyl JP, Maciejewski PK, O'Malley S, Innis RB (2001) Sex differences in [123I]beta-CIT SPECT measures of dopamine and serotonin transporter availability in healthy smokers and nonsmokers. Synapse 41:275–284

Stewart DE, Rolfe DE, Robertson E (2004) Depression, estrogen, and the Women's Health Initiative. Psychosomatics 45:445–447

Succari M, Foglietti MJ, Percheron F (1990) Microheterogeneity of alpha 1-acid glycoprotein: variation during the menstrual cycle in healthy women, and profile in women receiving estrogen-progestogen treatment. Clin Chim Acta 187:235–241

Swaab DF, Bao AM, Lucassen PJ (2005) The stress system in the human brain in depression and neurodegeneration. Ageing Res Rev 4:141–194

Thase ME, Frank E, Kornstein SG, Yonkers KA (2000) Gender differences in response to treatments of depression. In: Frank E (ed) Gender and its effects on psychopathology. American Psychiatric Press, Washington, pp 103–125

Thiels C, Linden M, Grieger F, Leonard J (2005) Gender differences in routine treatment of depressed outpatients with the selective serotonin reuptake inhibitor sertraline. Int Clin Psychopharmacol 20:1–7

Timmer CJ, Sitsen JM, Delbressine LP (2000) Clinical pharmacokinetics of mirtazapine. Clin Pharmacokinet 38:461–474

Vadakkadath Meethal S, Atwood CS (2005) The role of hypothalamic-pituitary-gonadal hormones in the normal structure and functioning of the brain. Cell Mol Life Sci 62:257–270

van Amelsvoort TA, Abel KM, Robertson DM, Daly E, Critchley H, Whitehead M, Murphy DG (2001) Prolactin response to d-fenfluramine in postmenopausal women on and off ERT: comparison with young women. Psychoneuroendocrinology 26:493–502

Vredenburg K, Krames L, Flett GL (1986) Sex differences in the clinical expression of depression. Sex Roles 14:37–49

Weissman MM, Klerman GL (1977) Sex differences and the epidemiology of depression. Arch Gen Psychiatry 34:98–111

Weissman MM, Bland R, Joyce PR, Newman S, Wells JE, Wittchen HU (1993) Sex differences in rates of depression: cross-national perspectives. J Affect Disord 29:77–84

Weissman MM, Bland RC, Canino GJ, Faravelli C, Greenwald S, Hwu HG, Joyce PR, Karam EG, Lee CK, Lellouch J, Lepine JP, Newman SC, Rubio-Stipec M, Wells JE, Wickramaratne PJ, Wittchen H, Yeh EK (1996) Cross-national epidemiology of major depression and bipolar disorder. JAMA 276:293–299

White TL, Justice AJ, de Wit H (2002) Differential subjective effects of D-amphetamine by gender, hormone levels and menstrual cycle phase. Pharmacol Biochem Behav 73:729–741

Wilhelm K, Roy K, Mitchell P, Brownhill S, Parker G (2002) Gender differences in depression risk and coping factors in a clinical sample. Acta Psychiatr Scand 106:45–53

Williams JB, Spitzer RL, Linzer M, Kroenke K, Hahn SR, deGruy FV, Lazev A (1995) Gender differences in depression in primary care. Am J Obstet Gynecol 173:654–659

Winokur G (1973) Depression in the menopause. Am J Psychiatry 130:92–93

Wohlfarth T, Storosum JG, Elferink AJ, van Zwieten BJ, Fouwels A, van den Brink W (2004) Response to tricyclic antidepressants: independent of gender? Am J Psychiatry 161:370–372

Xiao L, Becker JB (1994) Quantitative microdialysis determination of extracellular striatal dopamine concentration in male and female rats: effects of estrous cycle and gonadectomy. Neurosci Lett 180:155–158

Yonkers KA, Brawman-Mintzer O (2002) The pharmacologic treatment of depression: is gender a critical factor? J Clin Psychiatry 63:610–615

Yonkers KA, Kando JC, Cole JO, Blumenthal S (1992) Gender differences in pharmacokinetics and pharmacodynamics of psychotropic medication. Am J Psychiatry 149:587–595

Young EA (1995) The role of gonadal steroids in hypothalamic-pituitary-adrenal axis regulation. Crit Rev Neurobiol 9:371–381

Young SN, Gauthier S, Anderson GM, Purdy WC (1980) Tryptophan, 5-hydroxyindoleacetic acid and indoleacetic acid in human cerebrospinal fluid: interrelationships and the influence of age, sex, epilepsy and anticonvulsant drugs. J Neurol Neurosurg Psychiatry 43:438–445

Young MA, Scheftner WA, Fawcett J, Klerman GL (1990) Gender differences in the clinical features of unipolar major depressive disorder. J Nerv Ment Dis 178:200–203

Young EA, Midgley AR, Carlson NE, Brown MB (2000) Alteration in the hypothalamic-pituitary-ovarian axis in depressed women. Arch Gen Psychiatry 57:1157–1162

Young EA, Kornstein SG, Marcus SM, Harvey AT, Warden D, Wisniewski SR, Balasubramani GK, Fava M, Trivedi MH, John Rush A (2009) Sex differences in response to citalopram: a STAR*D report. J Psychiatr Res 43:503–511

Zanardi R, Rossini D, Magri L, Malaguti A, Colombo C, Smeraldi E (2007) Response to SSRIs and role of the hormonal therapy in post-menopausal depression. Eur Neuropsychopharmacol 17:400–405

Index

J.C. Neill and J. Kulkarni (eds.), *Biological Basis of Sex Differences in Psychopharmacology*, Current Topics in Behavioral Neurosciences 8, DOI 10.1007/7854_2010,

Printed by Publishers' Graphics LLC